handbook of noise and vibration control

6th Edition

by Antony Barber MSc, CEng, FIMechE, MRAeS, ACIArb

ELSEVIER ADVANCED TECHNOLOGY
MAYFIELD HOUSE, 256 BANBURY ROAD, OXFORD OX2 7DH, UK

ISBN 1 85617 079 9

Printed in UK by Professional Book Supplies, Abingdon, Oxon.

PREFACE FROM THE PUBLISHER

The Sixth Edition of the Handbook of Noise and Vibration Control has been completely revised, updated and extended to keep pace with the rapid development of the discipline.

Machinery is the principal cause of noise and vibration. The constant quest for more power, more capacity, more speed, more weight loading, to meet the demands of industry can easily lead to greater noise and vibration. Yet all industrialized countries have introduced stringent laws to combat noise in the factory, the office and the environment. To sell at home and abroad, machinery manufacturers must comply with the laws of the country of origin or importation. Also a quiet machine is an efficient machine. The ultimate in machine efficiency would eliminate noise and vibration completely.

For all those concerned with reducing noise and vibration, from machine designers to architects, from public health and environmental officers to factory managers, this handbook will prove an indispensable source of technical data and practical information. Cause, effect, measurement, acceptable levels, methods of control, materials, working data and sources of specialized assistance and equipment are all incorporated in this one volume.

G.S. Kitteringham
ELSEVIER ADVANCED TECHNOLOGY

ACKNOWLEDGMENTS

Accent Group Ltd
AIRO (Acoustic Investigation & Research Organisation)
Alpha dBK Ltd
Applied Acoustic Variables
Anthony Best Dynamics
Atlas Copco
Bersafe
Bilson
Boral Edenhall
British Gypsum Ltd
Bruel & Kjaer
Building Regulations
Burgess Industrial Silencing
Cabot Safety Ltd
Christie and Grey
Cirrus Research Ltd
CMT Dynamics
Compair Holman
Contranoise Ltd
Datarec
Econocruise
Endevco
Environmental Elements Corp
Farrat Machinery
Fawcett Christie Hydraulics
Fuji Ceramics
GEBR
GERB

Getzuer
Gleason Works
Hanson Transmissions
Health & Safety Executive
Hydrorane
IAC
Interwand Eibergen
IMV Corporation
James Walker
Kistler
Ling Electronics
Lloyd's Register
Lucus Industries Ltd
Mercury Electronics
MIRA Loughborough University
Monarch Instrument
Pafel
Pilkington Environmental Advisory Service
Pneurop/Cagi
Ricardo
Rockwool
Scientific Atlanta
SDRC
SKF
SPM Instrument
Sound Alternators Industrial
Stewart Hughes
Stevens
Wilcoxon Research

CONTENTS

SECTION 1

Chapter 1
Units and Nomenclature

Symbols

Unless otherwise specified in the text, the following symbols are used consistently throughout this book.

A = surface area (m^2)

c = velocity of sound in specified medium (m/s)

DI = directivity index (dB)

f = frequency (cycles/s)

I = sound intensity level (W/m^2)

L_I = sound intensity level expressed in decibels. The reference level is 1 pW/m^2 (picawatt/m^2)

L_p = sound pressure level expressed in decibels. (The reference level is 2 x 10^{-5} Pa)

L_W = sound power level expressed in decibels. (The reference level is 1 pW)

p = rms value of sound pressure (Pa)

Q = directivity factor

r = distance from noise source (m)

R = room constant (m^2)

S = room surface area (m^2)

t = time (s)

T = temperature (°C) also reverberation time (s)

V = volume (m^3)

W = sound power of source (w)

Greek symbols

α = absorption coefficient (dimensionless) also attenuation (dB/m)

ρ = density (kg/m^3)

λ = wavelength of sound (m/s)

γ = ratio of specific heat at constant pressure to that at constant temperature (Cp/Cv)

ζ = damping ratio

Acoustic Units

Bel – A unit for expressing the ratio of two values of power. It is defined as the logarithm (base 10) of the ratio of the two powers. The bel is considered too large for practical purposes, so a more common unit is the decibel (dB) which is one tenth of a bel. Because of the logarithmic scale,

1 dB = $10^{0.1}$ bel. When specifying absolute sound levels, the reference power is 1pW.

Neper – When the level difference is expressed as the *natural* logarithm of the ratio of the two quantities, the appropriate unit is the neper (Np) rather than the decibel. 1 Np = 8.686 dB.

Noy – Unit of perceived noisiness in which equal noisiness contours replace equal loudness contours. 1 noy is equal to the noisiness of a one-third octave band of noise centred on 1kHz and having a sound pressure level of 40 dB.

Phon – Unit of objective loudness on the sound level scale. It is numerically equal to the sound pressure level of a 1 kHz tone, which is judged by a reliable observer to be as loud as the unknown sound. This hypothetical observer can, for practical purposes, be replaced by a microphone and weighting network.

Rayl – A unit for the measurement of acoustic impedance, equivalent to 1 kg/(m^{2}s)

Sabin – A measure of sound absorption of a surface. It is numerically equal to the assumed total absorption (i.e. as for an open window) of 1 ft^2 area. By extension, the metric sabin is defined for a 1 m^2 area.

Sone – The ratio of the loudness of sound to that of a 1 kHz tone 40 db above the threshold of hearing. The loudness in sones is given by the formula

$$2^{0.1(L-40)}$$

SI units are used consistently throughout the text except for occasional reference to Imperial units, where necessary.

Prefixes

When dealing with very small powers, the standard prefixes are

Multiple	Prefix	Symbol
10^{-6}	micro	μ
10^{-9}	nano	n
10^{-12}	pico	p
10^{-15}	femto	f
10^{-18}	atto	a

Glossary of Terms

Absorption – The reduction of sound energy as it passes through an acoustic medium and its conversion into heat.

Absorption coefficient – The measurement of the ability of a medium to absorb sound. It is given by the ratio of the absorbed energy to that of the incident energy. (= 0 for total reflection, = 1 for total absorption). Reverberation absorption coefficient is the value obtained when the sound field is diffuse.

Accelerometer – A sensor whose output is proportional to acceleration.

Ambient noise – The total noise level in a specified environment.

Anechoic chamber – A room for noise testing which is so constructed that the boundary surfaces absorb all the incident sound.

Audibility threshold – The sound pressure level at a specific frequency which persons with normal hearing can just detect. It is conventionally quoted as 2 x 10^{-5} Pa rms at 1000 Hz, which is also the basis of the decibel scale of sound pressure.

Background noise – The prevailing noise level in a specified environment measured in the absence of the noise source being studied.

Charge amplifier – An amplifier whose output voltage is proportional to the input charge from a transducer. The voltage is unaffected by the length of the cable. See voltage amplifier.

Compliance – The displacement per unit of applied force; the reciprocal of stiffness

Damping – Method of dissipating vibration energy.

Damping ratio – The ratio of the actual damping in a system to the critical damping, at resonant frequency.

Directivity factor – The ratio of the sound intensity at a specified distance from the source of the sound in a specified direction to the sound intensity at the same distance, but averaged over all directions.

Directivity index – Ten times the logarithm of the directivity factor.

Free field – An environment in which there is no reflective surface.

Far field – That part of the sound field where the sound wave is spreading spherically, i.e. the sound decays at 6dB for a doubling of the distance from the source.

Hearing loss – An increase in the audibility threshold caused by disease, injury, age or exposure to a high level of noise. It may be temporary or permanent.

Impedance – Ratio of applied force to resulting velocity under harmonic excitation. Driving point impedance is when force and velocity are measured at the same point; transfer impedance if force and velocity are at different points.

Impedance, characteristic acoustic – The product of the density and the velocity of sound in a medium.

Infrasonic – Frequencies below the audible range, below 16 Hz

Integrating circuit – A frequency filter which converts a vibratory acceleration signal into one in which the amplitude is proportional to velocity or displacement.

Jerk – A vector quantity specifying rate of change of acceleration with respect to time.

Noise – Sound and Noise are often used interchangeably. When dealing with physical properties, 'sound' is preferred, because it does not carry any sense of subjectivity. It is better to restrict the use of 'noise' to unwanted sound, although this does not always produce an infallible distinction. One person's music is often another's noise.

Pink noise – Broadband noise whose energy content is inversely proportional to frequency (−3dB per octave or −10dB per decade).

Random noise – A noise whose instantaneous amplitude is not specified at any instant of time. Its character can only be defined statistically by a distribution function.

Reverberation – The persistence of sound in an enclosure after a sound source has been removed. The reverberation time is the time for the sound pressure, at a specific frequency, to decay by 60dB.

Reverberant chamber – A room designed for the study of acoustic properties of materials and for other purposes where a diffuse field is required. It has highly reflective walls, so that the average energy density is the same throughout the chamber: the converse of an anechoic chamber

Sound power level – The rms sound power quoted in decibels. See chapter on *Sound Propagation*.

Sound pressure level – The rms pressure quoted in decibels. See chapter on *Sound Propagation*.

Sound reduction index – The ratio of sound energy incident on a partition to that which is transmitted through the partition.

Threshold shift – See hearing loss.

Transmission coefficient – The ratio of the transmitted energy to the incident energy of a panel or structure.

Ultrasonic – Frequencies above the audible range, above 20 kHz.

Voltage amplifier – An amplifier which produces an output voltage proportional to input voltage from an accelerometer.

Weighting network – A series of filters in a sound level meter or signal processor, which approximates under defined conditions to the frequency response of the human ear. A variety of different weighting networks has been devised, the most common one being A–weighting. Sound meters frequently give A–, B– and C– weightings, which are then quoted as, e.g., dB(A). D–, E– and SI–weightings are also occasionally found.

White noise – Broadband noise having constant energy per unit of frequency.

See also BS 4727 Part 3 Group 08.

Decibel Scales

In the following chapters, sound is usually expressed in terms of the decibel scale, so a brief summary of the principles is included here. Decibel is the unit of power level difference and is used as a measure of response in all types of electrical circuit. Confusion can arise through the use of decibels for both sound pressure and sound power.

The question often arises as to why, when measuring sound pressure, the unit employed is not the standard SI unit of the Pascal. Certainly when studying the mechanical effects of sound power, one has to use the actual pressure in Pascal. A microphone, for example, is a mechanical device which only responds to a pressure change measured by a force per unit area. A structure will only move, or be affected by, a force acting on it and at that stage the decibel level has to be converted to pressure units.

The sound pressures and powers that are the concern of the acoustical engineer cover an enormous range – a ratio of $1 : 10^9$ for pressures and $1 : 10^{18}$ for power. It would be difficult to put meaning into such values quoted in Pascals or watts. The same range expressed in decibels is from 0 to 200, which is much more manageable.

Another merit in the decibel scale lies in the way the human ear responds. The lowest sound that the healthy human ear can detect is 2×10^{-5} Pa; at the other end of the scale, the threshold of pain is approximately 100 Pa – a range from 0 to 120 dB. The ear/brain system tends to respond logarithmically rather than linearly between the two. An illustration of the subjective effect of changes in noise levels expressed in decibels is Table 1.

TABLE 1 – SUBJECTIVE EFFECT OF CHANGES IN NOISE LEVEL

Change in level	Subjective effect
3	Just perceptible
5	Clearly perceptible
10	Twice as loud

Table 2 gives some typical examples of sound sources.

The fundamental unit is the bel (after Alexander Graham Bell), defined as the logarithm to the base 10 of the ratio of two powers. This is considered too large for practical purposes so the decibel (dB), defined as 0.1 bel, is preferred. For most purposes it is not necessary to use fractions of a decibel; indeed practical sound meters do not measure to a greater accuracy than about 0.5 dB, even if the sound level could be relied not to vary by more than that. One point to remember is that a level of 0 dB does not represent absence of sound, but rather that it is equal to the reference level.

Manipulation of values expressed in decibels

It is sometimes required to add two values expressed in decibels. If for example two or more machines, each characterised by sound pressure levels expressed in dB, are put together it may be required to determine the combined level. Only pressures or powers can be added directly, so if the individual pressure levels are (L_1, L_2 . . . etc) dB, the combined pressure level is L_{tot}, where:

$$L_{tot} = 10 \log \left(antilog \frac{L_1}{10} + antilog \frac{L_2}{10} \ldots \right)$$

For rapid solutions to this equation it is often convenient to use a graph such as Figure 1. For more than 2 values, the graph can be used repeatedly.

Some features of this graph are immediately apparent: two equal sound sources put together increase the level by 3 dB; if two sound sources differ by more than about 10 dB, the smaller one can be ignored.

It is meaningless to add together arithmetically two or more values expressed in decibels, but there is one exception: when determining an average sound pressure over a region, providing that the spread of the readings does not exceed 5 dB, the arithmetical average gives adequate accuracy for practical purposes; such a calculation is described in some test codes, although it must be remembered that it is not theoretically correct.

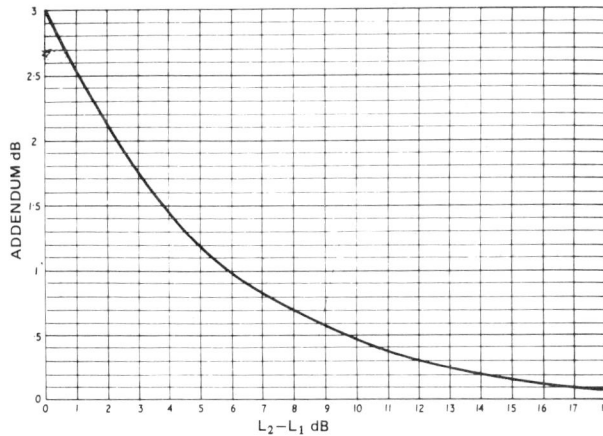

Figure 1

Chapter 2
Propagation of Sound

Velocity of Sound

The velocity of sound in a gas is given by

$$c = (\gamma P/\rho)^{0.5}$$

or alternatively

$$c = (\gamma RT/M)^{0.5}$$

where P is the ambient pressure, R the gas constant, T the absolute temperature and M the molecular weight.

It can be seen that c is proportional to $(T)^{0.5}$ only, and therefore independent of pressure. This is true for moderate pressure but at high pressure other factors affect the value.

The velocity in air, with which this volume is mainly concerned, depends on a number of other factors such as humidity and CO_2 content; it also varies with frequency. For most practical purposes the velocity in air is given by

$$c_{air} = 20.06 \, (T)^{0.5}$$
$$= 343 \text{ m/s at } 20°C$$

The velocity of sound in a liquid or solid is much higher. The subject is complex, so reference should be made to a specialist text, but for convenience some typical values are given in Table 1, which also gives values for attenuation where known.

Attenuation of Sound in Air

Various mechanisms attenuate sound in air; this subject is also complex. Viscosity and thermal losses (conversion of sound energy into heat) are small. Humidity has a more powerful effect. Attenuation varies with temperature, humidity and frequency. Table 2 gives some typical values for attenuation of plane waves in air. The Table demonstrates the common experience that 'low frequency sounds travel further through the air than high frequency'. Another common belief is that sound travels better in mist or fog. Although Table 2 does indicate a reduced attenuation under high humidity conditions, the observation is more likely to be caused by the still air conditions associated with fog – see the paragraph on Wind Effects below.

Attenuation in Other Media

Attenuation of sound waves occurs in all media. The value of for plane waves can be found from

$$\alpha = 1/2d \ \log (I_1/I_2) \ dB/m$$

where the initial intensity I_1 has decreased to I_2 after traversing a distance d. Attenuation in gases other than air is a complex study and practical values are not readily available. Some values are available for liquids and are given in Table 1. Note that for this part of the table the attenuation α is proportional to f^2. For solids, only a few values are known in the literature and these are given in the table.

Wavelength of Sound

The wavelength of sound is related to the velocity and frequency by the relationship:

$$\lambda = c/f$$

This applies to any material, using the appropriate velocity.

TABLE 1 – VELOCITY AND ATTENUATION OF SOUND IN VARIOUS MEDIA

Medium	Velocity	Attenuation
Gases at 0°C	**c (m/s)**	
Argon	308	
Carbon dioxide	259	
Carbon monoxide	337	
Helium	972	
Nitrogen	337	
Water (100°C)	405	
Liquids at 20°C		$\alpha /f^2 \ \lvert 10^{-18} \ dB/(m \ Hz^2)\rvert$
Kerosene	1315	0.95
Mercury	1454	0.0478
Nitrogen (−189°C)	745	0.528
Oil (castor)	1500	94.7
Oxygen (−186°C)	950	0.0747
Water, distilled	1482	0.217
Water, sea (Note 1)	1522	–
Solids (longitudinal waves)	**(Note 2)**	**(dB/m)**
Aluminium	6374	3.5
Brass	4372	–
Concrete	4250 – 5250	–
Glass	5660	17
Ice	3840	–
Iron, cast	4994	–
Lead	2160	–
Nylon	2680	100
Rubber, natural	1600	130
Steel, mild	5960	–
Steel, hardened tool	5874	43
Wood with grain	3600 – 4600	–
Wood across grain	1400	–

Notes: (1) Velocity and attenuation for sea water depend on depth, temperature, salinity and frequency. Attenuation is much greater than for distilled water. Refer to a specialist text for a fuller treatment, e.g. *Kaye & Labey, Tables of Physical and Chemical Constants.*

(2) Only values for longitudinal bulk waves are quoted. Rod waves, shear waves and surface waves can also occur. Rod waves occur in straight bars and thin tubes; the velocity of these waves can be found from $c_E = (E/\rho)^{0.5}$, where E is Young's modulus.

TABLE 2 – ATTENUATION OF SOUND IN AIR (dB/km)

Frequency (kHz)	Relative humidity (%)					
	0	10	20	40	60	80
1	0	12	6	4	4	3
2	1	40	19	10	9	8
4	3	110	68	32	23	20
16	41	280	460	380	270	210

Radiation of Sound

Sound is a form of radiant energy and as such behaves according to the law of radiation, i.e. its intensity at a distance from the source varies inversely with the square of the distance. In practice, sound does not radiate from a single point; it is more likely to come from a machine with several sources. There will however be some distance, large compared with the size of the machine (about 4 or 5 times its characteristic length), where the inverse square law will apply. This region is known as the 'far field'. The region close to the object is known as the 'near field'.

At a distance r from the source, the intensity of sound will be uniformly distributed over the surface of a hypothetical sphere, radius r. The intensity will be proportional to $1/r^2$. It follows that a doubling of the distance will reduce the sound intensity to $\frac{1}{4}$.

Expressed in decibels, doubling the distance reduces the intensity by 10 log 4 = 6 dB.

The acoustic intensity is found by dividing the sound power of the source in watts by the area of the sphere.

The sound power radiates in the form of pressure fluctuations in the air. The rms value of the

$$I = \frac{W}{4\pi r^2}$$

pressure fluctuations is found from:

$$I = \frac{p^2}{\rho c} \qquad \text{...(1)}$$

Equating the two gives the relationship

$$p^2 = \frac{W\rho c}{4\pi r^2} \qquad \text{...(2)}$$

This equation shows that the (acoustic pressure)^2 is inversely proportional to the (distance)^2 from the source.

The acoustic pressure is the quantity measured by a microphone. If the actual value of the sound power is required, it can be found by application of the above formulas.

Comparison between sound pressure in air and other media

It is of interest to calculate the sound pressure generated in water and compare it with that in air for the same power source. Equation 1 can be used to compare the intensity

$$\frac{\text{Iw}}{\text{Ia}} = \frac{(p^2/\rho c) \text{ for water}}{(p^2/\rho c) \text{ for air}}$$

using   c for water = 1000 x 1500 kg/(m² s)
and   c for air = 1.209 x 343 kg/(m² s)
and   Iw = Ia

$$p_w/p_a = 60$$

For the same intensity of sound, the pressure in water is 60 times as great as in air, equivalent to a difference of 36 dB.

Wind Effects

It is an observed experience that sound is carried by the wind, i.e. sound levels are higher downwind than upwind. Because air is viscous, a boundary layer is formed near the ground, in which the relative velocity at ground level is zero. The air velocity increases with height until it reaches that of the main air mass. Figure 1 demonstrates this behaviour; it can be seen that downwind the sound rays are closer together and therefore the intensity is increased. Upstream the reverse effect applies, with a shadow region where the attenuation can be as high as 30 dB. The magnitude of the attenuation due to wind cannot be easily predicted. When measuring sound, the best advice is always to wait for still air conditions. It cannot be assumed, when determining the total sound power of a source, that the increase of the downwind intensity is balanced by an equal reduction in the upstream intensity.

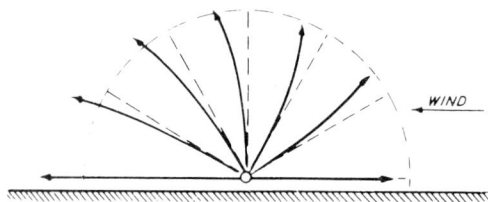

WIND   Figure 1

Directivity

Most practical sources of sound are directional, in that they radiate sound more in one direction than another. The actual pattern may be quite complex, and can be determined by performing an extensive survey of the sound pressure level in a 3-dimensional array around the equipment. With these measurements, contours of equal sound pressure level in the planes of interest can be made. Figure 2 shows a simplified plot of this kind. It is not only the character of the sound source which can cause directional effects, but also the nature of any reflective surface in the vicinity of the source. The presence of a reflective surface, such as a concrete base or a wall, will give rise to a directional quality to the sound even though it may be otherwise symmetrical.

The directivity of a source is characterised by its Directivity Factor, Q or Directivity Index, DI. They are of use in describing a highly directional source and are defined by:

$$Q = I_n/I_{av}$$
$$DI \text{ (dB)} = 10 \log (I_n/I_{av}) = 20 \log (p_n/p_{av}) = L_{pn} - L_{pav}$$

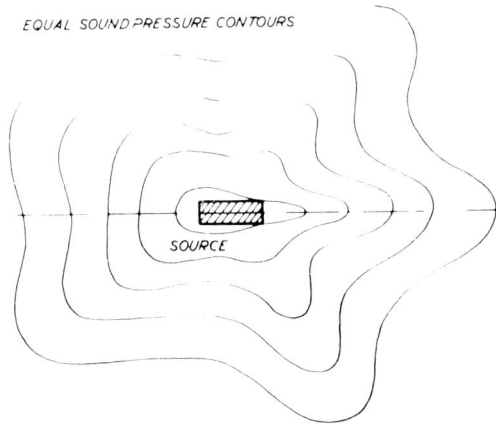

EQUAL SOUND PRESSURE CONTOURS

SOURCE

Figure 2

where the suffix n refers to values in the n th direction and the suffix av refers to the average over the whole of the measurement sphere. In general DI may vary with direction, distance from source and frequency. When only one value of DI is mentioned in a report, it can be assumed to be the maximum in any direction.

When measurements are made over a refective surface – a typical situation in many applications of practical interest, the intensity of the sound field is on average twice what would be expected for spherical radiation, since the total sound is now distributed over the surface of a hemisphere. In this case

$$DI \text{ (hemisphere)} = L_{pn} - L_{pav} + 3$$

3 is the decibel addition to take account of the doubling of the intensity.

Several other cases are of interest. Table 3 summarises the DI caused by surface effects for a uniform sound source positioned in a rectangular room.

Sound Pressure Level

The use of the word 'level' always indicates that the value is in decibels. The sound pressure level is defined by:

$$L_p = 10 \log (p^2/p^2_{ref}) \text{ dB}$$
$$= 20 \log (p/p_{ref}) \text{ dB} \qquad \qquad ...(3)$$

p is the rms value of the sound pressure at a particular point and $p_{ref} = 2 \times 10^{-5}$ Pa

TABLE 3

Location of Source	DI
Centre of room	0
Centre of one wall or floor	3
At junction of two room surfaces	6
At junction of three room surfaces	9

The reference level is chosen by convention to be the rms pressure that is equal to the limit of audibility in a healthy young adult at 1 kHz.

Suppose, as an example, that a machine generates a noise field such that at 7 m from its centre, the sound pressure level is 75 dB. Using equation 3 to find the pressure:

$$75 = 20 \log \frac{p}{2 \times 10^{-5}}$$

from which p = 0.1125 Pa = 1.125 x 10^{-6} bar

To find the sound power of the source, use equation 2 which can be rewritten as

$$W = \frac{4\pi r^2 p^2}{\rho c}$$

for ρ = 1.209 kg/m^3 and c = 343 m/s

W = 0.0188 watt

This calculation demonstrates that even for high sound pressure levels both the rms pressure and the magnitude of the acoustic power are small.

The total attenuation with distance is the sum of the losses due to the application of the inverse square law and the values obtained from Table 2.

Sound Power Level

Sound power refers to the acoustic power of the source. Confusion sometimes arises when the difference between sound power and sound pressure is not made clear; this confusion arises because both values are expressed in decibels. Sound pressure level should be looked upon as the value to be used when considering the effect of the sound on an observer. Sound power level is the value to be used when studying the total sound generated by a piece of equipment. Sound pressure is the value measured by a microphone. Sound power cannot be determined from a single measurement – it can only be derived by measuring the sound pressure at a number of points around the equipment (arranged for example on the surface of an imaginary sphere), and calculating the total sound power of the source, taking into account the area of the bounding surface.

One frequently sees that the noise output of an item of equipment is quoted as having a sound pressure level of X dB. This is a meaningless statement unless the distance and direction from the source and the kind of environment is also given.

Sound Power Level is given by the formula:

$$L_w = 10 \log (P/P_0)$$

P is the sound power of the source and P_0 is the reference sound power, which conventionally is given the value 1 pW (10^{-12} W).

The relationship between L_p and L_w is found from:

$$L_w = L_p + 10 \log (S/S_0)$$

S_0 is the reference area of 1 m^2 and S is the area of the measuring surface.

This relationship is only approximately correct, in that it depends upon a source of sound power of 1 pW, giving rise to a sound pressure of 2×10^{-5} Pa on the surface of a sphere of surface area 1 m². Using the values for and ρ quoted above, the actual sound pressure produced by a 1 pW source would be 2.036×10^{-5} Pa. The difference is negligible in this example, but care should be taken when either the velocity or density differ markedly from those assumed.

In terms of the distance r from the source the relationship between L_p and L_W is

$L_W = L_p + 20 \log r + 11$ in a free field or
$L_W = L_p + 20 \log r + 8$ on a reflecting plane

Sound Intensity Level

This quantity is also sometimes required. It represents the sound power transmitted per unit area in a given direction and is also expressed in decibels. It is given by:

$L_I = 10 \log(I/I_0)$

I is the sound intensity of the field and I_0 is the reference intensity of 1 pW/m².

Sound intensity is a vector quantity (it requires a direction to be specified). In a diffuse field such as might be generated by a source located in a reverberant chamber, the flow of energy is equal in all directions and the intensity loses its directional quality.

The distinction between pressure, power and intensity must always be kept in mind. Pressure is the easiest to measure – merely requiring a non-directional microphone. Power is a derived quantity, requiring several pressure measurements which have then to be combined. Intensity is a property of the sound at a particular point in space and in principle can be determined by a single measuring device, but such devices require careful design; this subject is dealt with in greater detail under Measurement Techniques. For most purposes, sound measurements are limited to meters giving sound pressure levels.

Typical Noise Levels

To give an indication of the sound pressure levels that might be encountered in typical situations, Table 4 has been prepared. There are wide variations within the ranges, so this table must be used with great care.

Table 5 gives the sound power levels of some typical noise sources. Once more, wide variations are possible between different examples of the sources quoted.

TABLE 4 – DECIBEL RATING AND EQUIVALENT ENERGY OF COMMON SOUNDS

Sound Pressure Level dB	Condition	Relative Energy Intensity	Sound Pressure	General Class
120	Threshold of pain	1 000 000 000 000	2 000 000	
110	Thunder, Artillery	100 000 000 000		
100	Steel riveter at 4.5m	10 000 000 000	200 000	Deafening
90	Noisy Factory	1 000 000 000		
80	Tube train (open window)	100 000 000	20 000	
70	Average factory	10 000 000		Distracting
60	Loud conversation	1 000 000	2000	
50	Average office	100 000	200	Range of Conversation
40	Average living room	10 000	200	
30	Private office	1 000		Extreme quiet
20	Whisper	100	20	
10	Soundproof room	10		
0	Threshold of audibility	1	2	Soundproof chambers

TABLE 5 – SOUND POWER OUTPUT OF SOME TYPICAL NOISE SOURCES

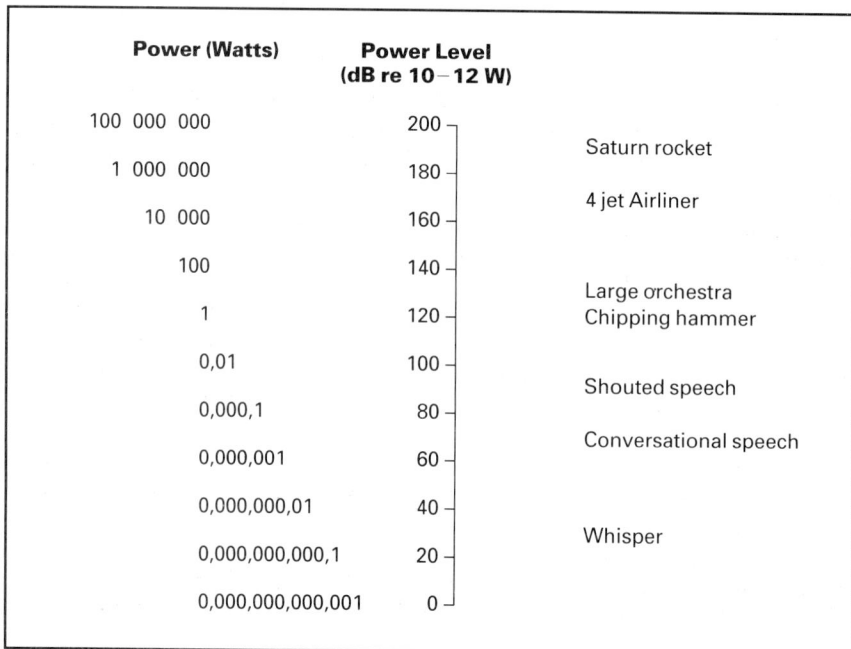

Power (Watts)	Power Level (dB re $10-12$ W)	
100 000 000	200	
		Saturn rocket
1 000 000	180	
		4 jet Airliner
10 000	160	
100	140	
		Large orchestra
1	120	Chipping hammer
0,01	100	
		Shouted speech
0,000,1	80	
		Conversational speech
0,000,001	60	
0,000,000,01	40	
		Whisper
0,000,000,000,1	20	
0,000,000,000,001	0	

Chapter 3
Measurement of Sound

When determining the sound pressure level to which a worker is exposed, measurements are taken in the actual circumstances at the time, but for analysis purposes it is necessary to know the environmental conditions (the kind of room or open air) in which the measurements are taken. A number of standard conditions can be defined:

Free field in a completely open space where there are no sound reflections or other modifying factors present.

Reverberant field where the sound energy at any point is the sum of that directly radiated from the source and sound levels reflected from adjacent surfaces. In a fully reverberant field all the sound energy striking the bounding surfaces is reflected without loss.

Semi-reverberant field where the prevailing conditions may be anywhere between free field and reverberant field conditions.

Anechoic field where all the sound measured comes directly from the source, equivalent to taking measurements in an acoustically dead room where all incident sound energy striking the walls is fully absorbed.

Semi-anechoic field where the source is mounted above a hard reflective surface.

Advantages and disadvantages of these different conditions are summarised in Table 1.

As a general rule measurements taken outdoors can be considered to approximate to free field conditions and show reasonable agreement with anechoic measurements, provided there are no reflective surfaces nearby.

Measurements taken *indoors* can be considered as approximating to diffuse field conditions and show reasonable agreement with reverberant field measurement. Average rooms, without sound deadening, can approximate closely to diffuse field conditions where the flow of sound energy in all directions is more or less equal. On the other hand, those with sound deadening treatment or with inherent sound deadening characteristics could approximate more closely to anechoic or free field conditions. Reverberant field measurements may provide a suitable approximation for general indoor use when the directional characteristics of the sound source are not important.

With both outdoor and indoor measurements considerable difference may be experienced between measured levels and predicted levels (assessed on the basis of anechoic or reverberant field measurements respectively), and between individual measurements taken under different site or enclosure conditions. This makes it difficult – or even impossible – to quote a specific noise level for any particular source as such a figure is unlikely to be valid when assessing the same source under a variety of different sets of practical conditions. Direct measurement under the

TABLE 1 – COMPARISON OF MEASUREMENT CONDITIONS

Condition	Advantages	Disadvantages
Free field	simplest technique simple distance relationship in far-fielded zone provides a basic reference level (for comparison or correction) can contain compound and directivity information	measurement can be dependent on position measurement is frequency dependent in near-field zone measurements may be modified by atmosphere true free field conditions may not hold in practice even in far-field zone
Reverberant field	good general standard for indoor measurement with hard (reflective) walls, floor and ceiling	readings may be modified by standing waves no directivity information
Semi-reverberant field	good general standard for measure ment in rooms or buildings	room characteristics must be correction to free field conditions is required. sound pressure has no simple relationship to distance from source
Anechoic field	true free field measurement	requires the use of a special anechoic chamber or anechoic room
Semi-anechoic field	free field over a reflective ground surface for equipment testing	requires a special chamber as above with a hard floor

actual conditions involved is the only reliable method of obtaining a true objective reading in semi-reverberant field. Measurements of the same source under free field conditions are not valueless, however, as these can serve as a very useful indication of sound reduction possibilities. Rendered in terms of the sound power level of the source, such data can be used to predict average sound pressure levels in a semi-reverberant field.

As a general rule, measurements taken outdoors in still air can be considered as approximating to free-field conditions, providing there are no reflective surfaces nearby. A hard concrete or asphalt base can be considered to be fully reflective; tests on equipment frequently have to be performed over such a surface.

Anechoic Chamber

No room can be entirely free of echoes, but it is possible to approach very closely to this condition. Special test rooms are expensive to construct and so their use is restricted to universities and research establishments; they are to be distinguished from rooms intended for seclusion of operators, which require much less in the way of special treatment. A common form of treatment to the walls is by lining them with anechoic wedges, made of foam or some similar material. A semi anechoic chamber has the walls and ceiling treated, but has a hard reflective floor.

Reverberant Chambers

Such chambers represent the other extreme, constructed so as to maximise the internal reflections. They are solidly made rooms, of concrete rather than brick, and surfaced with hard plaster.

The surfaces should have a sound absorption figure of less than 0.06. They have two main purposes: the diffuse field is ideal for determining the sound insulation properties of materials; and provided that the characteristics of the room are known, the sound power of a source can be determined.

No room can be perfectly reflective, otherwise the sound intensity would increase indefinitely. In practice what happens is that the sound increases until the absorption of the walls just equals the output of the source. The absorption characteristics can be determined:

(a)   by calculation from a knowledge of the properties of the surface

(b)   the reverberation time can be measured directly

(c)   a known reference noise source can be situated in the room and the unknown source compared with it.

This subject is dealt with in detail in the chapter on *Room Acoustics*.

Quantitative Measurements of Sound

A comprehensive description of the sound power emitted by a source would include a detailed frequency distribution spectrum over the audio range, as well as the directivity on a three-dimensional plot. If in addition the sound varied with time, that would also need to be specified. For many noise sources, the time-wise variation is not describable other than in statistical terms, so a probability function would also be required. In practice some simplification and averaging is employed in order to make the measurements more manageable.

Root mean square (rms) averaging

Since sound is a form of vibration energy, its proper measure is the energy content of the vibration, which is proportional to the (pressure)^2. When considering the effect of the pressure, it is not so much the peak pressure which is important as that average pressure which relates better to the energy content; this is known as the root mean square (rms) average. For a simple sine wave, the average pressure is of course zero, but the rms average is found from:

$$p_{rms} = \frac{1}{T}\left\{ \int_0^T p^2(t) \; dt \right\}^{0.5}$$

If $p(t) = p_{max} \sin \omega t$ and $T = 2\pi/\omega$
then $p_{rms} = 1/(2)^{0.5} p_{max} = 0.707 \, p_{max}$

So for a simple sine wave, and for any linear summation of sine waves, the rms average of the pressure signal is 0.707 x the peak pressure. Most sound level meters give rms values, except for special instruments intended for impulse noise or single events.

Octave and one–third octave bands

Some sound meters can measure the pressure in intervals as small as 1 Hz, but except where a pure tone predominates, it is sufficient to give values in octave bands or one-third octave bands, and most meters designed for noise measurements are limited to octave bands. Band values are obtained by passing the signal through a series of band-pass filters, for a standardised set of frequencies. These frequencies are given in Table 2 for the audio range.

Sound meters contain a time-averaging circuit which ensures that the readings are not affected by fluctuations that occur more rapidly than the characteristic value of the time constant - for a slow response this is about 1 second, and for a fast response about 0.125 seconds.

TABLE 2 ONE-THIRD OCTAVE AND OCTAVE PASSBANDS

Centre Frequency	Third Octave Passband	Octave Passband
16	14.1 – 17.8	11.2 – 22.4
20	17.8 – 22.4	
25	22.4 – 28.2	
31.5	28.2 – 35.5	22.4 – 44.7
40	35.5 – 44.7	
50	44.7 – 56.2	
63	56.2 – 70.8	44.7 – 89.1
80	70.8 – 89.1	
100	89.1 – 112	
125	112 – 141	89.1 – 178
160	141 – 178	
200	178 – 224	
250	224 – 282	178 – 355
315	282 – 355	
400	355 – 447	
500	447 – 562	355 – 708
630	562 – 708	
800	708 – 891	
1k	891 – 1.12k	708 – 1.41k
1.25k	1.12k – 1.41k	
1.6k	1.41k – 1.78k	
2k	1.78k – 2.24k	1.41k – 2.82k
2.5k	2.24k – 2.82k	
3.15k	2.82k – 3.55k	
4k	3.55k – 4.47k	2.82k – 5.62k
5k	4.47k – 5.62k	
6.3k	5.62k – 7.08k	
8k	7.08k – 8.91k	5.62k – 11.2k
10k	8.91k – 11.2k	
12.5k	11.2k – 14.1k	

Many noise sources, such as automobiles or power tools, are characterised by a number of single frequency tones plus a broad spectrum noise. These sources can cause odd behaviour in sound meters, particularly when the single frequency occurs near to the cut-off frequency of one of the octave bands, so while the use of octave or third-octave bands can be valuable, care must always be taken.

Weighting Scales

The human ear responds not only in a logarithmic way to fluctuations in sound pressure, but also differentially throughout the audio frequency spectrum; subjectively, high frequencies appear much louder than low frequencies. Figure 1 demonstrates this by reproducing the results of a large number of tests on healthy young subjects; the curves represent equal loudness contours for pure tones.

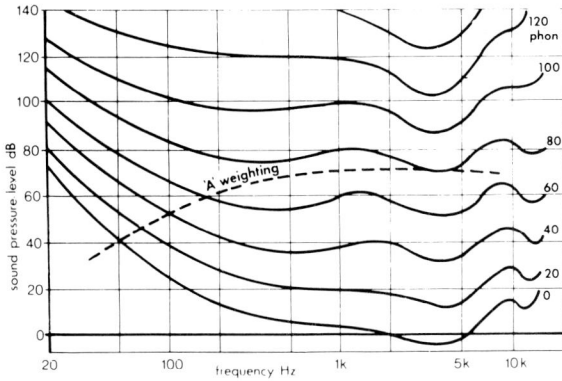

Figure 1 A-weighting compared with the
ear's response.

Sound level meters incorporate weighting filters which compensate for this subjective response. Although the weighting filters do not exactly follow the undulations of Figure 1, the approximation is considered adequate. It was originally intended that different weighting networks should apply to different levels, to match the differences exhibited in the Figure, but now for most purposes only one, the A weighting, is used. Strictly speeeaking, A-weighting is intended for levels below 55 dB, B-weighting between 55 dB and 85 and C-weighting above 85 dB. D-weighting follows a contour of perceived noisiness and was originally intended for aircraft noise measurements.

Figure 2 and Table 3 give the weighting in decibels to be added to the individual sound pressure level readings to obtain the final weighted result. Note that an A (or any other) weighting is a single value, found by adding together the separate frequency values. It is not necessary to actually perform this calculation, most modern sound level meters do it electronically.

C-weighting is likely to be used in the future for peak sound pressures. A European community directive (89/392/EEC) will make it mandatory for C-weighted information to be supplied for all

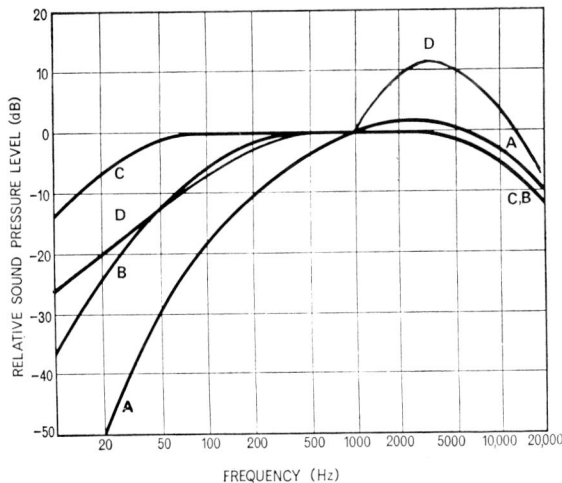

Figure 2 International standard A, B, C and
D weighting curves for sound levelmetres.

TABLE 3 – A, B, C AND D WEIGHTINGS COMPARED
(Weightings in decibels)

Weighting	Frequency Hz									
	16	31.5	63	125	250	500	1000	2000	4000	8000
A	−56.7	−39.4	−26.2	−16.1	−8.6	−3.2	0	1.2	1.0	−1.1
B	−28.5	−17.1	−9.3	−4.2	−1.3	−0.3	0	−0.1	−0.7	−2.9
C	−8.5	−3.0	−0.8	−0.2	0	0	0	−0.2	−0.8	−3.0
D	−22.5	−16.5	−11.0	−6.0	−2.0	0	0	8.0	11.0	−6.0

equipment where there is likely to be a sound pressure level in excess of 120dB. This will be in force by the end of 1992.

The weighted level is only of use in assessing the effect on the human ear. For investigating the sources of noise in machines, a full spectrum is required. The conversion of A-weighted sound pressure to A-weighted sound power is often done, particularly for the purpose of rating machines to meet legislative standards, but the theoretical justification for such a calculation is dubious.

The existence of single frequency tones may give rise to a subjective assessment differing from that of a meter giving an A-weighted value, so once again care should be taken.

An Example of A-weighting calculation

The following table gives specimen readings in octaves of a noise source and the calculations to derive A-weighting:

Frequency (Hz)	Reading dB(A)	A-Weighting Corrected dB(A)	Correction Values dB(A)	Antilog of Corrected Values
31.5	81	−39.4	42	$10^4 \times 1.58$
63	76	−26.2	50	$10^5 \times 1$
125	74	−16.1	58	$10^5 \times 6.31$
250	68	−8.6	59	$10^5 \times 7.94$
500	62	−3.2	59	$10^5 \times 7.94$
1000	68	0	68	$10^6 \times 6.31$
2000	71	1.2	72.2	$10^7 \times 1.58$
4000	69	1	70	$10^7 \times 1$
8000	70	−1.1	68.9	$10^6 \times 9.5$
		Total		$10^7 \times 4.39$
		$10 \times$ log(total)		76.4

The weighted value is therefore 76 dB(A).

It is usual to ignore fractional parts of a decibel.

Impulse Noise

Impulse noises as generated by guns, cartridge operated tools, piling hammers and explosions may not be sufficiently well measured by any of the ways discussed above. Subjectively, a single event of this kind may not seem to be as loud as if a sound of the same magnitude were repetitive; the shorter the duration of the sound the less loud it appears to be, although it may be more physically damaging. The peak value of the pressure is often required, particularly where noise exposure regulations specify a particular action which has to be taken to protect hearing. Special meters are available for this purpose. The peak is often quoted as an actual pressure (e.g. 200 Pa, rather than a level of 140 dB). Even if a conventional meter is set to 'fast' response it may still not be adequate for measuring impulse noise; a meter with a rise time of less than 50 microseconds as well as a peak hold facility may be needed.

Single Event Noise

All sound occurrences have a time history associated with them. The characteristic time varies according to the application — a week, for example in a factory where that is the period of a batch process, or a few seconds for an aircraft flyover or a vehicle drive-by. Some kind of integration of the total sound level may be required, particularly when evaluating the disturbance in the community or assessing the risk to a worker. Special meters are available for this purpose, where usually the A-weighted, or some other function is sampled and integrated over the appropriate period. These subjects are dealt with in greater detail in later chapters.

Sound Power Determination

A variety of techniques have been devised to convert sound pressure level readings into sound power levels, i.e. to calculate the sound power output of a noise source, from the information recorded by a sound meter. Most of these techniques have been brought together in a set of British and International Standards. Table 4 summarises these current standards; and Table 5 quotes the likely accuracy for the individual standards. They can be used as they stand for determination of sound power levels, but they are probably more often used as source documents for type testing, for which purpose refer to BS 4196: Part 0 (ISO 3740). Depending on the test environment (type of room used or outdoors) a correction may first have to be made to the measured sound pressure levels. Any such correction has to be calculated or measured, and its magnitude is heavily dependent on the refectivity or absorptive character of the place where measurements are taken. In general the greater the correction, the less reliable is the final value of the sound power.

The first correction which has always to be considered is that due to the background noise (caused by noise sources other than the one being studied). Readings are first taken with the machine turned off and if these prove to be more than about 10 dB less than with the machine operating they have to be subtracted logarithmically. Other corrections which may be necessary are described in the appropriate standard, or refer to the next chapter on *Room Acoustics*.

In deciding the appropriateness of a particular standard, these factors have to be considered:

Size of noise source

Test environment

Character of source

Frequency range of interest

Acoustic data required (sound power, directivity, temporal pattern)

Accuracy required

TABLE 4 – STANDARDS SPECIFYING VARIOUS METHODS FOR DETERMINING THE SOUND POWER LEVELS OF MACHINES AND EQUIPMENT

British Standard No.	International Standard No.	Classification of method	Test environment	Volume of source	Character of noise	Sound power levels obtainable	Optional information available
4196: Part 1	3741	Precision	Reverberation room meeting specified requirements		Steady, broad-band		
4196: Part 2	3742			Preferably less than 1% of test room volume	Steady, discrete-frequency or narrow-band	In one-third octave or octave bands	A-weighted sound power level
4196: Part 3	3743	Engineering	Special reverberation test room		Steady, broad-band, narrow-band, or discrete-frequency	A-weighted and in octave bands	Other weighted sound power levels
4196: Part 4	3744	Engineering	Outdoors or in large room	Greatest dimension less than 15 m	Any		Directivity information and sound pressure levels as a function of time; other weighted sound power levels
4196 Part 5	3745	Precision	Anechoic or semi-anechoic room	Preferably less than 0.5% of test room volume	Any	A-weighted and in one-third octave or octave bands	
4196 Part 6	3746	Survey	No special test environment	No restrictions: limited only by available test environment	Any	A-weighted	Sound pressure levels as a function of time; other weighted sound power levels
4196: Part 7	3747	Survey	No special test environment; source under test not movable	No restrictions	Steady, broad-band narrow-band, or discrete frequency	A-weighted	Sound power levels in octave bands
	3748 (Draft)	Engineering	Essentially free-field over a reflecting plane (indoors or outdoors)	Less than 1 m³	Any	A-weighted Sound power levels in octave bands	Sound pressure levels as a function of time; other weighted sound power levels

TABLE 5 – UNCERTAINTY IN DETERMINING SOUND POWER LEVELS, EXPRESSED AT THE LARGEST VALUE OF THE STANDARD DEVIATION IN DECIBELS

International Standard No.	Octave bands (Hz) 125 / 1/3 Octave bands (Hz) 100 to 160	250 / 200 to 315	500 / 400 to 630	1000 to 4000 / 800 to 5 000	8000 / 6 300 to 10 000	A-weighting
3741 3742	3	2	1,5		3	-
3743	5	3	2		3	2
3744	3	2		1,5	2,5	2
3745 (Anechoic room)	1	1		0,5	1	-
3745 (Semi-anechoic room)	1,5	1,5		1	1,5	-
3746	-	-	-	-	-	5
3747	-	-	-	-	-	5

The standards will repay careful study, but some general indications of their use are given here:

BS 4196: Part 1 is for broad–band noise, where the precision of a reverberation room is adequate. The qualities of the room are specified by volume (200 m³ or greater) and method of construction. Three microphone locations are used. Sound power is determined either by a direct method or a comparison method. The direct method requires corrections for room size, reverberation time, surface area, frequency of interest and barometric pressure. The comparison method requires the use of a reference sound source; no corrections for reverberation time etc., are then necessary. It is not possible with this or the next two standards to obtain information on the directivity of the source or the variation of the noise with time. It is not suitable for impact noise measurement.

BS 4196: Part 2 is to be used where the noise is characterised by the presence of discrete frequencies or narrow-band sources. The test conditions are the same as for Part 1 except that more microphone locations are required (up to 30) if high frequencies are of interest.

BS 4196: Part 3 is for use when a lower accuracy is adequate. The reverberation room design is specified but to a lower standard than in Parts 1 and 2. The volume can be as small as 70 m³ but should preferably be larger. Either a direct method (requiring corrections for reverberation time and room volume) or a comparison method may be used.

BS 4196: Part 4 is a useful standard for many engineering purposes. Sound pressure readings are taken on an imaginary measuring surface surrounding the noise source which can be a hemisphere (for equipment sited on a reflective surface) or a sphere or a rectangular parallelopiped. If testing is done in a room, an environmental correction for volume and reverberation time is to be applied.

BS 4196: Part 5 is a precision method for use in an anechoic or semi-anechoic room. It is the most accurate method. Provided that the test environment is as specified, no corrections need be applied.

BS 4196: Parts 6 and 7 are survey methods useful for obtaining weighted levels. Part 6 relies on direct measurement with corrections. Part 7 requires a reference sound source. Although no special test environment is specified, a correction may have to be made if testing is performed indoors. If this is large, the test may be vitiated.

ISO 3748, although it appears in Table 4, is at present only as draft and may not mature into a full standard.

Two-surface Method of Sound Power Determination

A method of sound power determination of equipment *in situ*, which has not so far become the subject of an International Standard, although it is favoured by workers in USA, is the two-surface technique. It is claimed to be particularly suitable when taking measurements on large installed machinery which cannot easily be moved, providing reasonable accuracy without having to go through the difficulties inherent in correcting for reverberation time, which can be particularly difficult if not impossible in the presence of high levels of background noise. The machine acts as its own reference sound source. The only equipment required is a sound level meter. One further claim made for the method is that it is applicable even when the sound pressure level of the source is lower than that of the ambient noise.

Readings are taken over two imaginary measurement surfaces, which can conveniently be rectangular parallelopipeds or some simple conforming surface. A reference surface is first defined; this is a parallelopiped just enclosing the machine. The first measurement surface, area S_1, is situated 1 m away from the reference surface. The second measurement surface, area S_2, is located at some greater distance, totally enclosing the first measurement surface; it should be as large as possible, without enclosing any other sources.

The procedure involves the taking of readings of octave band sound pressure levels over each measurement surface and determining the averages. These readings are then corrected if necessary for background noise.

The following sound pressure levels are defined:

L_p = average octave band levels over the area S_1 corrected for background and environment

L_{p1} = average octave band levels over the area S_2 corrected for background only

C = environment correction

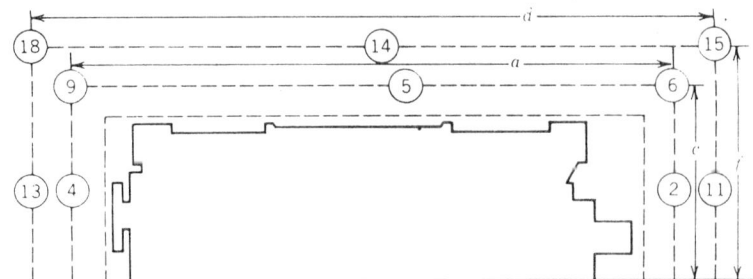

Figure 3 Plan view of Two-surface method of Sound Power Determination microphone locations

(a)

L_{p2} = average octave band levels over the area S_2 corrected for background only
Then

$$Lp = Lp_1 - C$$

where

$$C = 10 \log \left\{ \left[\frac{K}{K-1} \right] \quad \left[1 - \frac{S_1}{S_2} \right] \right\}$$

and $K = 10^{0.1(Lp1 - Lp2)}$

The sound power level of the machine is then given by

$$Lw = Lp + 10 \log \frac{(S_1)}{S_0}$$

$$S_0 = 1 \, m^2$$

C can be calculated directly or found from a set of curves such as Figure 4.

Sound Power Determination by Sound Intensity Measurements

The standard methods of sound power determination described above require a knowledge of the acoustic properties of the environment – either free-field conditions or the reverberation time of the test room. There is a technique which allows determination of the sound power without having to make any assumptions about the sound field; this technique involves the measurement of sound intensity. It will be recalled that intensity is a vector quantity, i.e. it requires both direction and magnitude to be fully defined. In Figure 5, the sound source is radiating energy. All this energy must pass through an area enclosing the source. Intensity is power per unit area so if we

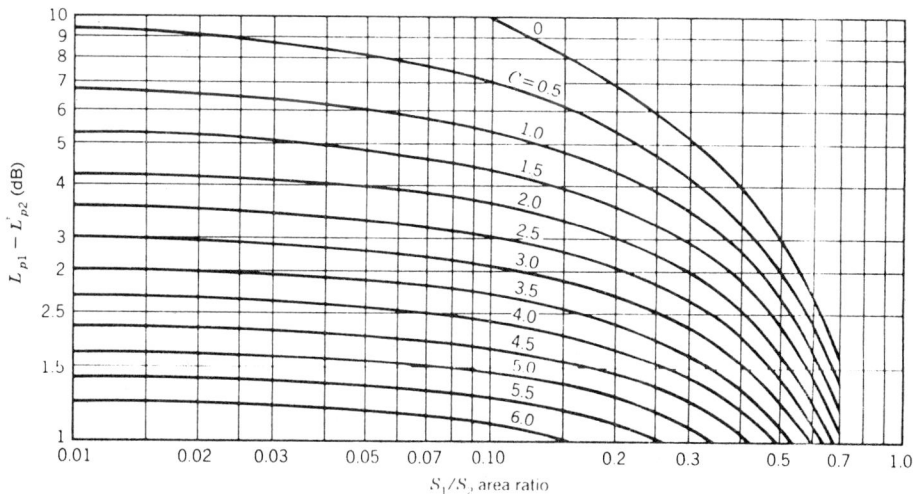

Figure 4 $S_1 S_2$ area ratio

can measure the intensity over the whole of the area then multiply its space average value by the area, the sound power of the source can be readily determined. The intensity measurements have to be taken in a direction normal to the measurement surface, and to do so requires a special kind of measurement device known as an intensity probe. Sound intensity requires the determination of both pressure and particle velocity.

Intensity = Pressure × Particle velocity

$$= \frac{\text{Force}}{\text{Area}} \times \frac{\text{Distance}}{\text{Time}} = \frac{\text{Energy}}{\text{Area} \times \text{Time}} = \frac{\text{Power}}{\text{Area}}$$

Pressure can be measured by an ordinary microphone without difficulty; particle velocity, being a vector quantity requires more care. Particle velocity can be related to the pressure gradient, i.e. the rate at which the instantaneous pressure varies with distance. The relationship is known as Euler's equation, which is a form of Newton's Second Law applied to fluids. For a fuller discussion of the mathematics behind this, it is best to refer to a specialised text, but it is sufficient to know at this stage that the intensity can be measured by means of two microphones separated by a solid spacer, built into a probe as shown in Figure 6. The signals from the two microphones require to be analysed in a intensity analyser, which is capable of giving the intensity in octave bands, 1/3 octave or narrow bands. The two microphones in the intensity probe have to be carefully matched in magnitude and phase and regularly calibrated, but aside from that the technique is easy to use.

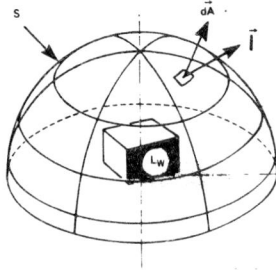

Figure 5 Calculation of sound power from Sound Intensity Measurements. S is the total area; dA and I are the incremental area and intensity respectively.

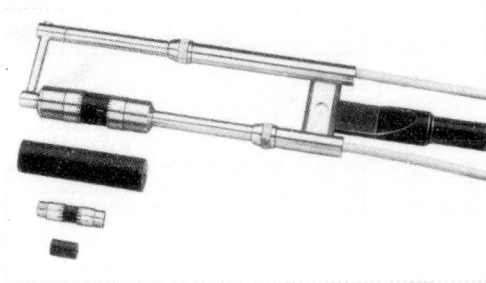

Figure 6 Sound Intensity Probe showing two microphones separated by a 12 mm spacer, suitable up to a frequency of 5 kHz. As an alternative a 6 mm spacer is suitable up to 10 kHz.

The advantages of intensity measurement as a means of sound power determination are:

Accurate results in the presence of high background noise – sound power can be measured to an accuracy of 1 dB even when the source produces a sound pressure as much as 10 dB lower than the background.

No particular acoustic environment is needed, eliminating the need for anechoic chambers etc.

Where a machine generates noise from a number of separate sources, each can be assessed. The contributions from individual components of an engine for example can be measured.

In building acoustics only one reverberation room is required.

There are disadvantages which may make the technique unsuitable in some applications:

The measurements have to be taken over the whole surface, which may require either a large number of probes or an automatic scanning technique.

If the background noise is at a high level it has to be constant over the scanning time.

Any particular probe has a low-frequency and a high frequency measurement limit, which depends on the microphone spacing – a large spacer is needed for low frequencies and a small spacer for high frequencies.

Probably the greatest disadvantage is that there are as yet no internationally adopted standards for the technique, so it tends to be limited to research laboratories. This is likely to change in the near future.

For further information on this technique refer to the chapter on *Measurement Techniques*.

Chapter 4
Noise Scales and Ratings

A large number of Noise Scales and Ratings have been introduced over the years in an attempt to give a qualitative assessment to the noise experienced by particular people in particular environments. The simplest descriptor of the character of a noise is a weighted value such as dB(A) for example, which is an instantaneous reading approximating to the response of the human ear. (See chapter on *Measurement*)

When assessing community response, the instantaneous weighted value has severe limitations: it cannot take account of the duration of the noise, whether it occurs during day or night, or the type of neighbourhood (e.g. a residential area, or near to a a hospital or school). The annoyance aspect of the noise is particularly important when deciding on the location of airports, motorways and factories. The community noise level is not usually so large as to affect the health of individuals, but it is nevertheless of great importance. In a factory or construction site, on the other hand, the health of the workers can be seriously affected in a measurable way, through permanent loss of hearing.

It would clearly be of benefit to be able to quote a single measure of annoyance or injury, particularly when legislation is under consideration. The task of producing such a measure is formidable and so it is not surprising that there is no single value available but rather a variety, each adapted to a specific purpose.

The A-weighted sound pressure level, Lp, forms the basic unit in most of the following measures. There may be instances in which a weighting other than A needs to be used, in which case it must be specified. A-weighting is to be assumed unless otherwise stated.

L_{eq} – Equivalent Continuous Sound Level

L_{eq} is one of the the simplest of the various derived measures originally developed for assessing environmental noise. It is the A-weighted energy of the sound level averaged over the specified measurement period. It can be defined as the continuous noise which would have the same acoustic power as the real measured noise over the same period. It has found much favour both as a means of assessing community noise as well as estimating hearing damage.

In most real situations where the sound level varies continuously, the use of a simple meter would give only a few spot values, so a special meter capable of integration over an extended time period has to be used.

L_{eq} can be defined mathematically as:

$$L_{eq} = 10 \log \left\{ 1/T \int_0^T (p_A(t)/p_0)^2 \, dt \right\}$$

where T is the total time, $p_A(t)$ is the instantaneous value of the sound pressure and p_0 is the reference pressure. In this form it is suitable for manipulation in a sound level meter.

If the overall sound during the time T can be adequately represented by a limited number of discrete levels, then

$$L_{eq} = 10 \log \frac{t_1 \log^{-1} \frac{L_1}{10} + t_2 \log^{-1} \frac{L_2}{10} + \ldots + t_n \log^{-1} \frac{L_n}{10}}{T}$$

where $L_1 \ldots$ etc are the A-weighted sound pressure levels and $t_1 \ldots$ etc are their durations. In this form it is suitable for manual manipulation.

Sound level meters are available which can integrate the A-weighted sound pressure level over periods of many hours (often limited only by their battery life). Noise dose meters, which measure the acumulated noise over the working day, also give a reading of L_{eq}.

L_E – Sound Exposure Level

This can be looked upon as an L_{eq}, normalised to a period of one second. In a noise environment which consists of a number different types of noise events, it is useful to be able to reduce each of them to the same basis of comparability. L_E is equivalent to L_{AX}, the Single Event Noise Exposure Level. It is defined as:

$$L_E = 10 \log \left\{ \frac{1}{T_o} \int_0^T (p_A(t)/p_0)^2 \, dt \right\}$$

T is the actual time duration of the event, T_0 is the reference time of 1 second. Usually the integration need only be taken over the period during which the level remains within 10 dB of the maximum.

There are two uses of this measure: the exposure level of any particular event or period can be assessed in relation to other events or periods; and where the noise environment consists of a number of sound exposure levels, the equivalent L_{eq} can be found by taking their logarithmic sum over the total time period.

The sound exposure level is related to the equivalent continuous sound level as follows:

$$L_E = L_{eq} + 10 \log (T/T_0)$$

$L_{EP,d}$ – Daily Personal Noise Exposure

This can be regarded as the total exposure of a worker to noise during a working day, taking account of the average noise levels in the working areas and the times spent in those areas. It takes no account of any hearing protection which may be worn. Its determination is an essential stage in assessing the need for protection and the degree of such protection.

It is formally defined as

$$L_{EP,d} = 10 \log \left\{ \frac{1}{T_0} \int_0^{T_e} \left[\frac{p_A(t)}{p_0} \right]^2 \, dt \right\}$$

where

T_e is the duration of a person's exposure to noise

T_0 = 8 hours

p_0 = 20 Pa and

$p_A(t)$ = the instantaneous A-weighted sound pressure (Pa) to which a person is exposed.

In measuring $p_A(t)$, readings may be taken either in the work place in the absence of the worker, or by a microphone situated adjacent to the worker's ear (conveniently attached to his helmet).

$L_{EP,w}$ – Weekly average of daily personal noise exposure

Can be used to extend the applicability of $L_{EP,d}$, by taking the logarithmic average of $L_{EP,d}$ over a working week of 5 days

L_{DN} – Day-Night Average Sound Level

This is an L_{eq} with an extra 10 dB weighting for noise occurring during the night time period from 10 p.m. to 7 a.m. It accounts for the extra annoyance caused to communities by night time noise.

L_{PN} – Perceived Noise Level

L_{PN} (also designated PNL) is a noise rating primarily used for jet aircraft flyovers, although it has been adapted for other purposes. Its origin is in a subjective assessment derived from experiments in which listeners were asked to judge between a reference sound and given sound. For practical purposes, instead of having to rely on a panel of listeners, it is now calculated as an approximation by a complex procedure to be found in BS 5727 (ISO 3981). The sound pressure level in one-third octave bands is converted to a perceived noisiness by use of tables; the summation of these bands gives the noisiness in noys, which are then converted into decibels. Refer to the standard for full details of the procedure.

L_{TPN} – Tone Corrected Perceived Noise Level

Where the noise spectrum of an aircraft flyover contains pure tones or other marked irregularities caused by propellers, compressors, turbines or fans, these features cause an increased annoyance; the L_{PN} can be corrected to account for them by a further procedure described in BS 5727. The result is called the Tone Corrected Perceived Noise Level.

L_{EPN} – Effective Perceived Noise Level

The total subjective effect of an aircraft flyover depends not only on the maximum value of L_{PN} (or L_{TPN}) but also on the variation of the noise with time, so L_{EPN} was introduced to allow for this; it involves integration over the time interval of the flyover. By definition it is the level in decibels of the time integral of the antilog of one-tenth of the Perceived Noise Level; the reference duration is 10 s.

$$L_{EPN} = 10 \log \left(\frac{1}{T_0} \int 10^{0.1 L_{TPN}} \, dt \right)$$

T_0 is the reference time of 10 s. The actual integration time is the total time of the flyover

NNI – Noise and Number Index

The index is an early attempt to evaluate the annoyance caused to residents in the vicinity of London Airport (Heathrow). It derives from a suggestion made in the Wilson Committee Report on Noise in 1963 (Cmnd 2056) after the analysis of a social survey made around the airport. The committee concluded that the average annoyance is a function both of the average L_{PN} and the number of aircraft flights. The definition of NNI is:

$$NNI = L_{PNav} + 15 \log N - 80$$

where L_{PNav} is the logarithmic average of all the N flyovers in a day. The value 80 is introduced, because it was found that there was zero annoyance for a L_{PN} less than 80. The index is purely empirical, but was sufficiently well received by the authorities as to form the basis of compensation paid to local residents for the cost of noise insulation (double glazing etc.).

Subjective Noise Assessment

Many attempts have been made to establish subjective loudness scales for sound, a number of which have been widely quoted in the past and still continue to be used in certain applications. these are based on *subjective* units of sound, i.e.

phon – a measure of 'loudness level'.

sone – a measure of 'subjective loudness'.

noy – a measure of 'perceived noise'.

The *phon* is a measurement of 'loudness level'. Loudness levels are derived as contours of sound pressure levels of simple tones over a range of frequencies, each contour representing a constant loudness. Each contour is designated by a number, representing the loudness level in phons. At a specific frequency (1000 Hz) this number is defined by the band sound pressure level in dB. The contours express the frequency dependency of subjective response.

Since these data are essentially subjective different contours can be derived from different techniques or from different sampling methods, although the ISO equal loudness contours are generally accepted.

Loudness levels may be determined from octave band analysis when values may be expressed as phons (OD); or phons (GF) for loudness levels in a free field, and phons (GD) for loudness levels in a diffuse field.

Loudness Scale

A loudness scale can be derived as a transfer function from equal loudness contours, the unit for such a scale being the sone. As an arbitrary starting point a loudness of 1 sone is equivalent to a loudness level of 40 phons. The complete relationship is then defined as:

$$\log_{10} N = 0.03 (L_N - 40)$$
where
N = loudness in sones
L_N = loudness in phons or the second pressure level of a 1000 Hz pure tone (which is assessed to be equal in loudness to the sound under comparison).

The Sone

The *sone* as a unit has the attractive feature that numerical values are substantially linear in expressing loudness levels, e.g. a reduction in sound level to one half of its original value is equivalent to halving the number of sones. Similarly, well separated components of sound are additive on the sone scale, e.g. a loudness of 51 sones from one source combined with a loudness of 52 sones from a second source gives a combined loudness of 51 + 52. In the case of two sounds of

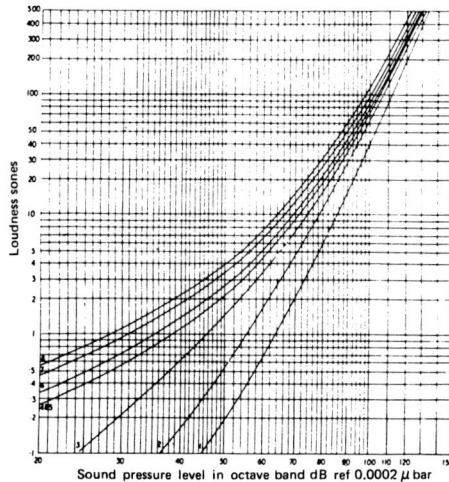

Figure 1 Octave Band Analysis

equal loudness 'S', the combined effect will be equal to 2 × S. By comparison, with loudness *level* two sources of equal loudness will give a theoretical increase of 10 *phons*.

Loudness scale values in sones can be obtained directly from octave-band analysis by reference to tables or charts (e.g. see Figure 1). In the case of charts, the procedure is as follows:

(i)   Tabulate loudness values (sone equivalent) for the measured sound pressure levels in each band.

(ii)   Find the highest value given in this tabulation.

(iii)   Multiply the loudness value of all *other* bands by 0.3.

(iv)   Determine the sum of (ii) and all the other corrected values (iii).

        This will give the total loudness of the noise in sones.

(v)   The total loudness in sones can then be converted to Loudness Level (L_N).

A similar procedure is adopted in the case of one-third band analysis, the appropriate charts being shown in Figure 2. The only difference (apart from the greater number of bands) is the correction factor value used in step (iii). This is 0.15.

Determination of loudness in this manner is applicable only to a diffuse sound field and is accurate only for steady noise with a broad band spectrum.

Loudness scale values derived from octave band analysis may be referred to as sones (OD) and sones (GF) for loudness in a free field or sones (GD) for loudness in a diffuse field.

Loudness Index (LI)

The *loudness index* (LI) is an alternative method of expressing loudness levels in the form of a series of curves established with a slope of −3 dB per octave through the 1000 Hz reference frequency. The curves are thus rendered in straight line form; although above 9000 Hz the slope changes to a constant 12 dB per octave and below a certain frequency each curve changes to a slope of −6 dB per octave – see Figure 3. The diagonal line which defines this lower frequency change point passes through the 10 dB pressure level at 1000 Hz.

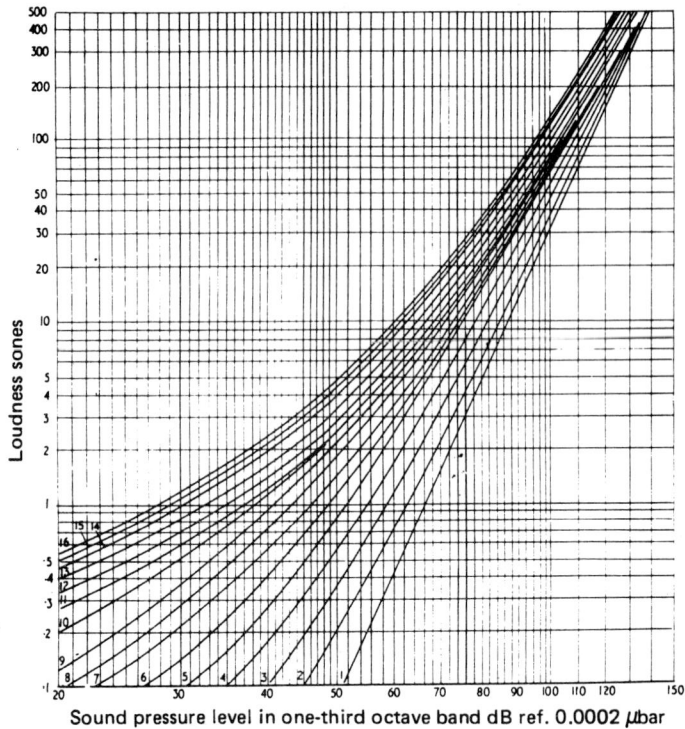

Figure 2 One-third Band Analysis

Again loudness index can be determined from octave band analysis. The procedure is:

(i)   Determine the LI for each band.

(ii)   Calculate the total loudness (in terms of LI value) from the formula–

$$LI(total) = LI_{max} + 0.3 (LI_{sum} - LI_{max})$$

where

LI_{max} is the greatest numerical value of LI found in (i)

LI_{sum} is the sum of all the values found in (i)

Noise Criteria (NC) Curves

NC curves are octave band contours, each curve being given a number numerically equal to the sound pressure level in the 1200 – 2400 Hz octave – Figure 4. Each curve has a loudness level in phons of 22 units greater than speech interference level. Basically they are 'smoothed' equal loudness contours, originated in America in the early 1950s.

NCA curves are another similar family with the main difference that they designate a loudness level in phons of 30 units above speech interference level (see Figure 5). They are somewhat more realistic than NC curves where there is a predominance of low frequency content in the sound spectrum.

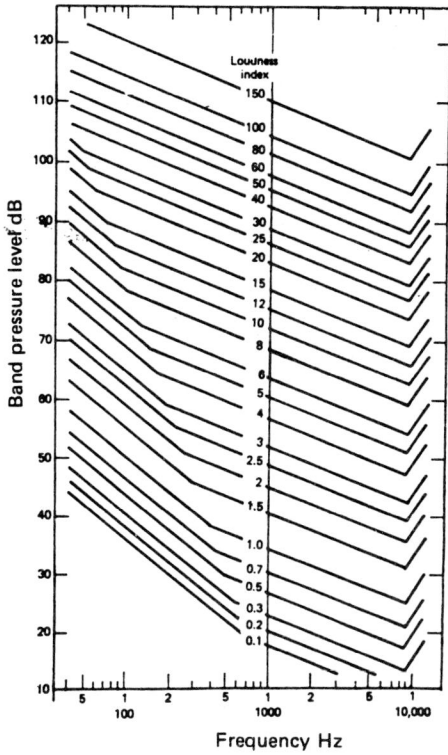

Figure 3 Loudness index curves

Figure 4 Noise Criteria Curves

Figure 5 NCA curves

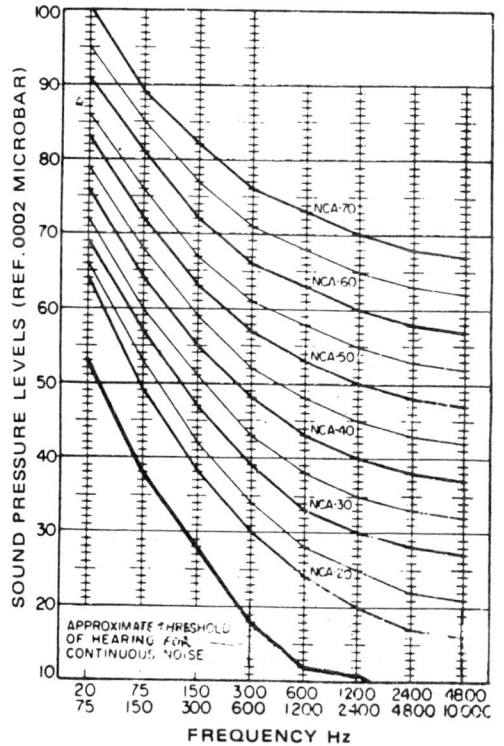

Noise Rating Curves (N or NR)

Noise rating curves are subjective data comprising a series of octave contours, similar to equal loudness curves but 'smoothed'. The number given to each curve is the sound pressure level at 1000 Hz – see Figure 6.

Noise Rating (NR) curves are similar to concept to NC curves, the numerical value of each curve being read as the sound pressure level at 1000 Hz.

The noise rating of a particular sound is established as the next number above the *highest* NR equivalent of the sound pressure level measurement in dB in any octave band. That is, after converting all band dB measurements into equivalent NR numbers, the highest value of NR appearing in the tabulation is taken and 1 added to it to determine the final figure noise rating in NR.

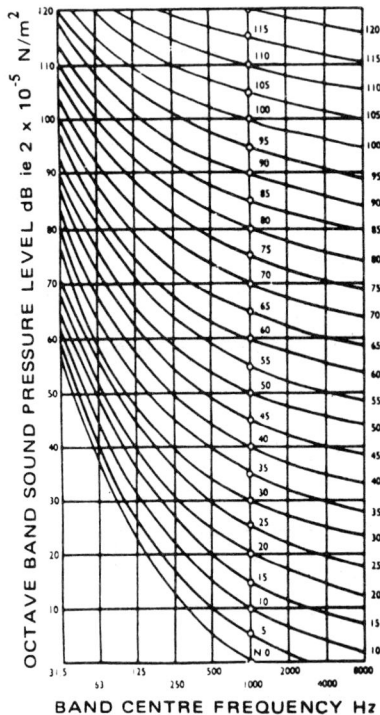

Figure 6 Noise rating curves

Chapter 5
Room Acoustics

An acoustic room is a room or enclosure intended for a specific purpose in relation to measurement of sound or for the isolation of the interior from external sources. As already mentioned in the chapter on Measurement of Sound, there are several kinds of room used for measurement purposes. These are classified in Table 1 of the chapter on *Measurement of Sound*.

There are other situations in which special acoustic rooms are required, for example in a control room where the emphasis is on the isolation of the interior from external noise and vibration. Another type of acoustic treatment is needed in recording studios and in laboratories intended for audiometry. In these there is a requirement for a degree of absorption of the interior wall surface with a lower specification than in a true anechoic chamber. These may be called 'silent' rooms. A true anechoic chamber is expensive to construct, and for that reason the use of these rooms is restricted to special purposes, where a true free-field condition is required. If the test requirements specify a free-field, it is clearly cheaper to carry out testing in the open air, but the number of days in a year when it is possible to satisfy the proper test conditions (temperature, wind speed etc.) may force the construction of an anechoic chamber.

Any room, whose acoustic properties are known or can be accurately determined, can be used for the determination of sound power of equipment; but in practice an anechoic or semi-anechoic chamber will give the most accurate results. A reverberation room is the next favoured choice. The use of a room with no special characteristics can be used for survey purposes. However such a room is likely to have odd reflections with variable characteristics, standing waves or some other unpredictable behaviour.

Design Considerations for Anechoic Chambers

A well constructed anechoic chamber must possess noise attenuating properties such that resulting ambient sound levels will not interfere with acoustic measurement. This will require exceptional treatment to the interior wall surfaces to reduce the sound reflection and may also require the use of a double wall structure, with the interior wall isolated from the exterior wall by a rubber or air spring suspension system. Depending on the quality of wall treatment, true free field conditions will exist only in the central region of the room volume; close to the boundaries, the specification of free-field may not be satisfied.

The structure of the chamber should be of concrete or steel. Brick is not to be recommended because of its unpredictability.

A properly constructed anechoic chamber should have walls, ceiling and floor possessing a sound absorption coefficient of at least 0.99 throughout the frequency range of interest (usually 125 Hz upwards) for normal incidence waves, and approaching that value for grazing angles of

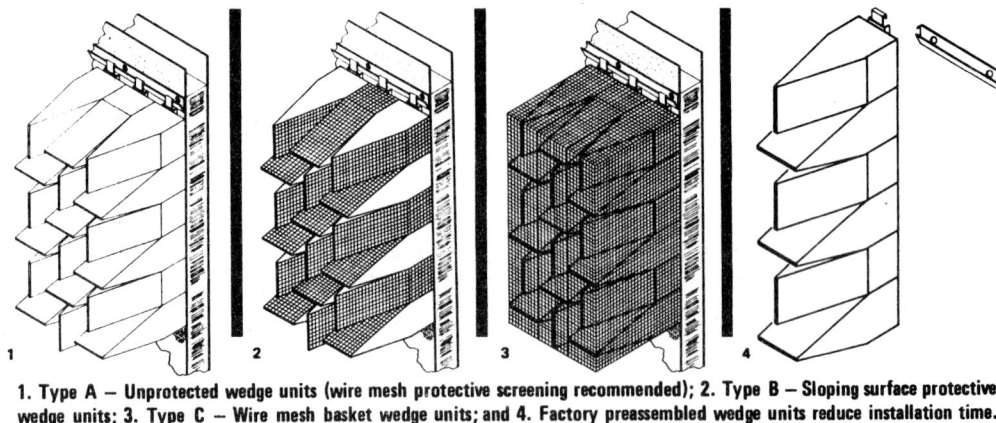

1. Type A – Unprotected wedge units (wire mesh protective screening recommended); 2. Type B – Sloping surface protective wedge units; 3. Type C – Wire mesh basket wedge units; and 4. Factory preassembled wedge units reduce installation time.

Figure 1   Types of Acoustic Wedges used in Anechoic Chambers

incidence. It is not possible to achieve this condition with a plane homogeneous layer applied to the walls. As an example, a surface treatment consisting of 30 mm thick perforated fibreboard tiles, which would be excellent treatment for an office or control room, has a maximum absorption coefficient of about 0.8 at 500 Hz, falling to 0.2 at 125 Hz.

The usual wall treatment for anechoic chambers consists of acoustic wedges made of foam or fibre glass, as illustrated in Figure 1. These wedges are in effect an approximation to a series of acoustic cones, tuned to a specific frequency. The length of the wedges is determined by the desired lowest frequency of interest, the cut–off frequency. The wavelength of that frequency is approximately 3 × (the length of the wedge). Thus for a cut-off frequency of 125 Hz, the wedge length has to be greater than about 0.9 m. To reduce the cut-off frequency to 80 Hz would increase the length to 1.4 m. The frequency for a given wedge length can be reduced by mounting them on a series of cavity resonators, consisting of air spaces behind the wedges; it is possible by this means to reduce the frequency by about 20%.

In deciding on the size and shape of a chamber, these principles should be followed:

It should have a volume at least 200 times greater than the the volume of the largest noise source to be tested.

Measurements should be taken no closer to the walls than one quarter of the wavelength of the lowest frequency encountered.

Many test codes require measurements on a spherical or hemispherical surface, so the chamber must be large enough to accommodate the largest spherical surface to be employed.

Support for the equipment under test must be provided. In a semi-anechoic chamber, the test component will be mounted on the floor, but in a fully anechoic chamber, with absorption treatment applied to the floor, it is usual to provide a grid or a spring suspended cable floor strong enough to carry the component. Special non-reflective cables are preferred to reduce any danger of interference; for heavy loading a stronger grid may be required. See Figure 2 and refer to BS 4196 (ISO 3745)

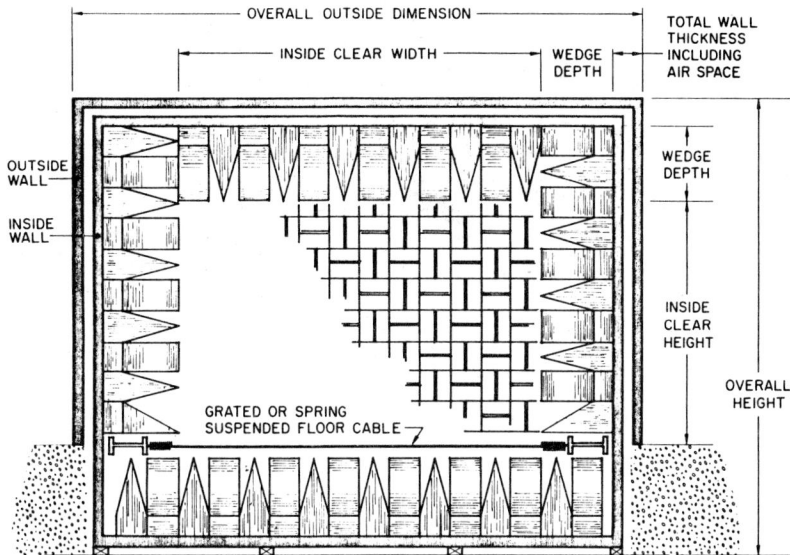

Figure 2 Construction of Anechoic Chamber

In the detailed design of an anechoic chamber, the following further points need to be considered:

Provision of ventilation and removal of exhaust from the equipment.

Lighting.

Temperature and humidity control.

Size of access doors.

Sealing of service inlet conduits.

Provision of remote control and observation.

Mounting of equipment.

The construction of anechoic chambers can be very expensive. There are companies that specialise in this area, and it is recommended that they be approached by anyone contemplating installing one.

Semi-anechoic chambers

There are many test conditions which require a room with a reflective floor surface. For construction equipment, cars, tractors and compressors, which are always used on a hard surface, a semi-anechoic chamber is preferred. A chamber designed to this condition will have the same quality of wall treatment as the fully anechoic chamber but will have a hard, solid floor with a sound absorption coefficient less than 0.06. It should be possible to vary the absorption coefficient of the floor so as to simulate concrete, tarmac, grass etc.

Reverberant chambers

These chambers are the converse of anechoic chambers, in that the interior wall surfaces are intended to be as reflective as possible, with the object of producing a uniform diffuse field in the interior. They may be used for noise testing of equipment, but are more commonly used for evaluating the sound absorbent characteristics of furnishings and other items to be found in interiors. In these rooms it is possible to determine the effect of adding chairs, curtains, acoustic screens and human beings, and to assess the advantages of various kinds of surface treatments.

When a source of sound operates in an enclosed space, the extent to which the sound level builds up and its subsequent decay when the source is switched off are governed by the characteristics of the boundary of the space. A fully reverberant room is a practical impossibility (in such a room the sound level would build up infinitely and would never decay). If the sound intensity is independent of location and has a random incidence, the field is known as diffuse. This is the ideal condition for test purposes.

A reverberant chamber has to meet certain design conditions to give satisfactory results (see for example BS 3638 (ISO 354)).

It can be made of walls which are neither rectangular or vertical, or it can be rectangular with no two dimensions in the ratio of small numbers. A commonly found set of ratios is 3:7:13.

Its volume should be at least 150 m^3 but preferably 200m^3

The longest straight line in the room (i.e. the line joining the corners furthest apart), l_{max}, is given by the inequality: $l_{max} < 1.9\ V^{1/3}$

It is generally required that sound diffusers shall be suspended from the roof; these can be reflective sheets of varying sizes and of random orientation, or they may be rotating vanes, taking care that the fan frequency does not coincide with a frequency of interest.

Two factors describe the sound properties of an reverberant chamber: the reverberation time which is the time taken for the sound pressure level to fall by 60 dB (i.e. one millionth of its value) after the source of sound is switched off; and the equivalent sound absorption area which is the hypothetical area of a totally absorbent surface which would give the same reverberation time as the actual room under consideration.

In performing tests on specimens placed in such a room, it is convenient to measure the reverberation time both with and without the specimen, and calculate the absorption characteristics from the measurements.

If T_1 is the reverberation time in the empty room, T_2 is the reverberation time of the room with the test specimen, V is the volume of the room and c is the velocity of sound in air, then

$$A_1 = \frac{55.3\ V}{cT_1} \qquad \text{and}$$

$$A_2 = \frac{55.3\ V}{cT_2}$$

where A_1 is the sound absorption area in the empty room and A_2 is the area in the room with the specimen. The equivalent sound absorption area of the specimen, A, is $A_2 - A_1$.

These relationships are forms of the Sabine equation, so-called because it was determined experimentally by Sabine at the turn of the century. It is more usually stated as:

$$A = 0.161 \; V/T$$

The coefficient 0.161 is correct for a sound velocity of 343 m/s. Note that this is an empirical relationship (although it can be justified theoretically) and is correct in this form only when V is in m^3 and A in m^2.

When measuring the effectiveness of an acoustic panel, or a portion of an absorbent surface, the specimen should be mounted in a reflective frame so as to minimise edge effects. In this case the effectiveness of the panel is assessed by dividing the equivalent sound absorption area by the actual area of the panel to obtain the unit sound absorption area. For further information refer to BS 3638 (ISO 354).

Failing a precise measurement of the reverberation time, the mean acoustic absorption coefficient of a room can be estimated from:

$$A = \alpha S_v$$

where

α = the mean absorption coefficient
S_v = total surface area of the room (walls, floor and ceiling)

The appropriate value of α can be estimated by use of Table 1.

Reverberant Chambers used for Assessment of Sound Insulation

A special arrangement of reverberant chambers is necessary when measuring the transmission loss of building elements such as acoustic screens, different kinds of wall construction, windows and doors. In these cases the parameter to be measured is the transmission loss of the element. The test arrangement consists of two side-by-side reverberant chambers, one containing the noise source and the other the measuring equipment, with an opening between them to carry a specimen of the construction panel under investigation.

The reverberant chambers are necessary to produce diffuse fields, although the quality of diffusivity need not be as good as for a room intended for absorption measurements. BS 2750 (ISO 140) requires each of the rooms to have a volume greater than 50 m^3 with no more than a 10% difference between them. It is preferred that the test opening should cover the whole of the side of the common wall and have an area of at least 10 m^2. The reverberation time should be less than 2 seconds.

The sound reduction index or transmission loss is given by:

$$R = 10 \log \frac{W_1}{W_2}, \text{ where } W_1 \text{ is the incident sound power}$$

and W_2 is the transmitted sound power.

In terms of sound pressure:

$$R = L_1 - L_2 + 10 \log \frac{S}{A}$$

where L_1 and L_2 are the sound pressure levels in the source and receiving rooms respectively, S is the area of the test specimen and A is the absorption area of the receiving room.

The main problem in carrying out this kind of test is the avoidance of sound transmission by a direct path between the two rooms rather than through the specimen. This may involve the complete structural isolation of the two rooms and the provision of some kind of sound deadening treatment to the exteriors. The specimen must be well sealed in the opening so that there is no chance of stray sound passing through.

Noise Measurements on Equipment Installed Indoors

The ideal way of measuring the noise emitted by a piece of machinery is by placing it in an anechoic chamber or in a free field outdoors, where there is no conflict with other sources and no unwanted reflections. It may not be possible to secure this ideal, so some attempt has to be made to take measurements when the equipment under investigation is installed in its final location in an industrial site under actual operating conditions, which may be among other noise sources. The location is likely to be semi-reverberant: the walls are neither absorbent nor reflective, but between the two. The parameter which describes the characteristics of the site is the room constant, and various techniques are available to determine its value, which are discussed below.

If all that is required is an idea of the *in situ* sound pressure levels so as to determine the exposure of an operator, then a noise scan need only be taken at the positions of interest. If, however, a value is needed for the overall sound power level, then measurements have to be taken on a predetermined measurement surface totally enclosing the test item. This surface may be a hemisphere, a rectangular parallelopiped or a conformal surface. The last is a parallelopiped with the corners and edges composed of cylinders and portions of spheres. For most purposes the rectangular parallelopiped is adequate, which makes the location of the microphones easier, but a hemisphere is theoretically better.

The relationship between sound power and sound pressure in a semi-reverberant enclosure is:

$$L_W = L_P + 10 \log \left(\frac{1}{4\pi r^2} + \frac{4}{R} \right)$$

where Q is the directivity factor, r the distance from the noise source and R the room constant. The two terms inside the brackets represent the direct field ($Q/4\pi r^2$) and the reverberant field ($4/R$).

For hemispherical radiation (the source is placed on a reflecting plane), $Q = 2$ and the above relation becomes:

$$L_W = L_P + 10 \log \left(\frac{1}{2\pi r^2} + \frac{4}{R} \right)$$

This can only be expected to apply in the far field, i.e. when the distance r is large in comparison with the size of the noise source.

$L_W - L_P$ is plotted in Figure 3, for $Q = 1$. Similar plots can be drawn for other values of Q.

Note that the units of R are m^2, commonly called sabin (or strictly metric sabin, as originally sabin was defined in ft^2 units). One sabin is equal to the total absorption (as in an open window) of $1 m^2$.

Another method of assessing the sound power output of a piece of machinery involves

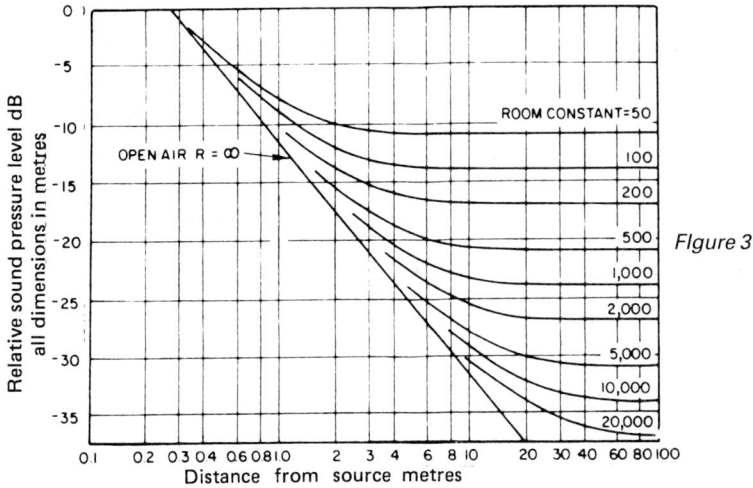

Figure 3

replacing it by a calibrated noise source of known characteristics. The comparison between the sound pressure levels generated by the equipment and that of the calibrated source enables the sound power to be calculated. This has its difficulties, particularly if the machinery cannot readily be moved.

Determination of the room constant R

The room constant is defined as

$$R = \frac{S}{1 - \alpha}$$

S is the total surface area of the room and α is the average absorption coefficient. Note that α and consequently R are frequency dependent, so the absorption data needs to be available or measurable throughout the frequency range of interest; R then becomes a plot against frequency rather than a single parameter. In some rooms, where the absorption coefficients of the surfaces are known, R may be calculated, but in many other cases, it may prove very difficult, particularly where there are other objects present possessing unknown characteristics.

The room constant can be found indirectly by measuring the reverberation time (the time taken for the sound pressure level to decay by 60 dB). Special equipment is required for accurate results (although Sabine himself managed with nothing more sophisticated than a stop-watch). If the reverberation time can be determined, R is found from:

$$R = \frac{V}{6.25\,T - V/A}$$

where V is the room volume in m³, A is the total area of the surface in m² and T is the reverberation time in seconds.

To find R by calculation, it is necessary to assemble the separate absorption coefficients of the various surfaces, not only the floors, walls and ceiling but also the other equipment in the room. In a simple room this may give reasonable accuracy, but in many cases this method is either inadequate or impossible.

Absorption Coefficients

The preferred method of determining the acoustic properties of a room is by measurements performed in the room after construction, but some attempt at predictive calculation may be needed at the planning or design stage or when the question is raised as to the best way of altering the measured characteristics of a room.

There are several standard methods of determining the absorption coefficient of surfaces. One such is the American Society of Testing Materials ASTM C384, a laboratory test measuring normal incidence coefficients by use of an acoustic tube (Kundt's tube); it consists of a tube containing a loudspeaker, microphone and a specimen of the material under test. It is a cheap and easy test where only small samples are available (100 mm dia. or less). An example of commercial equipment used for this purpose is the Bruel and Kjaer Standing Wave Apparatus Type 4002.

The results of such tests are not as useful for practical applications as those obtained from tests done in a reverberation room, where random incidence coefficients are measured. Such a test is BS 3638 (ISO 354), see paragraph above on Reverberant Chambers. Comprehensive test results to this standard are not readily available, except where manufacturers have done their own tests on proprietary products.

Coefficients measured to different standards are likely to be reasonably comparable provided that the general basis of the tests is similar. As a rough guide, acoustic tube results are likely to be slightly lower than those for a reverberation room at low frequencies but about 50% lower at high frequencies.

It may happen, for highly absorbent materials, that the measured coefficient exceeds unity at some frequencies. It is difficult to give a theoretical justification for this and so it is usually better to limit the value for calculation purposes to 0.95.

Table 1 gives some widely quoted values, based on testing done in a reverberant chamber at the National Physical Laboratory. The qualities of this chamber do not exactly meet the requirements of BS 3638, but the test values on over 280 different materials still represent the most comprehensive set of data available. For further information refer to *Sound Absorbing Materials* by Evans and Bazley, 1960 (HMSO), from which Table 1 is extracted.

As an example of a highly absorbent material, Table 2 gives data supplied by Vitafoam for Pyrosorb-S, an acoustic foam suitable for applications requiring suppression of internal refelections. As well as giving typical values for porous material, it demonstrates the difference between the two forms of test.

Of the pair of values given for each frequency, the top value is for test to ASTM C384; the bottom value is for test to BS 3638. These data are examples only. Considerable variations are possible with different formulations of foam and facings; refer to manufacturers.

TABLE 1 – APPROXIMATE VALUES OF THE MEAN
ACOUSTIC ABSORPTION COEFFICIENT α

Mean acoustic absoprtion coefficient α	Description of room
0,05	Nearly empty room with smooth hard walls made of concrete, brick, plaster or tile
0,1	Partly empty room, room with smooth walls
0,15	Room with furniture, rectangular machinery room, rectangular industrial room
0,2	Irregularly shaped room with furniture, irregularly shaped machinery room or industrial room
0,25	Room with upholstered furniture, machinery or industrial room with a small amount of acoustical material (for example, partially absorptive ceiling) on ceiling or walls
0,35	Room with acoustical materials on both ceiling and walls
0,5	Room with large amounts of acoustical materials on ceiling and walls

TABLE 2 – ACOUSTIC PROPERTIES OF FOAM
LINING – PYROSORB–S

Thickness (mm)	Frequency (Hz)					
	125	250	500	1000	2000	4000
12	–	0.04	0.14	0.15	0.29	0.54
	0.06	0.09	0.18	0.29	0.38	0.58
25	–	0.07	0.14	0.32	0.80	0.71
	0.09	0.22	0.38	0.52	0.63	0.73
50	0.11	0.11	0.50	0.86	0.84	0.90
	0.24	0.46	0.71	0.84	0.87	1.02

Chapter 6
Human Response and Hearing Protection

Noise-induced Hearing Loss

It is estimated by the Health and Safety Executive that 1.7 million workers are exposed to noise over 85 dB(A) and 630 000 to noise over 90 dB(A). Although hearing loss is not life threatening, the result of such exposure is that a significant number of workers suffer hearing loss which affects their work and leisure activities. In addition to hearing loss, tinnitus (noises in the head) may be a result of noise exposure; 4 million people in UK are estimated to suffer from it. Exposure to noise doubles the risk of contracting tinnitus, particularly for people under 40 years. Tinnitus is not curable, whereas hearing loss can be alleviated to some extent by the use of hearing aids.

General Personal Protection

Personal protective equipment is defined as any device or appliance designed to be worn or held by an individual for protection against one or more safety and health hazards in the execution of the user's activities. This definition includes protection against the harmful effect of noise and vibration. The broad principles are contained in the Personal Protective Directive of the European Community (Directive 89/686/EEC), which will come into force on 1 July 1992; implementing provisions are to be adopted and published by 31 December 1991.

From 1 July 1992, all protective equipment must:

satisfy wide ranging safety requirements

be subject to type-examination by an approved body or satisfy approved methods of manufacture

carry a CE mark (see Figure 1)

provide instructions in the language of the country where used

$$C \in$$

Figure 1 The mark which must be affixed to personal protective equipment. CE must be followed by the last two figures of the year in which it is affixed and, if an approved body has been involved, its distinguishing number

At present it is likely that for hearing protection, a type-test procedure will be adopted by the UK Government. A list of approved laboratories for this purpose will be available from NAMAS (National Measurement Accreditation Service), operated by the National Physical Laboratory. Other European countries have their own procedure, but equipment approval in one country will imply approval in the others. Failure to comply with the requirements of the Directive will be made a criminal offence. The Directive will be implemented in the UK by regulation and enforced by the Trading Standards Departments of local authorities and the Health and Safety Executive.

It is desirable that the ambient noise level in the work place shall be so low that there is no need for the extra protection provided by muffs or plugs, but that is clearly not possible in the present state of the art of noise control. It is always desirable however to make an effort to reduce the primary noise sources. If it can be done, it may make the wearing of protection unnecessary for some or all of the employees, or it may allow the use of a low attenuation device, e.g. plugs instead of muffs. Even the best hearing protection has some drawbacks:

> spoken messages may be impossible to pass on (unless the muff has a built-in microphone)
>
> alarm signals or unusual behaviour of equipment may not be heard
>
> muffs and ear plugs may be uncomfortable or hot
>
> earplugs may cause infection in susceptible users – medical advice should be sought by anyone with a history of ear disease.

In any particular situation, hearing protection must be capable of attenuating the ambient noise at least to the level laid down by Council Directive 86/188/EEC (implemented in UK law as the Noise at Work Regulations SI 1989/1790). The correct choice of hearing protection depends therefore upon a knowledge of the noise experienced during a working day by every affected worker. It is first necessary to find the value of $L_{EP,d}$ (as defined in the chapter on *Noise Scales and Rating*, and determined by the method described in the chapter on *Hearing Conservation*). The attenuation required will dictate the choice of protection.

Types of Hearing Protector

There are two general types of protector in use: ear muffs and ear plugs. It is desirable that whichever are chosen they shall be manufactured in accordance with the appropriate standard. British Standard BS 6344 sets basic requirements for construction, performance, methods of test, weight, size and marking (Part 1 for ear muffs, Part 2 for ear plugs). BS 5108 (ISO 4869) describes the test procedure for determining the attenuation of the protector.

BS 6344 should be studied by the manufacturer of muffs. The purchaser should satisfy himself that those he buys meet the standard. In the UK the BS Kitemark scheme is a guarantee of quality.

Function of a Hearing Protector

No ear protector working on the basis of an insertion loss between the source of sound and the ear drum can be entirely effective at excluding sound since sound vibrations are transmitted through the skull by bone conduction as well as through the auditory canal. An ear protector modifies both the air conduction and to a lesser extent the bone conduction. Four distinct sound paths can be identified in Figure 2.

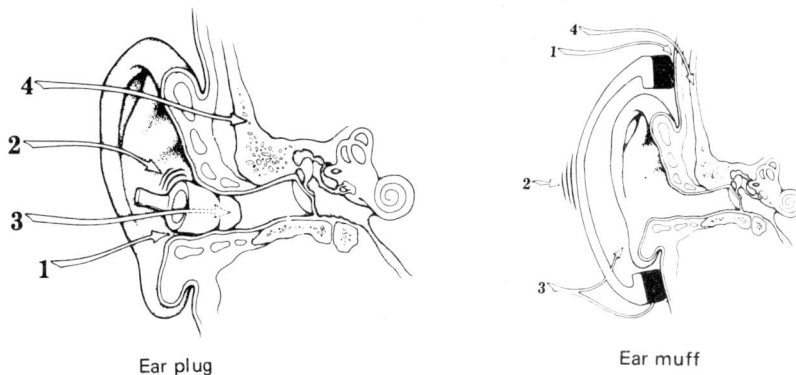

Ear plug   Ear muff

*Figure 2 Illustrations of the four paths by
which sound reaches the occluded ear.*

1 Air leaks.

For maximum protection the device must make an airtight seal with the canal or the side of the head. Plugs must be flexible enough to fit the contours of the canal, and cushions must accurately adjust themselves to the contours of the external ear. Either foam or liquid filled cushions may be used.

2 Vibration of the air protector.

Plugs may themselves vibrate in the canal, which limits the low frequency attenuation. Similarly, muffs may resonate aginst the side of the head, the stiffness being governed by the volume of the air and the stiffness of the surrounding flesh.

3 Transmission through the material of the protector.

The use of a soft material such as a modern plastic foam should ensure that must plugs are effective barriers to this path. Muffs, being of a larger area could have significant transmission unless the cushions are soft and the material of the cup chosen to have a high damping.

4 Bone conduction.

This is the main channel remaining. Plugs and muffs are primarily designed to reduce the air transmission. Unless helmets which completely surround the head are employed, there is a little than can be done to eliminate bone conduction. Muffs are better than plugs in this respect.

BS 5108 (ISO 4869) defines the test method for determining the the attenuation given by muffs. The test is basically a *subjective* test of the hearing threshold shift in healthy subjects. An alternative *objective* test, using a fixture and a microphone is described in BS 6344 may be used for quality control purposes, but the subjective test is to be preferred when quoting the performance of hearing protection devices. The standard prescribes the recording of the average attenuation and the statistical variation during the tests, caused not only by differences in construction but also by differences in the human subjects and their ability to detect low level sounds. Tests are performed throughout the frequency range of speech (63 Hz to 8 kHz) and results quoted for all the frequency bands.

The Assumed Protection Value (APV) given by the ear protection (muffs and plugs) is the mean attenuation less one standard deviation at each test frequency. Note that a normal distribution with one standard deviation on either side of the mean includes 68% of the population, so 16% of the users of the muffs will not enjoy a protection equivalent to the APV. Using two standard deviations will increase the proportion of the users who will be protected, so that only 2.5% will not receive the APV. This factor may need to be taken into account in critical cases.

A minimum attenuation is quoted in the standards as Table 1

TABLE 1 – MINIMUM ASSUMED PROTECTION VALUES (APV) FOR HEARING PROTECTORS AS QUOTED IN BS 5108, ISO 4869 AND EN 352 (DRAFT)

f (Hz)	125	250	500	1000	2000	4000	8000
APV (dB)	5	8	10	12	12	12	12

Most ear muffs on the market are better than those quoted in Table 1, although some older types do not meet the low frequency requirements.

It follows that the acoustic characteristics of ear muffs are described by a table of attenuation in each frequency band. It is not at present acceptable in most countries to define a single number to replace that table, although in USA, protectors carry a Noise Reduction Rating (NRR) which is an attempt to do that. The appropriate American standards are ANSI S3.19 and S12.6. Other similar systems are the Australian SLC 80 and the German Z value. It is not possible to obtain any of these single value numbers from a knowledge of the attenuation spectrum of BS 5108.

There is in hand a proposal in Europe that simpler figures to describe performance should be acceptable. These are ENR (a single number to indicate performance) and H-M-L (a set of three numbers). They require the measurement of dB(A) and dB(C), which may cause difficulties, since most noise meters do not possess a C-weighting, however we can expect manufacturers to respond to a need.

It is desirable to have a choice of protectors available for any workshop situation. Personal preference is an important factor when promoting their use; some people will prefer the comfort of one type against another. However it is confusing to have available ear muffs with different degrees of attenuation for different work cycles or locations within a factory.

In choosing which ear muff to use, the factors additional to attenuation which will determine the choice are weight, headband force, cushion pressure, range of cup rotation and resistance to damage. All these factors are measured during test to the appropriate standards and should be available on demand from the suppliers. Cost is also a factor; at 1991 prices, ear plugs vary from about 10p to £3.50 per pair and ear muffs from £4.00 to £30.00 or even higher for electronic types.

Some authorities suggest the use of ear plugs and ear muffs in combination, in the expectation that the two together will be better than just one. Use of the two in combination is likely to increse the attenuation of the muffs by about 5 dB; it is certainly not possible to state that the APV of the combination is equal to the sum of the individual APVs. The only sure way of determining the attenuation of the combination is by a test performed according to BS 5108. If the attenuation required is insufficient for those muffs to hand, a model giving a higher protection type should be chosen.

Types of Ear Muff

A basic ear muff consists of a band passing over the head, joining together two cups. The band may also be worn behind the head and under the chin. The sealing of the cup against the head is obtained by a soft cushion ring filled with foam or liquid, or a combination of the two. The ear shell is filled with an absorbent foam. It is sometimes claimed that a liquid seal could fill the ear with the liquid if the cushion breaks, but this rarely is a problem with modern muffs. The effectiveness tends to increase with the weight of the muffs, so for comfort the lightest muff providing the desired attenuation should be chosen. The muffs should be tried with any spectacles that need to be worn. The head band should be comfortably padded. Commercial ear muffs are available in a variety of sizes.

Ear muffs may fit under a safety helmet, or be attached to it. The advantage of the attached type is that they can conveniently be parked on the helmet when not in use.

Passive and Active Ear Muffs

Most ear muffs behave in a linear way, that is, whatever the sound pressure at the ear may be, they reduce it by the attenuation value; these are known as *passive* muffs. In this type, low level sounds are reduced as much as high level sounds, and this can be a disadvantage to the user who wishes to hear normal conversation or where there is only occasional exposure to high levels, as for example on a firing range. Several attempts have been made to address this problem by the use of *active* muffs. These are designed to cut off entirely the damaging high levels but attenuate the low levels; they incorporate either an electronic circuit with microphone and speaker or an acoustic valve which has much the same effect. They are heavier and more expensive than the passive type, but they may be useful in special situations. Ear plugs have also been made with an acoustic valve, but these are less satisfactory. There is no standard method of measuring the behaviour of active devices.

Ear Plugs

An ear plug is defined as a hearing protector worn either within the external ear canal (aural) or in the concha against the external ear canal (semi-aural). They can be classified into several groups, following the definitions of European Standard (draft) EN 352-2:

> Disposable
> Reusable
> Custom moulded
> Banded

Disposable plugs are made of low cost materials that are intended to be kept in place throughout a shift and then disposed of. A typical construction is the Bilsom type, which consists of a combination of mineral fibres and plastic foam in a protective plastic film. Another type is made of foam which is intended to be compressed by the fingers, inserted in the ear canal and allowed to expand. At one time a crude form of plug consisting of cotton wool or some similar material was used. These are not really satisfactory today for two main reasons: the attenuation is uncertain, and there is a danger that the plug will be left in the ear or even driven further down onto the ear drum. Some of the disposable or semi-disposable plugs, particularly those based on a foam which require manipulation, could lead to ear infection or irritation if handled with dirty hands; and if this is likely to happen, instructions should be given that they are to be used once only.

Reusable plugs may be made of foam or moulded from a thermoplastic elastomer and are washable in warm water. They should be discarded after about 5 or 6 uses. Some types are joined together in pairs by a cord.

Custom moulded plugs are individually made to match the contours of the individual's ear canal. This 'personalised' aspect may make it easier to encourage an employee to wear them.

Banded plugs are joined as a pair by a semi-rigid interconnection, which helps to retain them in the ear in the presence of vibration and ensures that they are less easily lost when taken off temporarily. They are available in a variety of band sizes.

For applications in the food industry, plugs containing a permanently embedded steel ball are available making them metal-detectable.

Choice of Hearing Protector

If it is established that some kind of hearing protector is needed, either because a survey done under the Noise at Work Regulations requires it or it is likely to be conducive to good working practices, the next step is to choose the most suitable type and manufacturer.

It is first necessary to obtain values of assumed protection for the various muffs or plugs under consideration. A selection of those currently available in the UK with their Assumed Protection Values are to be found in Tables 2 and 3. Remember that the assumed protection equals the mean of the test results less one standard deviation. Any supplier should be prepared to supply attenuation data tested to BS 5108 (ISO 4869). If the protectors themselves comply with BS 6344 (or EN 352 when this becomes current) which sets requirements for size, weight and durability, this data will be available. If the hearing protectors do not comply with this Standard, such data as are available will need to be interpreted by someone familiar with test procedures and their limitation. It should not be assumed that because a hearing protector does not comply with a standard it is unsuitable. So far there is no standard covering special types such as those which are amplitude sensitive or have electronic limiting circuits; they are for particular applications and extra care needs to be taken in their choice.

The correct procedure to determine whether a muff or ear plug can give adequate attenuation is as follows:

A spectrum analysis of the measured noise is needed. Usually an octave band analysis will suffice. If protection is needed against just one noise environment, this can be readily found from a noise meter incorporating octave band filters. If however the situation is more complex, in that an employee needs protection against differing noise regimes during the day, then an octave-band daily-personal-exposure value has to be obtained. This not the quite the same as $L_{EP,d}$ which by definition applies to A-weighting only. A meter giving an L_{eq} in octave bands is required, or failing that a survey taking sample values throughout the day in the various locations may suffice.

Calculation of Protection Available from Ear Muffs

Suppose that a noise with an A-weighted level of 112 dB(A) has octave band pressure levels as in Table 4.

The Assumed Protection is obtained by taking the mean of the test results less one standard deviation for the particular protector being considered. Note that the data provided will include attenuation values for 3.15 kHz and 6.3 kHz, which can be ignored for this exercise. The Assumed Protected Level (APL) is the band pressure level less the Assumed Protection. To convert the band APL back to a dB(A) value for calculating the daily personal exposure, the method described in the chapter on the *Measurement of Sound* can be used. This results in an A-weighted APL of 88 dB(A). The determination of the new $L_{EP,d}$ can now proceed as previously described.

In this example, if the value of 88 dB(A) is the actual daily personal exposure, it falls between the First and Second Action Level, which requires hearing protection to be made available, but it is already being worn! A better protector should be sought with increased attenuation in the 500 Hz to 1 kHz range.

**TABLE 2 – ASSUMED PROTECTION VALUES FOR EAR MUFFS TESTED IN ACCORDANCE
WITH BS 5108 (ISO 4869). ATTENUATION IN dB**

	63	125	250	500	1k	2k	31.5k	4k	6.3k	8k	Weight (gm)	Notes
Muffs with Headbands												
Bilsom Loton	8.4	7.6	9.2	18.1	26.6	27.8	29.2	27.8	27.2	30.9	129	1
Blue	7.3	6.6	11	18.3	27.1	30.3	28.9	30.5	33	32.8	158	1
Yellow	10.9	8.3	11.1	17.6	26.3	28.8	33.4	34.3	31.7	30.7	206	2
Comfort	11.1	6.7	12.8	19.4	27.1	29.1	32.7	36.4	35	35.3	172	1
Compact	7.1	6	8.8	16	23.5	27.9	30.7	34.6	31.1	32.9	125	1, 3
Viking	13	10.9	16.4	26.8	31.4	30.1	33.4	39.2	38	40	211	1
Pocket	11.1	9	12.4	20.9	30.1	30.4	33.4	35.5	34	32.6		1, 3
Hellberg Popular	6	4.5	7	14.5	20.5	25	30.5	26.5	23	19.5	120	1
Mark I	9.5	7.5	11.5	20	29.5	28	28.5	25.5	27.5	24	150	1
Mark V	15	15	19.5	26	34	36	33	32	33.5	33.5	230	1
Mark X	12	14.5	19.5	26	34.5	36.5	37	36.5	38	35	280	1
North ED 2000	11.8	9.2	12.7	20.4	31.2	29.8	28.5	27.3	27.1	24.0	161	1
ED 2010	7.4	7.0	11.0	15.6	26.7	28.0	28.5	27.3	27.1	24.0	127	1
ED 2020	15.2	14.4	17.1	26.0	33.7	35.9	32.4	31.1	33.7	33.6	166	1
ED 2030	15.6	15.4	21.0	28.1	37.5	37.2	34.2	36.8	40.1	41.6	194	1
Racal Classic 1	8.4	6.6	12.9	20.1	29.4	30.7	34.4	35.1	33.9	27.1	181	1
Classic 2	10.2	8	17.7	23	30	34.9	40.7	34.3	34.6	30.5	195	1
Classic 3	11.7	13.4	20.1	25.8	35.2	34.7	41.3	39.6	39.7	35.2	267	2
Ultramuff	10	5	12	20	29	29	29	28	23	24	195	1
Supamuff	5.2	4.3	8.1	16.1	20.3	24.4	34.4	33.6	27.4	27.8	169	1, 3
Sonomuff	12.7	14.5	16.7	23	31.8	31.9	34.9	35.3	30.7	31.1	250	1, 3
Sonogard	8.2	9.6	17.3	22.6	31.2	31.5	35.7	29.7	26.9	19.2	373	2
Safeline QT	6.7	5.6	9.8	19	27.3	22.7	34.9	31.8	33.7	30.5	149	1
E.A.R 1000	14.1	11.1	12.3	21.4	28.6	31.5	34.6	33.5	34	36.5	210	1
4000	7.6	7.7	12.3	18.8	21.4	30.2	36	34.7	34.7	32.2	210	1
9000	9.2	7.2	15.4	21.3	20.5	22.1	26.7	23.6	21.5	20.5	220	1
MSA Hilo I	9.4	6.0	12.1	22.6	23.2	22.3	24.4	26.7	28.0	27.5	165	1
Hilo II	11.5	9.8	15.7	25.9	30.5	30.1	26.7	29.1	31.3	29.3	185	1
Hilo III1	2.0	10.1	16.8	25.5	33.7	32.3	32.6	33.8	33.7	32.8	210	1
Helmet Muffs												
Bilsom Blue	8.4	5.7	10.1	19.1	27	30	34.5	36.4	35	37.2	–	1
Comfort	10.6	7.6	12.1	18.8	24.7	31.3	33	34.9	33.4	30	–	1
Viking		9.5	16.3	27.1	31.9	30.8	34.5	38.3	33.3	33.9	–	1
Hellberg Mark I	12	11	14	20.5	30	30	31.5	29	31	34	–	1
Mark V	15	15	18	23	28	33	32	33	33	35	–	1
Mark X	15	16	19	28.5	29	34.5	35.5	37.5	40	37.5	–	1
North ED 2015	6.9	7.2	9.2	20.2	20.6	25.1	29.3	31.3	29.8	28.5	–	1
ED 2035	8.4	8.9	14.1	19.8	27.8	29.5	30.8	28.9	30.5	27.2	–	1

Notes: 1. Foam filled ear cushion
2. Liquid cushion
3. Slim design intended to fit under visors and helmets

Special Types – Non-linear and electronic. Tests to BS 5108 not appropriate.

Bilsom Impact Viking.	Electronic; user has normal hearing, but high level impact noises are blocked.
Bilsom Impact.	As above
Hellberg Active.	As above; maximum level inside the cups is 82 dB(A)
E.A.R 9000.	Non-electronic, acoustic valve operates above 120 dB. Data in table above is for linear region

TABLE 3 – ASSUMED PROTECTION VALUES FOR EAR PLUGS TESTED ACCORDING TO BS5108. ATTENUATION IN dB

	63	125	250	500	1k	2k	3.15k	4k	6.3k	8k	Type
E.A.R Foam	17.5	18.3	19.3	21.9	24.5	27.9	38.2	38.6	39.4	38.9	4, 8
Cabocord	17.5	18.3	19.3	21.9	24.5	27.9	38.2	38.6	39.4	38.9	4, 8, 11
Ultrafit	17	13.7	14	14.8	16.3	20.6	23	20.3	27.3	30.4	6, 8, 11
Model 200	18.1	17.6	16.7	16.7	17.9	29.8	36	35.4	37.7	32.9	4, 8, 9
Model 600	18.6	17.7	18.5	16.9	19.1	22.5	27.3	30.4	30.4	28.9	5, 8, 9
Hexafit	7.7	10.5	14.3	18	20.5	28.5	37.3	38.7	39	36.4	4, 7
Tracers	17	13.7	14	14.8	16.3	20.6	23	20.3	27.3	30.4	6, 8, 10, 11
Bilsom P.O.P	7.5	9.3	10.6	12	13.3	20.2	24.7	24.9	25.6	28.6	4, 7
Premier	18.4	17.7	17	17.2	19.3	24.8	29.7	29.7	30.3	30.1	5, 7
Soft	12.3	12.1	13.6	17.3	20.6	26.4	34.3	36.7	37	34.9	5, 7
Perfit *		25.3	25.2	27.8	29.9	34.5	40	34.9	34.4	31.9	6, 8, 11
North ED 1770	18.9	18.4	17.2	19.1	18.3	30.3	36.7	36.5	37.8	27.6	4, 7
Racal DBA	14	14	13	14	17	26	34	36	34	29	6, 8, 11
Airsoft	12	11	9	12	12	19	25	23	21	26	6, 8, 11

Racal Gunfenders Special type for impulsive noise incorporating an acoustic valve

* Attenuation values according to ANSI S3.19 –1974. Not directly comparable with BS 5108

Notes: 4. Foam
5. Foam in elastomer cover
6. Premoulded
7. Disposable, single use
8. Washable
9. Banded
10. Metal detectable
11. In pairs with cord

TABLE 4

Octave band frequency (Hz)	63	125	250	500	1k	2k	4k	8k
Band pressure level dB	89	91	95	100	103	107	105	98
Assumed Protection dB	2	2	9	14	21	26	31	25
Octave band APL dB	87	89	86	86	82	81	74	73

Chapter 7
Legislation – Noise Regulations and Hearing Protection

European Legislation on Noise

Within the European Community, personal protective equipment in use in industry must be capable of attenuating noise so that the perceived noise levels do not exceed the daily limit laid down by the Directive 86/188/EEC on the protection of workers from the risks relating to noise at work. This Directive is implemented in different ways by the Community members; in the UK the appropriate legislative instrument is the Noise at Work Regulations (SI 1989/1790) which came into force on 1 January 1990. The HSE (Health and Safety Executive) Code of Practice for Reducing the Exposure of Employed Persons to Noise is now obsolete. Employers have a legal duty to prevent damage to hearing. Noise levels in excess of 85 dB(A) require the employer to take action. The HSE have published a series of Noise Guides, which should be studied by those who need a more detailed coverage of the material in this chapter.

The fundamental physical value of noise exposure is the daily personal exposure to noise, $L_{EP,d}$. This is the same as L_{eq} over the working day. It can be determined from an integrating sound level meter complying with BS 6698 (IEC 804). A simpler meter complying with BS 5969 or BS 4197 might also be suitable but requires rather more skill in its use. A personal dosemeter worn by the worker, complying with BS 6402, can be used; this is suitable where an particular employee works for different parts of the day in several different noise environments. An integrating meter giving a direct reading of L_{eq} is the easiest to use, but failing that a calculation has to be done using a sampling method. Provided that the noise level is reasonably constant or follows a clear pattern through the day, the following procedure can be used, making use of the nomogram in Figure 1 or alternatively by use of the formula

$$f = 0.125 \, t \, \text{antilog} \, [0.1(L - 90)]$$

where f is the fractional exposure to a noise level L for a time t. A value of f is calculated for each time duration t until the full working day of eight hours (or more) is accounted for. The summation of all the values of f is known as f_{tot}. Then

$$L_{EP,d} = 10 \log f_{tot} + 90 \, \text{dB(A)}$$

This procedure is necessary only where an integrating sound level meter is not available

The value of $L_{EP,d}$ can be obtained from the nomogram in Figure 1 by drawing a straight line connecting the measured level on the L scale with the exposure duration on the t- scale, and reading either the value of f or $L_{EP,d}$ on the centre scale.

L

125

120

115

110

105

dB(A)

100

95

90

85

L_eq   f

120 ┬ 1000

      ┤ 500
      ┤ 400
      ┤ 300
      ┤ 200

110 ┼ 100

      ┤ 50
      ┤ 40
      ┤ 30
      ┤ 20

100 ┼ 10

      ┤ 5
      ┤ 4
      ┤ 3
      ┤ 2

90 ┼ 1

      ┤ 0·5
      ┤ 0·4
      ┤ 0·3
      ┤ 0·2

80 ┼ 0·1

      ┤ 0·05
      ┤ 0·04
      ┤ 0·03
      ┤ 0·02

70 ┴ 0·01

t

8
6
4

Hours

2

80
1   60
40

20

10
8
6

Minutes

4

2

80
60   1
40

Seconds

20

From:

$$f = \frac{t}{8} \text{ antilog } [0.1 \ (L - 90)]$$

where t is in hours

Also:

$$L_{EP,d} = \frac{\log f}{0.1} + 90$$

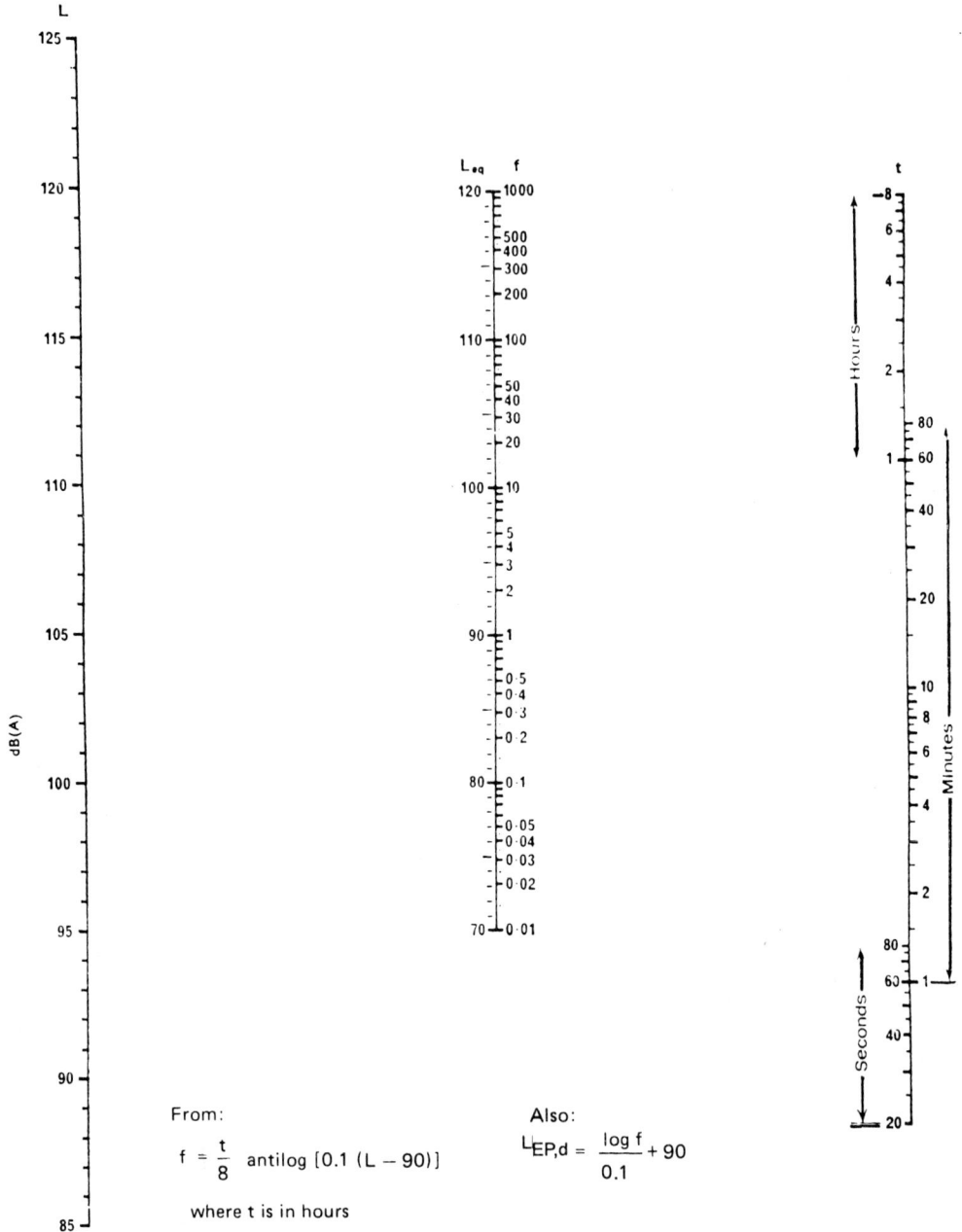

(1)   For each exposure connect sound level dB(A) with exposure duration t and read fractional exposure f on centre scale.

(2)   Add together values of f received during one day to obtain total value of f.

(3)   Read equivalent continuous sound level $L_{EP,d}$ opposite total value of f.

Figure 1 Nomogram for Calculation of Daily Personal Noise Exposure

Example 1 of the use of the nomogram.

A person is exposed to a sound level of 102 dB(A) for 2¼ hours per day. During the rest of the day, the level is below 75 dB(A) which can be ignored. From Figure 1 $L_{EP,d}$ = 96 dB(A)

Example 2 of the use of the nomogram

A person is subject to a pattern of exposure as given in the first two columns of the table below. The third column shows the values of f determined from the nomogram and the total, f_{tot}

Sound level dB(A)	Duration of exposure	f
114	10 min	5.2
105	45 min	3.0
92	10 hours	2.0
		$f_{tot} = \overline{10.2}$

From Figure 1 (central scale) $L_{EP,d}$ = 100 dB(A)

In both these examples $L_{EP,d}$ is clearly in excess of 85 dB(A) so action is required. It has to be admitted that there will be be many borderline cases, which can be cured by taking some elementary measures. Before arranging for a full scale survey, which may prove to be expensive, a few spot check may be done during the course of the day to see if 85 dB(A) is exceeded for any significant time. If there seems to be a problem, a full scale survey should be undertaken, which can be done either in-house or by a company specialising in this work. The HSE offers as a rough guide that wherever people have to shout or have difficulty in being heard clearly by someone 2 metres away, there is a potential problem and noise levels should be measured.

If the survey is to be done, the employer must ensure that it is undertaken by a competent person (CP) who needs to have undertaken a course of training. The level of training required will depend upon the situation. An appreciation of general principles rather than an advanced knowledge of acoustics is required, as well as an ability to use sound level meters and analyse the results. A number of companies offer courses ranging from 2½ to 5 days which are adequate for the purpose. These may be of high quality, but are often quite understandably directed towards the use of that company's equipment for measurement or protection. The Institute of Acoustics has published a list of Accredited Centres which meet their requirements for the training of a competent person, Table 1. These courses last 5 days, and may be taken in one block or in daily modules.

Action Levels

Under the new regulations action is required at a lower level than was previously the case. There are three action levels defined in the regulations:

> The First Action Level when $L_{EP,d}$ = 85 dB(A).
> The Second Action Level when $L_{EP,d}$ = 90 dB(A).
> The Peak Action Level when there is a peak sound pressure of 200 pascals (140 dB).

There is a quantifiable risk of hearing damage between the first and second action levels and a small risk below the first action level, so an employer will need to consider whether it is reasonably practical to reduce the exposure when it is below the first action level; the closer to the first action level the more attention should be given. Above the first action level there are certain steps that he must take.

TABLE 1 – LIST OF ACCREDITED CENTRES FOR TRAINING IN CERTIFICATE OF COMPETENCE IN WORK PLACE NOISE ASSESSMENT

Bristol Polytechnic, Ashley Down Road, Bristol BS7 9BU (Mr N J Pittams)

Colchester Institute, Sheepen Road Colchester CO3 3LL (Mr D Bull)

Cornwall College, Trevenson Road, Redruth TR15 3RD (Mr M Latham)

Derbyshire College of Higher Education, Kedleston Road, Derby DE3 1GB (Dr M Fillery)

Glasgow College, Cowcaddens Rd, Glasgow G1 0BA (Mr L Mair)

Heriot-Watt University, Riccarton Campus, Edinburgh EH14 4AS (Mr C Fleming)

Leeds Polytechnic, Brunswick Terrace, Leeds LS2 8BU (Mr J Bickerdike)

Liverpool University, PO Box 147, Liverpool L69 3BX (Dr J G Goodchild)

Newcastle upon Tyne Polytechnic, Ellison Building, Ellison Place, Newcastle upon Tyne NE1 8ST (Dr J Llewellyn)

College of North East London, All Saints College, White Hart Lane, London N17 8HR (Mr J P Seller)

North East Surrey College of Technology, Reigate Road, Ewell, Surrey KT17 3DS (Dr R J Peters)

Salford University, Dept of Applied Acoustics, Salford M5 4WT (Dr T I Hempstock)

Sheffield City Polytechnic, Pond Street, Sheffield S1 1WB (Dr R B W Heng)

South Bank Polytechnic, Borough Road, London SE1 0AA (Professor J P Roberts)

South Glamorgan Institute of Higher Education, LLandaff Centre, Western Avenue, Llandaff, Cardiff CV5 2YB (Mrs T Noble)

Staffordshire Polytechnic, College Road, Stoke on Trent ST4 2DE (Mr J B Leyland)

University of Ulster, Shore Road Newtownabbey, Co Antrim B37 0QB (Dr G C McCullagh)

Where employees are exposed between the first and second action levels, the regulations require the employer to provide protectors to all who ask for them.

Where exposure is above the second action level or the peak action level, the employer has to provide protectors to all workers and ensure that they are used. Note that occasionally a worker may be exposed to the peak action level, but still remain below the first or second action levels; this could occur in a workplace where there is a very occasional explosion or impact noise. A sound level meter, capable of recording peak levels must be used.

In addition to the provision of protectors, the employer will also be required to provide instruction and information about the risk of damage, the precautions that need to be taken to minimise the risk and where the protectors can be obtained.

Wherever reasonably practical, the employer will need to mark ear protection zones with signs showing where ear protectors are required.

Measurement of Peak Pressures

A rough check on peak pressures can be made by the use of a simple sound level meter set to 'fast' response. If the reading exceeds 125 dB(A) it should be assumed that a more accurate measurement is required. A meter complying with BS 5969 Type 1 set to Peak Hold will meet the requirement, or any meter with a rise time of 100 microsec or less. The equipment should have an unweighted response to audible frequencies and cut off the high and low frequencies; C-weighting is suitable for this. Although many meters currently available do not have C-weighting incorporated, they are likely to become more common as the European Community Directive on the safety of machinery (89/392/EEC) comes into force at the end of 1992, which will require C-weighted sound pressure levels to be quoted where they exceed 63 Pa (equivalent to 130 dB).

A useful summary of the Noise at Work Regulations 1989 has been prepared by the HSE. References to the various regulations should be followed through by anyone likely to be affected by them. HSE publish some useful Noise Guides which give much more information.

Chapter 8
Community Standards

Community Annoyance

In considering the effect that noise has on our lives, there are two aspects which need to be considered. The first is the potential for a direct physiological worsening of the hearing of those workers in proximity of the source of the noise, against which precautions can be taken. These will include such means as the wearing of hearing protectors, the isolation of the worker from the noisy equipment in a soundproof booth and the limitation of the time of exposure.

The other aspect is the affect that noise has on us as members of the community, by having to live in proximity of a noisy factory near to busy traffic, under the flight path of the local airport or having to suffer from roadworks outside our windows. This kind of noise we know to be annoying, but the effect on us is not usually measurable by scientific instruments. If it can be assessed at all it is by subjective methods such as observing the numbers of complaints from those living near the noisy factory. The precautions that can be taken are more limited than for the exposed worker. One can hardly supply hearing protectors to the neighbourhood, and while undertaking double glazing in one's home may be effective at cutting down the noise indoors, it does nothing to improve the situation in the garden. The only entirely satisfactory place to reduce environmental noise is at its source in the factory. This can be done by replacing a noisy process with one which is quieter, by silencing the equipment itself or by enclosing the workshop with a soundproof barrier.

In an attempt to assess the relative importance of various kinds of noise nuisance, a survey was carried out in the Netherlands, comparing the years 1977 and 1987. About 4000 people were interviewed about their experience of noise nuisance, with the following result:

TABLE 1 – NOISE NUISANCE IN THE NETHERLANDS

Percentages experiencing some nuisance and serious nuisance				
Noise Source	Some (%)		Serious (%)	
	1977	1987	1977	1987
Road traffic	49	60	20	19
Rail traffic	4	6	1	2
Construction equipment		31		12
Industry	9	15	3	3
Recreation	6	14	2	4
Residential		83		40

Blanks in the table for 1977 indicate absence of data.

These figures indicate that the two most important sources were road traffic and residential noise. It is probable that similar results would be found for many industrial societies.

In assessing the annoyance likely to be caused by a specific noise issuing from a factory or some other source, the procedure is first to establish the sound power level of the noise, which can be done using one of the techniques discussed elsewhere in the text, and secondly to determine the local sound pressure level at a place where complaints have been or are likely to be made.

Determination of the Sound Pressure Level

If the sound power level of the noise source is known, the sound pressure level at a distant point can be found from:

$$L_{pA} = L_w - 20 \log r - 8 - A$$

L_{pA} is the A-weighted sound pressure level, L_w is the sound power, r is the distance from the source and A is the attenuation. It is assumed that the noise is radiated uniformly in all directions over a flat ground surface. The factor A can include attenuation due to the atmosphere, wind, meteorological effects such as temperature inversion, the ground surface and screening. Atmospheric attenuation is important for large distances, but for localised factories and workshops it can often be ignored. If it is of importance, its magnitude can be assessed (see Table 2 in chapter on the *Propagation of Sound*); it depends on humidity and frequency.

Attenuation caused by the ground effects, depends on the nature of the surface. Concrete, tarmac and water are considered to be fully reflective, so no correction is needed; the attenuation due to other kinds of ground such as snow cover, gravel and a range of growing crops or types of lawn is in general fairly difficult to predict. The attenuation depends strongly on the height of the source and the measurement point; usually the effect is restricted to shallow grazing angles.

At the planning stage of a new factory or of a new process to be added to an existing factory, it is important that a good assessment of the environmental impact should be made before installation. This requires from the suppliers an accurate prediction of the sound power of their equipment.

Measurement of Environmental Noise

The basic method for determining noise levels outside buildings is given in BS 4142 (ISO 1996); the latest edition of this standard must be used. In addition to outlining methods of measurement it describes how to assess whether the measured noise issuing from industrial premises is likely to give rise to complaints from people living in the neighbourhood, so it is applicable to mixed industrial and residential areas. It presents a number of typical case studies.

The philosophy of this standard is that a noise is liable to provoke complaints whenever its level exceeds the background noise by a certain magnitude. The method of rating the noise depends on comparing the specific noise level, corrected to take account of its character, with the background noise level.

In determining the noise level from a specific noise source, the primary parameter to be measured is the equivalent continuous sound level, $L_{eq,T}$ (defined in the chapter on *Noise Scales and Ratings*), which is measurable directly by an integrating sound level meter. BS 4142 allows, in the case where the noise is steady, the use of a simple non-integrating meter. This relies on the operator doing his own visual averaging, and the technique is not to be recommended. A meter to BS 6698 (IEC 804) type 2 is to be preferred. An A-weighted value of the noise level may be inadequate if the particular item of equipment responsible for the complaint is not readily identified. It will then be necessary to perform a more detailed frequency analysis (either octave, 1/3 octave or closer), using the appropriate filter set.

The background noise level is also to be measured. This is defined as the A-weighted sound pressure level of the residual noise (i.e. the noise remaining when the specific noise source is suppressed) which is exceeded for 90% of the time, $L_{90,T}$. This requires the use of an environmental noise analyser complying with BS 5969 (IEC 651) type 2 set to Fast weighting. Note that it is not necessary to have two instruments to measure both parameters; there are sound level meters which satisfy both requirements.

The time interval, T, over which L_{eq} and L_{90} are determined is 1 hour during the day and 5 minutes during the night; the night period is that period during which adults are sleeping or preparing for sleep. It is not more closely defined in the standard, although earlier standards suggest that the period 10 p.m. to 7 a.m. would appear to be appropriate for the UK. The actual measurements may be taken over a longer period, particularly where the noise being assessed is cyclic with a longer period than T.

Measurements have to be taken at positions outside buildings in which people are likely to be affected or where complaints have arisen.

Interpretation of Results

The difference between $L_{eq,T}$ and $L_{90,T}$ is L_{diff}

$$L_{diff} = L_{eq,T} - L_{90,T}$$

If L_{diff} is 10 dB or greater, no correction is necessary to $L_{eq,T}$. If L_{diff} is less than 10 dB a correction to $L_{eq,T}$ has to be made in accordance with Table 2.

TABLE 2 – CORRECTIONS FOR BACKGROUND

L_{diff} (dB)	Subtract from $L_{eq,T}$ (dB)
6 to 9	1
4 to 5	2
3	3
< 3	Not applicable; move measurement position nearer to the source and calculate the equivalent level at the desired position.

If L_{diff} is small because the background noise fluctuates, the measurements should be taken during the time the background is a minimum.

A further correction has to be made if the specific noise level contains a distinguishable discrete frequency or if it is so irregular in character as to attract attention; 5 dB has to be added in that case.

The corrected value of L_{diff} (i.e. the difference between the corrected $L_{eq,T}$ and $L_{90,T}$) has then to be assessed as to the likelihood of the noise provoking complaints, as in Table 3.

TABLE 3 – LIKELIHOOD OF COMPLAINTS

L_{diff} (corrected)	Complaints
> 10 dB	Likely, increasing as value increases
about 5 dB	Marginal significance
< 5 dB	The lower the value, the less likelihood
− 10 dB	Positive indication that complaints are unlikely

A Typical Case Study

A factory influences the background noise at the nearest dwelling. Typical background pressure levels measured at the nearest dwelling are 33 dB at night and 46 dB during the day. The factory wishes to install a new machine with two phases lasting an hour – the washing phase lasting 14 minutes and the drying phase lasting 46 minutes. The predicted sound pressure level during the drying phase is a steady 35 dB, and during the washing phase it is 41 dB with irregular clunks and clicks.

(a) Night time with T = 5 min

$L_{eq,5\,min}$ = 41 dB which applies during the washing phase.

Because of the irregular noise 5 dB has to be added, giving a corrected L_{diff} of 41 + 5 − 33 = 13 dB. It is predicted that complaints will be very likely.

(b) Day time with T = 60 min

In this case the value of L_{eq} has to take account of both phases to cover the 1 hour time interval.

$$L_{eq,\,60\,min} = 10\log\left(\frac{46 \quad 10^{3.5} + 14 \quad 10^{4.1}}{60}\right) = 37\text{ dB}$$

Because of the irregular noise 5 dB has to be added, giving a corrected value of L_{diff} = 37 + 5 − 46 = − 4 dB. Complaints are unlikely.

L_{eq} has been found in this example by calculation. In cases where the measurements are made on an existing installation an integrating meter would be used which, with a reference time interval of 1 hour, would give a direct reading of L_{eq}.

In order to forestall complaints, operation of the equipment should be limited to the day time, or the noise source should be shielded or suppressed in some way. Some of the possible techniques for noise suppression will be found in other chapters.

Vehicular Noise

As indicated in Table 1, road traffic noise is a significant noise nuisance in the Netherlands and, it may be presumed, in other industrial communities. The way to tackle it is by suppression at source. Legislation has proved to be the most effective method, and there exists a large corpus of laws restricting the permissible sound level of the range of vehicles from a moped to the heaviest of commercial vehicles. In this area, a great deal of effort has been spent on reducing noise from all vehicular sources, and this effort has resulted in the placing of increasingly stringent requirements on manufacturers.

Tables 4 and 5 give the latest sound pressure levels specified by the EEC that have to be satisfied by new vehicles coming on to the UK market in April 1991. For the list of applicable Directives, refer to the chapter on *Standards, Codes of Practice and Publications*.

Note that vehicular noise is specified in terms of *sound* pressure level, where most other items of equipment employ sound *power*. The sound pressure from a vehicle has to be measured in a particular way as specified in the EEC Directive – it normally involves the measurement of the noise as the vehicle accelerates to a measurement zone and passes through the zone and the microphones have to be placed at a specified distance from the passing vehicle. See also BS 3425 (ISO 362).

TABLE 4 – SOUND PRESSURE LEVELS FOR MOTOR VEHICLES

Vehicle	SPL dB(A)
Intended for carriage of passengers with not more than 9 seats including driver	77*
Intended for carriage of passengers with more than 9 seats including driver, weight more than 3.5 tonne and:	
an engine power < 150 kW	80
an engine power > 150 kW	83
Intended for carriage of passengers, equipped with more than 9 seats including driver; intended for carriage of goods:	
max weight < 2 tonne	78*
max weight > 2 tonne, < 3.5 tonne	79*
Intended for carriage of goods, weight > 3.5 tonne	
an engine power < 75 kW	81
an engine power > 75 kW, < 150 kW	83
an engine power > 150 kW	84

* values increased by 1 dB(A) for vehicles with a direct injection diesel engine.

TABLE 5 – SOUND PRESSURE LEVELS FOR MOTOR CYCLES AND MOPEDS

Vehicle		SPL dB(A)
Moped		73
Motorcycle (capacity in cm³)	1st stage (current)	2nd stage
< 80	77	75 from 1 Oct 93
> 80 < 175	79	77 from 31 Dec 94
> 175	82	80 from 1 Oct 93

Different values of sound power apply to vehicles placed on the market earlier than 1991. For the determination of the specific values which apply to vehicles which do not fit into the categories listed, refer to the relevant legislation, in the UK in particular to Statutory Instrument SI 1986/1078. The subject is a complex one not easily covered by a text such as this.

It is not easy to calculate the sound level resulting from a mixture of vehicle and other machinery noise. One approach is to calculate the sound pressure level from those components whose sound power is known and then add in the vehicular noise from the tables. The tabled values are typical of what might be experienced outside a building close to a road.

Noise From Construction Sites

The main control applied to construction sites is by reduction of the sound power from commonly used equipment – compressors, concrete breakers and picks, power generators, tower cranes, tractors and welding generators. These items are considered in detail in a later chapter on *Construction Equipment*. But many processes used on a construction site do not lend themselves to legislative control through prescriptive restrictions on sound power, so a more general approach to noise reduction is called for.

BS 5228 (the latest issue, 1984 and 1986) is a helpful reference for techniques of noise control on construction and demolition sites. Separate parts deal with road construction, demolition, surface coal extraction and piling.

Construction sites are subject to the Control of Pollution Act 1974 in the UK; they can be controlled by the Local Authority through the conditions attached to planning consent for the work. Such conditions may specify noise levels on the site boundary, hours of working, equipment to be used etc. In particular, they may require the use of equipment which meets the latest legislation issuing from EEC. The authority must be bound by the state of technology at the time and must have due regard for the relevant codes of practice. The act itself does not set noise limits. It does however, through the approved codes of practice, recommend that the parameter to be used in specifying levels shall be L_{eq}, thus allowing averaging of the levels through the working day. The averaging time is best set at 12 hours for daytime operation. For those local residents who are troubled by noise from construction sites, there is some consolation in it being temporary, so the controls that are reasonable to apply should not be as restrictive as those from permanently established factories.

Street Noise

Except for restrictions on the use of noisy equipment and the general requirements of the Control of Pollution Act 1974, as mentioned above, there is little more that can be done by current legislation to reduce the noise in streets. Road construction is treated as a construction site. There are codes of practice on noise from ice cream van chimes, intruder alarms and model aircraft. There is a general ban in the Act on the use of loudspeakers at night or for the purpose of advertising except in emergency.

Chapter 9
Hearing Conservation

The human ear (Figure 1) is a delicate mechanism which accepts sound waves along its auditory canal and through to the brain where it translates them into pitch and volume. Specifically, the auditory canal is confined to the external ear, separated from the middle ear by the tympanic membrane (eardrum). The middle ear is a cavity in the temporal bone containing three articulated bones (ossicles) transmitting the motion of the eardrum to an oval 'window' opening into the inner ear, but covered by a membrane. The inner ear is filled with fluid and contains the cochlea or actual 'hearing' portion of the ear which is essentially a system of coiled tubes also filled with fluid and separated by membranes. The basilar membrane separates the media from another tube, the scala tympani, and contains on its surface the hair cells which are the sound receptors. Motion of the basilar membrane causes bending of the hair cells, this movement grenerating impulses in the cochlea nerve endings which encircle the cells. These impulses are then transmitted through a series of lower auditory structures to the brain.

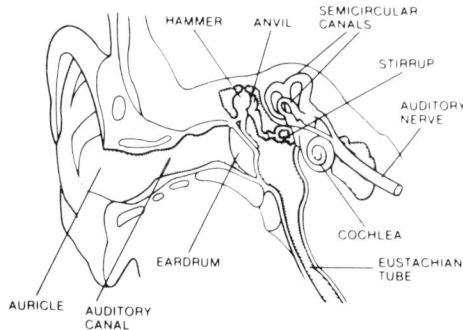

Figure 1

Apart from disorders or disease (or congenital defects), the two main causes of hearing loss are the deteriorating effects of age and damage to the delicate hair cells by exposure to excessive noise. (Damage to hair cells can also be caused by ototoxic drugs).

Hearing Impairment

Traditionally, hearing Impairment is looked upon as an inability to understand speech communication. On this basis to hear major speech components can be assessed as the average hearing levels at frequencies of 500, 1000 and 2000 Hz. An average loss of up to 25 dB represents no actual handicap, but each decibel of average loss above 25 dB represents a 1.5 per cent handicap.

More realistically, hearing handicap is assessed over a wider frequency range particularly as, when damaged by noise, the frequency affected first is 4000 Hz (and higher frequencies). On This basis, recommended criteria representing hearing loss are:

>   More than 15 dB at frequencies of 500, 1000 or 2000 Hz.
>   More than 20 dB at 3000 Hz.
>   More than 30 dB at 4000 and 6000 Hz.

Presbyacusis

This ageing effect, known as presbyacusis, normally starts at an age as early as 25 years, and with advancing age there is an ancillary loss of hearing, particularly with the higher speech frequencies. Figure 2 shows typical average patterns for healthy male and female subjects, with no history of exposure to excessive noise.

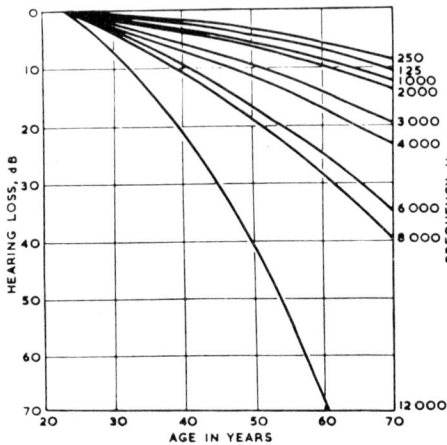

Figure 2 Presbyacusis curves for men and women with no history of exposure to excessive noise

Damaging Noise Levels

Noise within the sound pressure range 30 dB to 70 dB is generally accepted as 'safe'. That is, it has no adverse physiological effects on the human ear. The degree of tolerance within this range is, however, very variable. The higher figure represents what is probably the *logical* upper limit for tolerance provided this is based on A-weightings, i.e. sound pressure level measurement in dB(A). From the practical point of *achieving* such levels in typical 'noisy' atmospheres 80 dB or even 90 dB may be argued as 'acceptable'.

Continual exposure to sound pressure levels above 80 dB is potentially damaging. Short term exposure to sound pressure levels up to 100 dB is likely to lead to a temporary shift in the hearing threshold, generally with complete recovery. Longer term exposure presents the hazard that recovery of hearing will not be complete and a permanent loss of hearing will result. Above 100 dB the damage risk is drastically increased and only very short term exposure can be accepted if permanent loss of hearing is to be avoided. A sound pressure level of 120 dB also represents a *discomfort level*, i.e. a level of sound at which extreme discomfort is felt. At around 140-150 dB, discomfort is experienced as actual pain. At sound pressure levels with peaks of 160–180 dB there is the possibility of immediate mechanical damage to the ear – a burst eardrum. At lower levels, down into the 'risk' range, the damaging effect of sound is cumulative.

Noise has three main effects on hearing – acoustic trauma, temporary threshold shift and permanent threshold shift. It is important to realize that this type of deafness is 'perceptive' and, unlike 'conductive deafness', cannot be cured by either medical or surgical intervention. hence the serious nature of hearing damage caused in this way.

Acoustic trauma describes the immediate injury caused by exposure to intense sounds like blasts, explosions and gunfire. The perceptive deafness resulting from such sounds can be considerable and it is imperative that personnel exposed to noise levels in excess of 120 dB, no matter how short the duration, should take adequate precautions to protect their ears.

It is a general characteristic of both a temporary and a permanent threshold shift that the maximum loss of hearing occurs at around 4 kHz. In the case of permanent threshold shift the loss can be of the order of 60 dB at this frequency, with a substantial loss of hearing over the whole of the normal speech frequency range of 500 Hz to 4 kHz. It also follows that signs of permanent threshold shift are first associated with loss of acuity at the higher frequencies (e.g. the sharp consonants in speech), and progressive loss of acuity throughout the remainder of the speech frequency range. Such hearing loss is also accentuated by presbyacusis.

Frequency Response

The lowest frequency which the average person with normal hearing can detect is about 20 Hz. This does not alter with health or age. The highest frequency which the ear can detect is about 20 kHz but the ability to hear higher frequencies is very dependent on the individual's inherited acuity, age, health and history of exposure to noise. A typical normal limit for a younger person would be about 16–18 kHz, and for an older person could be very much less (e.g. 10–12 kHz). It is

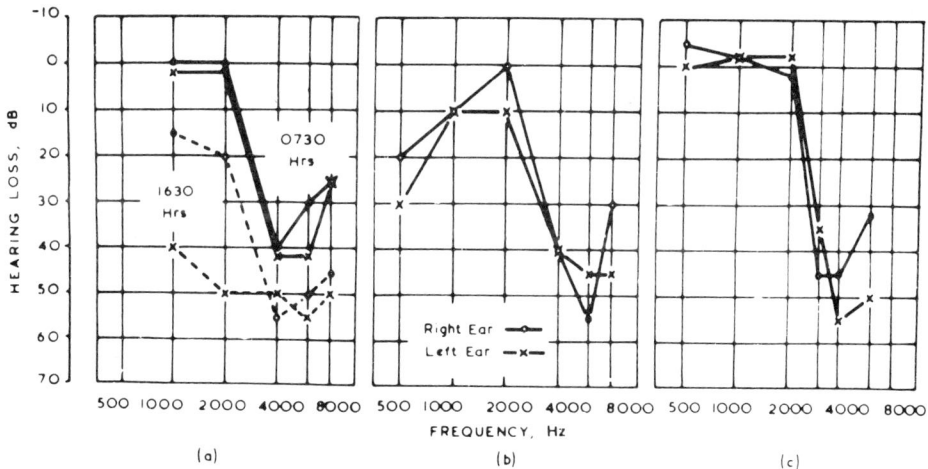

Figure 3 Pure audiogram showing temporary and permanent threshold shift caused by industrial noise:

(a) *Temporary threshold shift after one day's exposure to a sound field of 104 dB(A).*

(b) *Permanent threshold shift in a 28 year old man after 8 years' exposure to the automatic hammering of copper boilers.*

(c) *Permanent threshold shift in a 23 year old man showing the effect of working in a rolling mill and a machine shop.*

appropriate to comment here that 'Hi-Fi' performance calls for a minimum frequency range of 40 Hz to 12.5 kHz, or for complete 'Hi-Fi' capable of accommodating the full dynamic range of musical instruments, a range from 20 Hz to more than 15 kHz.

The human hearing mechanism is also more responsive to high frequency sounds and relatively insensitive to lower frequencies. In other words, listening to a high frequency sound and a low frequency sound, both at the same sound pressure level (or having the same sound energy), the high frequency sound would be judged 'louder'. This frequency content of a sound plays a large part in the psychological effect of noise at normal (below risk) levels.

Since assessment of 'loudness' is necessarily subjective, the relative loudness of different frequencies can only be measured by sampling – i.e. by a series of subjective tests. Moreover, as the tests are subjective, age will also influence the pattern established. As a general rule age difference is only noticeable to any extent at frequencies above 2 kHz.

Specifically, for the average person, a sound at 20 Hz would need to have a sound pressure level as much as 50 dB higher than a sound at 1 kHz to be judged equally as loud at low to moderate sound pressure levels. This would appear to indicate a relative insensitivity to low frequency sounds. However at higher sound pressure levels the difference to give 'equal loudness' response decreases.

Temporary threshold shift occurs when a person has been exposed for a few hours to noise levels of about 80 dB and above. These often leave a ringing in the ears for some time afterwards. The greater part of the hearing loss occurs soon after exposure and, similarly, recovery occurs largely in the 30 minutes following removal from the noise. Persons exposed to continuous noise at a level of 100 dB(A) for an eight hour working day could show a temporary threshold shift up to 40 dB in that part of the spectrum most affected (see Fig 3). However, this could overlap a degree of hearing loss already present; an older man with 30 dB presbyacutic loss, or an individual with a similar permanent hearing loss, would not be significantly affected.

Permanent threshold shift. When the ear is subjected to high intensity noise day after day, more lasting damage will take place. It is possible that a person may not recover full hearing between exposures, and what started as a temporary threshold shift eventually becomes permanent (see Figures 3b and 3c).

In the first stages, the pure tone audiogram shows a hearing loss in either the 3000 or 4000 Hz range, followed by further loss in these and higher ranges. At this stage the individual is not aware of his loss, but may complain of ringing in the ears. This is because frequencies between 3000 and 6000 Hz contribute little to the intelligibility of speech, although they are important to music. Gradually the condition progresses, extending down the frequency scale to affect speech frequencies (500–2000 Hz) until the sufferer is irreversibly deaf.

Speech Communication

The major source of intelligibility in speech communication is in the audio frequency range of 1000 – 4000 Hz, although maximum sound energy (loudness) of speech is normally generated at around 500 Hz; and the most significant frequency range for understanding speech is 500 – 2000 Hz. Equally, intelligibility will depend on the sound levels generated by individual noises, as well as articulation and phonetic balance of the spoken words. Typical sound levels for various bands of speech at a distance of 1 metre are given in Table 1. In general a typical female voice is about 5 dB lower than a male voice, unless deliberately raised, although sharing a similar peak pressure level at different frequencies – Figure 4.

TABLE 1 – SOUND PRESSURE LEVELS FOR SPEECH AT DISTANCE OF 1 METRE

Type of Speech	Male Voice		Female Voice	
	Level	Maximum Sound Energy	Level	Maximum Sound Energy
Whisper	20–30 dB(A) ⎫		20–25 dB(A) ⎫	
Low voice	50–60 dB(A) ⎪		45–55 dB(A) ⎬ At 900 Hz	
Normal voice	65–68 dB(A) ⎬ At 400 Hz		60–65 dB(A) ⎪	
Raised voice	70–75 dB(A) ⎪		68–70 dB(A) ⎭	
Loud voice	75–80 dB(A) ⎭		70–75 dB(A) ⎫ At	
Shouting	85 dB(A) and over		75–85 dB(A) ⎭ 1000–1100 Hz	

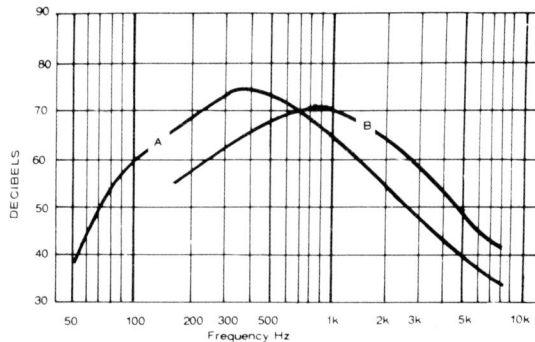

Figure 4 Generalised sound spectra of human speech. A – male B – female

Speech Interference Level

Background noise will inevitably have a masking effect on speech and the higher the background noise level the more the voice will have to be raised for satisfactory speech communication. Intelligibility will depend on the distance between speaker and listener.

Speech Interference Levels are a quantitative assessment of these parameters, expressing the masking effect of background noise which has to be overcome to establish reasonably reliable conversation – Table 2. Background noise levels are given in both dB and dB(A) in these tables. Original levels were based on averaging the actual sound levels preseing (in dB) in the three octaves above 600 Hz (i.e. 600 – 1200 Hz, 1200 – 2400 Hz and 2400 – 4800 Hz). The equivalent numerical values for dB(A) measurement, taken directly with a sound level meter with A-weighting, are also given in the table.

Examples of use:

(i)   At a distance of 1.5 metres (5 feet) a speaker using a normal voice should be intelligible provided the background noise level does not exceed 58 dB(A).

(ii)   If the background noise is 70 dB(A) and the distance between speaker and listener is 3 metres (10 feet), the speaker will have to shout to be heard.

TABLE 2 – SPEECH INTERFERENCE LEVELS

Distance From Speaker	Whisper		Low		Normal		Voice levels Raised		Very Loud		Shouting	
metres	dB	dB(A)	dB	dB(A)	dB	dB(A)	dB	dB(A)	dB	dB(A)	dB	dB(A)
0.25	41	47	56	62	67	74	73	79	79	86	85	90
0.50	35	41	50	56	61	68	67	73	73	80	79	84
0.75	32	38	47	53	57	64	63	69	69	76	75	82
1.0	29	35	44	50	55	62	61	67	67	74	73	78
1.5	26	32	41	47	51	58	57	63	63	70	69	76
2	–	–	38	44	49	56	55	61	61	68	67	72
3	–	–	–	–	45	52	51	57	57	64	63	70
4	–	–	–	–	43	50	49	55	55	62	61	68
5	–	–	–	–	41	48	47	53	53	60	59	66
6	–	–	–	–	39	46	45	51	51	58	57	64

Note: These figures are approximate

Alternatively, to use a normal voice the distance between speaker and listener would have to be reduced to about 0.3 metres (1 foot).

(iii) To be able to make conversation using a normal voice with a background level of 60 dB(A) the distance between speaker and listener will have to be 1.2 metres (4 feet) or less.

Values are based on the loudness of an average male voice and with the listener suffering no hearing loss. They are minima for barely reliable conversation (i.e. not less than 75% of the speech heard correctly). If the listener suffers from hearing loss, (usually expressed in dB), then this figure must be added to the background noise level.

Limits are as follows:

(i) Below 40 dB (45 dB(A)) background noise will be negligible.

(ii) If the background level exceeds 90 dB (97 dB(A)), intelligible speech communication is virtually impossible and can only be carried out satisfactorily with a headphone communications set.

Audiometry

Audiometry is the measurement of hearing, an audiometer is the instrument used to take measurements and an audiogram is a presentation of the results of those measurements in the form of a chart or graph. The technique used in audiometry is to determine the hearing threshold, i.e the minimum sound pressure level which a person can detect at a specified frequency.

In respect of Industry, European Directive 86/88/EEC requires audiometry facilities to be made available to those people at a risk of hearing loss, which is taken to mean at noisy sites, but the UK Noise at Work Regulations are silent on this point.

The argument for not including any such requirement in the Regulations is apparently that audiometry testing is available under the National Health Service and can be requested by any worker who feels the need. There is however no obligation on the patient's general practitioner to accede to his request unless he considers it appropriate. There are several drawbacks inherent in this attitude. Firstly, a careful employer may wish to have an audiogram of each worker at the commencement of his employment so that a change, if any, can be monitored; a GP may not think it wise to use Health Service resources for a patient who is apparently in good health and so may refuse to recommend a test on that ground. Secondly, even if an audiogram is called for, it may take several months if not years before the tests can take place. Thirdly, medical records are confidential so that any trend in hearing loss among workers in a particular area will only come to the employer's attention when it is too late for effective action. And fourthly, under this philosophy hearing tests cannot be made a condition of employment.

Many employers may wish to establish an audiometric test facility even though they are not obliged to do so by the regulations. There are also companies who offer mobile testing – usually those who supply ear protectors or audiometric equipment. It may be convenient for the person appointed to be a competent person (CP), as defined in the regulations, for carrying out a noise survey to have further training in audiometry. A number of companies and colleges offer suitable courses.

The following training schedule for training of industrial audiometricians has been recommended by the British Society of Audiology

Theoretical

1 Basic acoustics (45 minutes)
2 The ear and threshold of hearing (1 hour)
3 Noise-induced hearing loss and social handicap (1 hour 15 minutes)
4 Monitoring audiometry (1 hour 15 minutes)
5 Techniques of air-conduction audiometry without masking (1 hour 30 minutes)
6 Audiogram and categorisation (1 hour)
7 Organisation of audiometric programme (1 hour)
8 Personal hearing protection (1 hour)

Practical

1 Otoscopic examination (5 hours)
2 Tutorial: assessment of validity and categorisation of audiograms (1 hour)

The complete training programme takes 15 hours, but in practice it may take longer. It is not an arduous programme and could well be undertaken by someone who could combine it with another skill such as nursing or as a competent person (CP) for performing noise surveys. It would be desirable that a person appointed to be an audiometrician should work under the general direction of a medical practitioner.

The current recommendations in the UK for audiometric examination are to be found in the HSE Document 'Audiometry in Industry', which is currently out of print and due for revision.

An initial test on entry should identify those employees who have hearing peculiarities or are particularly sensitive and unsuitable for certain kinds of employment. Further audiograms at regular intervals should form part of the employee's record and can thus form part of the employer's defence, if one were needed, against a claim for industrial deafness.

An audiogram is taken by exposing the subject to a slowly varying sound level at several fixed frequencies, and a note made of the level that he can just hear. The results are compared with the normal threshold determined from a series of tests performed on healthy young adults. The audiometer must be capable of generating a set of pure tones at fixed frequencies of 0.5, 1, 2, 3, 4,

6 and 8 kHz, which are generated in a pair of earphones. The pure tone can be either continuous or pulsed.

It is essential that the audiometer must be capable of calibration to maintain the standard threshold, so that a comparison is valid. BS 2497 (ISO 389) prescribes the threshold and the accuracy required in various kinds of laboratory.

Indications for an Audiometric Programme

The main purposes of industrial audiometry are to establish the hearing status of an individual and monitor hearing during the period of employment. It will provide a test for the success of measures taken to reduce occupational hearing loss.

Routine examination of workers whose exposure does not exceed 85 dB(A) is not normally necessary, but should be instituted for all those working in a noise environment above 105 dB(A). As the noise level increases between the two values, the desirability for an audiometric programme increases. This comment should be read in conjunction with the Noise at Work Regulations 1989, regarding the provision of hearing protectors above 85 dB(A). Even if hearing protectors are available and properly used, this should not remove the desirability of having an audiometric programme.

The programme should consist of a pre-employment test followed by further tests at regular intervals. The interval may be adapted to the level of exposure. As a minimum, it is recommended to test annually for the first two years and thereafter at 3-year intervals.

Action following audiometric testing

HSE recommends the following action. The hearing threshold level otherwise known as the hearing level (HL) is first determined in dB. The HL values are summed over the low frequencies, 0.5, 1, 2 kHz (HL_1) and the high frequencies 3, 4, 6 kHz (HL_2).

If either sum shows an increase of 30 dB compared with the previous audiometric examination or 45 dB if the examination was performed more than 3 years earlier, the case is categorised 1.

If the difference of the sums between the two ears exceeds 45 dB for low frequencies or 60 dB for high frequencies, the case is 2.

The sums should be compared with those in Table 3. If the sum for either ear for either low or high frequencies exceeds the referral level, the case is 3.

TABLE 3 – HEARING LEVELS (dB) FOR CASES 3 AND 4

Age (years)	HL₁ (dB)		HL₂ (dB)	
	Warning level	Referral level	Warning level	Referral level
20 – 24	45	60	45	75
25 – 29	45	66	45	87
30 – 34	45	72	45	99
35 – 39	48	78	54	111
40 – 44	51	84	60	123
45 – 49	54	90	66	135
50 – 54	57	90	75	144
55 – 59	60	90	87	144
60 – 64	65	90	100	144
65 –	70	90	115	144

If the sum exceeds the warning level for either ear for either high or low frequencies, but in no case exceeds the referal level, the case is 4.

All other cases are case 5.

The designated medical practitioner will be responsible for deciding on the appropriate action to be taken in the case of persons categorised in any of the cases 1 to 4. In category 4 all persons should be told of the warning status and recommended to take precautions to preserve their hearing.

Audiometer Booths

When carrying out audiometric testing, unless an area can be found where the ambient noise is low, it will be necessary to use an audiometer booth. The maximum ambient noise level in the region of the subject's head should not exceed the values in the following table (ISO 6189)

Frequency Hz	125	250	500	1k	2k	4k	8k
Sound pressure level dB	37	35	33	25	21	5	43

A booth constructed to satisfy this condition will need to be better than a simple sound shelter, though not as good as an anechoic chamber. A suitable booth is illustrated in Figure 5. This is the Bilsom Audiometric Cabin. It is delivered dismantled and can be easily erected on site. It has attenuation characteristics:

Frequency Hz	25	250	500	1k	2k	4k	8k
Attenuation dB	25	33	47	51	51	56	61

It should not be difficult to find a location where the actual noise outside the booth was less than the sum of the values in these two tables.

Figure 5 Bilsom Audiometric Cabin

An alternative would be the use of a mobile facility, as illustrated in Figure 6. This contains all the equipment needed for a survey of the work force.

A number of recommendations for the quality of audiometric booths have been published, but ISO 6189 is likely to be embodied in national standards, and should be used as the basis for choice.

If an audiometric booth is used, it is important that its attenuation should be checked regularly, particularly if modifications to the structure have been made. Even a small hole for the purpose of admitting an electric cable, for example, can drastically alter the attenuation characteristics.

Figure 6 Mobile audiometric unit in use for on-site testing

SECTION 2

Chapter 1
Linear Vibration Theory

Basic Vibration Theory

The simplest form of mechanical vibration to consider is that based on a linear theory. This means that displacement, velocity and acceleration bear a proportional relationship to each other and to the mechanical stiffness of the system. This is clearly true of a coiled mechanical spring, where the displacement is proportional to the applied force, but is not generally true where the spring is an elastomer or a pneumatic or hydraulic cylinder. Not all springs based on mechanical deflection have a linear relationship between force and deflection, e.g. a spring based on a buckling strut, or when the metal of the spring works in the non-proportional region of the stress strain curve.

Linear systems are the easiest to analyse mathematically, and so engineers, in the main, consider only such systems and go to great lengths to approximate the real system in front of them by one which can be dealt with using simple mathematical techniques. Almost any structure, however non-linear, can be approximated by a linear relationship if it vibrates with a small displacement about a mean, perhaps very large, static deflection. When we deal with vibrations in this text, linear vibrations will be assumed unless otherwise stated, but it must be borne in mind that such a treatment is very often an approximation to the real situation.

Degrees of freedom

The number of independant coordinates needed to describe the motion of a system is known as the degrees of freedom of the system. Thus, in general, a particle has three degrees of freedom (in the $x-$, $y-$ and $z-$directions) and a rigid body has six (as well as the three directional coordinates it can also rotate about the same three axes). For the body to vibrate in one of the directions there must also be a spring force associated with that direction. It is usual to try to reduce a complex system to one consisting of a series of masses and springs. For rotational vibration the corresponding values are the moments of inertia and the torsional stiffness. It should always be understood that every practical component has both mass and flexibility, and part of the skill of the analyst is to find a system where the masses and the flexibilities can be lumped together.

There are several important components where it is not possible to lump the masses and springs together in this way, as for example in beams and plates which have an infinite number of degrees of freedom. For uniform beams and plates, supported in a simple way, the solutions have been worked out. For other structures, it is usual to try and lump together the parameters and solve the vibration problem by computer analysis.

Equations of Motion of a simple system

It is usual to start the study of vibration by finding the equation of motion of a mass on the end of a coiled spring, which can be set in motion either by a reciprocating force applied to the mass or by a displacment of the support structure, see Figure 1.

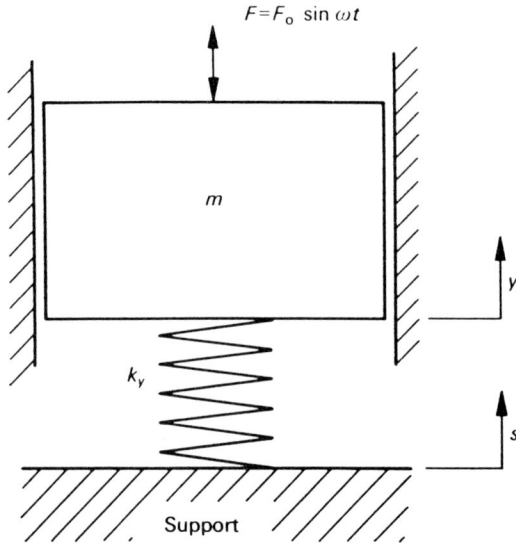

Figure 1 Idealised single degree of freedom system

Vibration caused by a force acting on the mass

The force is applied at an angular velocity of radians per second, where the relation between angular velocity and frequency is

$$\omega = f/2\pi$$

   f is angular velocity in radians per second

   ω is the frequency in cycles per second

The differential equation of motion is:

$$m \frac{d_2 y}{dt^2} = F_0 \sin \omega t - ky$$

The solution of this equation (ignoring the transients) is

$$y = \frac{F_0 \sin t}{k \{1 - (\omega/\omega_n)^2\}}$$

ω_n is the natural angular velocity of the system, given by

$$\omega_n = \sqrt{km} \text{ radians per second}$$

The maximum displacement y_0 is given by

$$y_0 = \frac{F_0}{k\{1 - (\omega/\omega_n)^2\}}$$

The natural frequency, i.e. the frequency with which the mass will oscillate if the force is removed is $f_n = \omega_n/2\pi$ cycles per second.

Equation (1) is a fundamental equation for forced harmonic motion.

It represents a sine wave with a repetition frequency of $\omega/2\pi$ per second, and as such it has a peak or maximum value of

$$y = y_0$$

similarly the velocity v, being the differential of displacement with respect to time, is given by

$$v = \frac{dy}{dt} = \omega\, y_0 \cos \omega\, t = v_0 \cos \omega\, t$$

and the acceleration being the differential of velocity or the second differential of displacement

$$a = \frac{d^2y}{dt^2} = -\omega\, y_0 \sin \omega\, t = a_0 \sin \omega\, t$$

v_0 and a_0 are the peak velocity and acceleration. It can be seen that displacement, velocity and acceleration each can be represented by a sine wave (a cosine is of course a sine wave displaced by $\pi/2$)

The motion which these equations describe is also known as simple harmonic motion.

Vibration caused by a displacement of the support

The mass can be set into motion by a displacement of the spring support. This situation is particularly relevant when studying methods of vibration isolation. If for example it is desired to isolate an item of equipment, represented by the mass, from a vibration of the support, it is important to study the conditions which make this possible.

Suppose, in reference to Figure 1, that the displacement of the support is

$$s = s_0 \sin \omega t$$

The differential equation of motion is

$$m\frac{d_2y}{dt^2} = k\,(s - y)$$

The solution of this equation is

$$y = \frac{s_0 \sin t}{1 - (\omega/\omega_n)^2}$$

The maximum displacement of the mass, y_0 is now given by

$$y_0 = \frac{s_0}{1 - (\omega/\omega_n)^2}$$

Transmissibility

When investigating the performance of a vibration isolation system it is sometimes helpful to know the ratio of the force applied to the mass to the force applied to the support. This is called the force transmissibility, T_F. The force applied to the support is $k\,y$ and so

$$T_F = k\,y/F = \frac{1}{1 - (\omega/\omega_n)^2}$$

The displacement transmissibility T_D is the ratio of the displacement of the mass to the displacement of the support.

$$T_D = y_0/s_0 = \frac{1}{1 - (\omega/\omega_n)^2}$$

In this case $T_F = T_D$. The transmissibilities as described by the above equations are illustrated in Figure 2.

A number of points can be made from a study of these equations: the peak values of velocity and acceleration are proportional to the peak displacement; the repetition frequency of displacement, velocity and acceleration is the same as that of the force. The displacement is in phase with the force, the velocity leads the displacement by 90° ($\pi/2$) and the acceleration leads the velocity by a further 90°. The acceleration is in the opposite direction from the displacement (180° out of phase).

Figure 3 shows the sine wave representation of a vibration. Although one should keep in mind the simplicity of the assumptions that have been made in deriving the formulas, more complex vibrations may have other descriptive quantities applied to them, and these can conveniently be

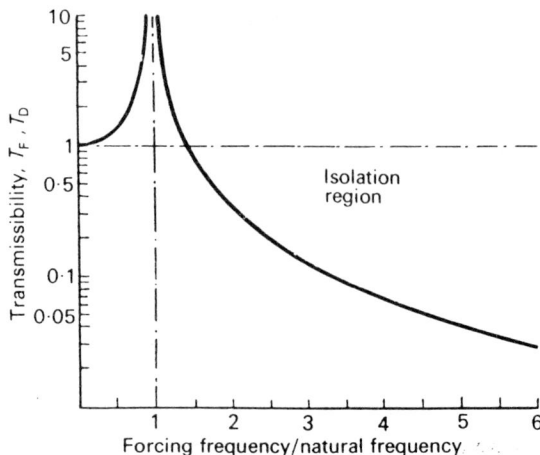

Figure 2 Transmissibilities of an undamped system

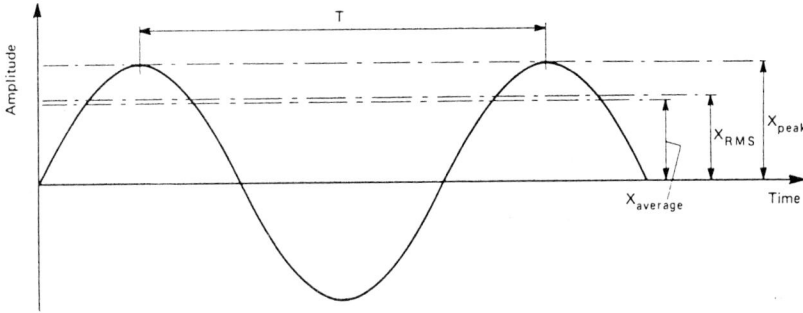

Figure 3 Example of a harmonic vibration signal with indication of the peak, the RMS and the average absolute value

discussed by referring to this figure. Certain definitions can be introduced at this point. The amplitude is the peak displacement, y_0, measured from the time axis, and the total excursion is the peak to peak measurement.

The peak value describes the vibration in terms of the instantaneous magnitude regardless of its time history. Other quantities do take the time history into account. The average absolute value is found from

$$y_{av} = 1/T \int_0^T |y| \, dt, \text{ which for a sine wave}$$

$$= 2y_0/\pi = 0.637 \, y_0$$

The average is rarely used, because of its limited theoretical application and the difficulty in taking the absolute value in a measuring circuit. A more commonly used descriptor of the magnitude, particularly of a complex vibration, is the root mean square (rms), primarily because one of the most important characteristics of a vibration is its energy content. The energy is proportional to the (amplitude) ^{2}, so a mean based on the second power will give a better idea of the power content when comparing vibrations.

$$y_{rms} = \sqrt{1/T \int_0^T y^2 \, dt} \quad \text{which for a sine wave}$$

$$= y_0/\sqrt{2} = 0.707 \, y_0$$

It may seem a trivial exercise to perform these calculations for a simple vibration represented by a sine wave, but rms is of great importance when dealing with more complex vibration regimes. At this stage it should be noted that it will not be true in general that the rms value can be calculated by multiplying the peak by 0.707.

Two other parameters are the crest factor, F_c, and form factor, F_f, where for simple harmonic motion,

$$F_c = y_0 / y_{rms} = \frac{2}{\sqrt{2}} = 1.414 \, (= 3 \, dB)$$

$$F_f = y_{rms} / y_{av} = \frac{\pi}{2\sqrt{2}} = 1.11 \, (= 1 \, dB)$$

Most of the vibrations encountered in practice are more complex than those described here. A vibration may be regular and periodic in the sense that each single wave form is the same as its predecessor and its sucessor in time, but may not be harmonic. A typical example is shown in Figure 4 which describes the piston acceleration of a combustion engine. By determining the peak, average and rms values of this vibration some useful information can be obtained, but a more precise analysis may be desired, particularly if the effect of the vibration on a structure is wanted. One useful method of interpreting the wave form illustrated in Figure 4 is by the use of Fourier analysis.

Fourier Analysis

This technique was originally developed by Joseph Fourier early in the 19th century to solve problems of heat flow. It is based on the theorem that any regularly periodic wave form, however complex can be broken down into a series of simple sinusoidal curves with harmonically related frequencies. So the vibration of Figure 4 can be looked at as the sum of two sine waves y_1 and y_2, where

$$y_1 = a_1 \sin \omega t \text{ and}$$
$$y_2 = a_2 \sin 2\omega t$$
$$y = y_1 + y_2$$

In general, any vibration has a form

$$y = a_0 + \sum_{k=1}^{k=n} a_k \sin k\omega_t + \phi_k$$

where ϕ_k is the appropriate phase angle difference that applies to the curve defined by the index k.

Another way of expressing the Fourier equation is:

$$y = a_0 = \sum_{k=1}^{k=n} a_k \sin k\omega_t + b_k \cos k\omega_t$$

Very often just the first few terms of the series are adequate for analysis purposes. For a square wave as shown in Figure 5, the most difficult point to approximate by a Fourier series is the corner of the square. By taking sufficient terms the point at the corner can be defined as closely as required but it can never be exactly defined.

Using Fourier analysis, the vibration as pictured in the time domain by the waveform can be represented in the frequency domain by the spectrum lines. The frequency domain concept of expressing vibration can be particularly useful when studying the effect of vibration on a structure. A further extension of the frequency domain is found in random vibrations where the line spectrum becomes a continuous curve. Fourier analysis then becomes a Fourier integral. This is dealt with in detail later.

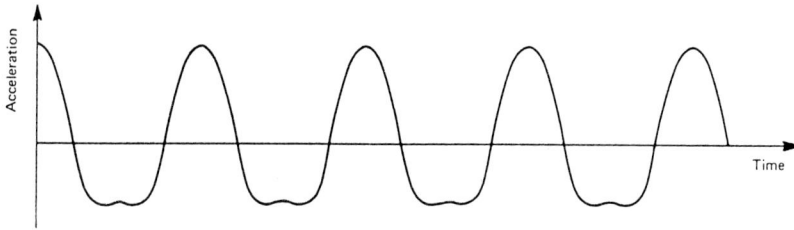

Example of a non-harmonic periodic motion (piston acceleration of a combustion engine)

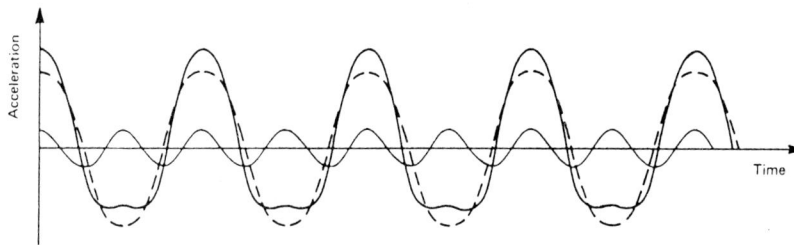

Illustration of how the waveform can be "broken up" into a sum of harmonically related sinewaves

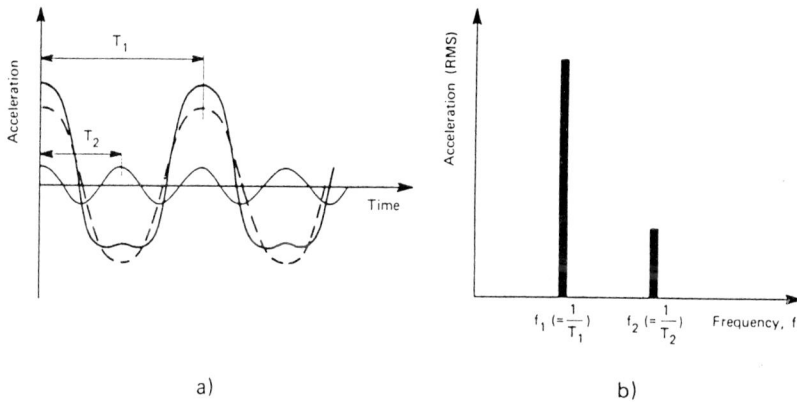

a)

b)

Illustration of how the signal, can be described in terms of a frequency spectrum
a) Description in the time domain
b) Description in the frequency domain

Figure 4

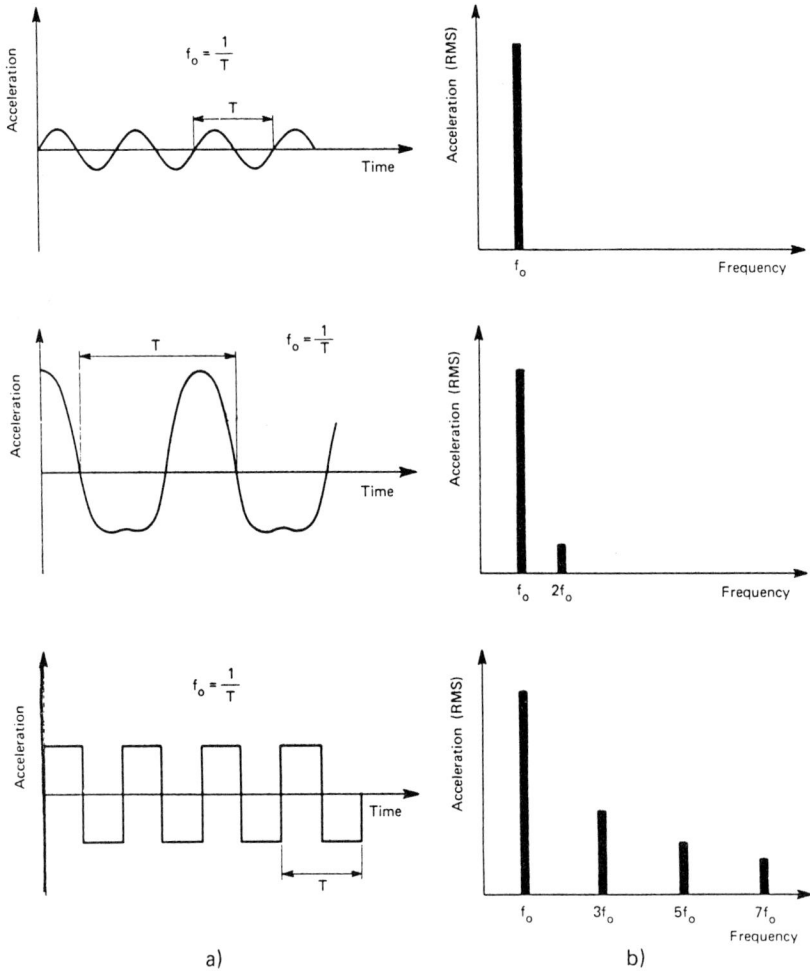

Figure 5 Examples of periodic signals and their frequency spectra
        a) Descriptions in the time domain
        b) Descriptions in the frequency domain

Damping

So far the analysis has been restricted to vibrations in which the only forces acting are caused by the acceleration of the mass and by the extension of the spring. If such a system were set into motion it would continue to vibrate indefinitely at its natural frequency, but in practice other forces come into play which act to stop it. All vibrating elements are subject to these damping forces which act to remove energy from the system. One of the most common forms of damping is viscous damping, not only because it is fairly commonly encountered in practice but also because it lends itself fairly readily to mathematical treatment, requiring only a modification to the equation of motion by the introduction of a term proportional to velocity. Viscous damping is found in a dashpot (a hydraulic device which by control of fluid flow through a restricted hole produces an output force proportional to the velocity of input).

Other forms of damping are to be found. Friction between dry surfaces, in which the force is dependent only on the normal force acting between two surfaces and the coefficient of friction, is known as Coulomb damping. Damping in which the force is in phase with the velocity but proportional to the displacement is known as hysteretic or structural damping; it is produced by internal damping of the material of the structure. Quadratic or velocity squared damping is encountered in turbulent flow of a fluid.

If a theoretical analysis of damping is required it is usual to approximate whatever is the true situation by the equivalent form of viscous damping, but it is worth bearing in mind that the approximation involved may have to be a gross one. The usual approach to a solution is to equate the energy dissipated by viscous damping to that dissipated by the non-viscous damping, assuming harmonic motion.

Figure 6 shows the same system as before with the addition of a dashpot. The differential equation of motion in the absence of an exciting force is

$$m\frac{d_2y}{dt^2} = -ky - c\frac{dy}{dt}$$

c is the damping coefficient, with units of Ns/m

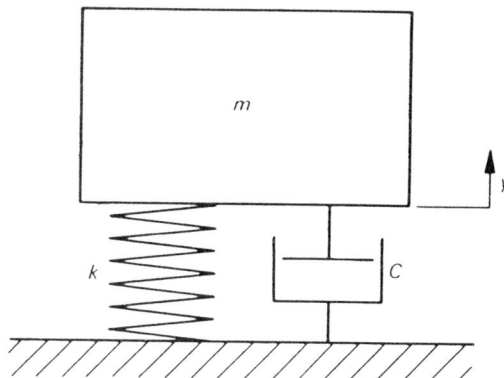

Figure 6 Idealised single degree of freedom system with viscous damping

This equation is conveniently rewritten

$$\frac{d_2 y}{dt^2} = -\frac{k}{m} y - \frac{c}{m} \frac{dy}{dt}$$

$$= -\omega^2 y - 2n \frac{dy}{dt}$$

The value $2n$ has replaced c/m for convenience.

The solution of this equation is

$$y = e^{-nt} (A_1 \sin \omega_1 t + A_2 \cos \omega_1 t)$$

where $\omega_1^2 = \omega^2 - n^2$ and A_1 and A_2 are constants which depend upon the initial conditions. For the details of the solution, the reader is referred to a specialist text. For the purposes of this chapter it is sufficient to realise that the solution of a damped vibrating system is the product of two functions, one represents the vibratory component and the other represents the exponential decay of the amplitude. This is shown in Figure 7. The dotted line represents the decay and the full line the vibration.

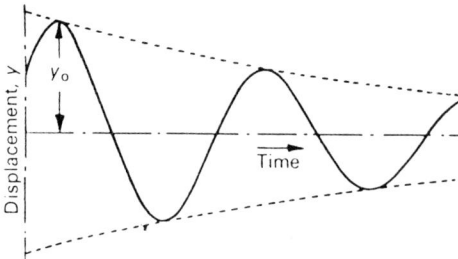

Figure 7 Free vibration of single degree of freedom system with viscous damping

If the damping is large, there is no vibratory component, or at most just one cycle, with the mass returning to its rest position asymptotically. When the damping is just large enough to satisfy this condition, the damping is said to be critical and the critical damping coeficient c_C is then equal to $2\sqrt{k/m}$. Most engineering systems possess less than critical damping, but the parameter c_C is a useful one. The damping ratio is defined as c/c_C.

The damping ratio can be found by measuring the ratio of successive peaks of the decaying vibration and determining the its natural logarithm. This is known as the logarithmic decrement.

$$\delta = \log_n (y_1/y_2)$$

In terms of the damping ratio

$$\delta = 2\pi \frac{c/c_C}{\sqrt{1 - (c/c_C)^2}}$$

If c/c_C is small, as it usually is,

$$\delta = 2\pi c/c_C$$

If the ratio of amplitudes in successive cycles is difficult to obtain accurately, the damping ratio can be obtained by averaging over a number of cycles of the decaying vibration. Thus if the first amplitude is y_1, the final amplitude is y_n and the number of cycles is n, then

$$c/c_C = \frac{\log_n (y_1/y_n)}{2 \quad n}$$

The presence of damping in a system clearly affects the amplitude of the vibration, but it also changes the natural frequency to a lesser extent. The relation between damped and undamped natural frequencies is found from

$$\frac{\omega \quad damped}{\omega \quad undamped} = \sqrt{1-(c/c_o)^2}$$

Forced Vibration with Damping
Forced vibration, as opposed to free vibration, is excited by the application of a periodically varying disturbance. The additional energy supplied to the system causes the amplitude of vibration to be maintained. Significant parameters are:

$$\text{the undamped natural frequency } \omega_n = \frac{kg}{W} = \sqrt{\frac{k}{m}}$$

$$\text{the static deflection of the system } d_0 = \frac{F_0}{k}$$

where
F_0 is the deflecting force
ω = weight

The relative response or *magnification factor* (R) is then given by:

$$R = \frac{d}{d_0}$$

$$= \frac{1}{\sqrt{1 - \left(\frac{\omega}{\omega_n}\right)^2 + 2\,(c/c_c)\,\frac{\omega}{\omega_n}\right)^2}}$$

Graphical analysis for single-degree-of-freedom systems is shown in Figure 8 rendered in terms of the dimensionless parameters - damping-ratio (c/c_c) and frequency-ratio (forcing frequency to undamped natural frequency, f_n). Such curves show the displacement amplitude generated by forced vibration of the system. Similar curves can be derived to show the transmissibility, or ratio of the resultant displacement to the applied displacement. Graphical solutions are also preferred for solutions of forced vibration applied to multi-degree-of-freedom systems. In all such cases, practical control of vibration is concerned with the design of a suitable elastic system.

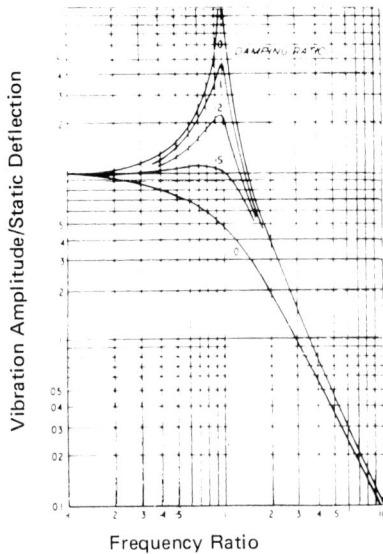

Figure 8

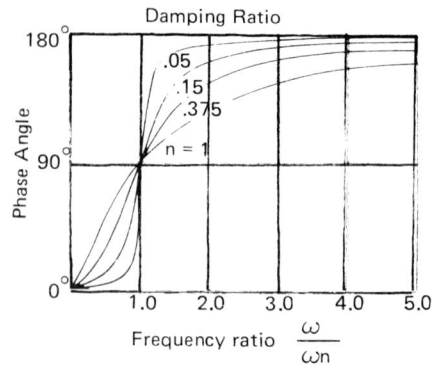

Figure 9

The phase angle is similarly defined by:

$$\text{the undamped natural frequency } \omega_n = \frac{kg}{W} = \sqrt{\frac{k}{m}}$$

This relationship is shown graphically related to frequency ratio – Figure 9.

The curves of Figures 8 and 9 show that both the vibration amplitude and the phase angle (between the forcing function and the resultant displacement of mass) are strongly affected by the damping ratio and the frequency of vibration. In general, as damping becomes smaller, amplitude of vibration at resonance becomes larger. Small values of damping are said to produce high 'Q' resonance. (In theory, with zero damping the vibration amplitude could become infinitely large). Also, for small damping the phase angle shifts more rapidly from 0° to 180°. For lightly damped systems, one technique for finding resonance is to determine the frequency of 90° phase shift.

Coulomb Damping

If the damping is primarily caused by friction, the decrement curve, instead of being exponential, is a straight line as in Figure 10. This gives an instant method for detecting a system with friction, and its magnitude. Thus in Figure 10 the value of the frictional force, F, is given by

$$F = \frac{k}{4}(y_1 - y_2)$$

For the purpose of analysis it is often convenient to express Coulomb damping in terms of the equivalent viscous damping coefficient; this is valuable when assessing the limiting amplitude of vibration. Because Coulomb damping introduces a non-linear component in the equation of

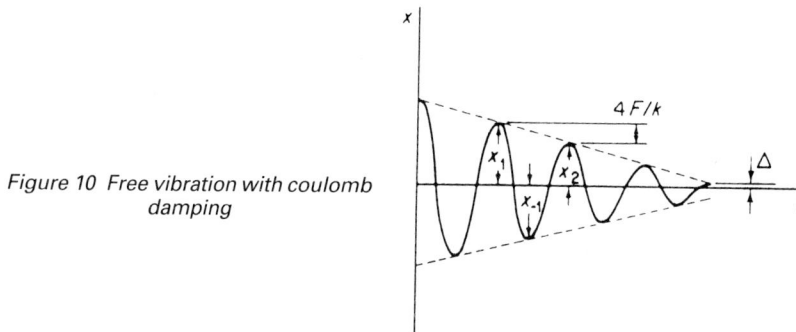

Figure 10 Free vibration with coulomb damping

motion, the equivalent coefficient depends on the amplitude of the vibration.

$$c_{eq} = \frac{4F}{\pi \omega y}$$

c_{eq} is the equivalent viscous coefficient and y is the amplitude at which this value applies.

The oscillation terminates when the amplitude reduces to Δ at which point the spring force is incapable of overcoming friction. The frequency of oscillation is unaffected by the presence of friction, i.e. $\omega = \sqrt{k/m}$.

Torsional Vibrations

A single degree of freedom system in torsion consists of a torsional spring (usually a flexible shaft) and a body (a wheel for example) possessing rotational inertia. The formulas developed above for vibrations of a mass on the end of a helical spring are exactly analogous to torsional vibrations where the mass is replaced by the moment of inertia of the body and the helical spring by the torsional stiffness of the shaft.

So if the stiffness of the torsional spring is K (in N m/rad) and the inertia of the body is J (kg m^2), the natural angular frequency of torsional vibrations is

$$= \sqrt{K/J} \text{ rad/s}$$

Torsional vibrations are important when studying the behaviour of engine stability and the coupling of engines with the machines they are used to drive. The subject will be dealt with in detail in subsequent chapters.

Spring Stiffness

The key to the calculation of vibrations lies in the calculation of the spring stiffness. The mass is usually easily determined.

The table gives a number of standard cases

TABLE 1 – SPRING STIFFNESSES

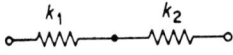

$$k = \frac{1}{1/k_1 + 1/k_2}$$

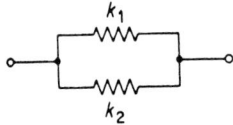

$$k = k_1 + k_2$$

$$k = \frac{EI}{l}, \qquad I = \text{moment of inertia of cross-sectional area}$$
$$l = \text{total length}$$

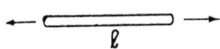

$$k = \frac{EA}{l} \qquad A = \text{cross-sectional area}$$

$$k = \frac{GJ}{l} \qquad J = \text{torsion constant of cross section}$$

$$k = \frac{Gd^4}{64nR^3} \qquad n = \text{number of turns}$$

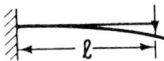

$$k = \frac{3EI}{l^3} \qquad k \text{ at position of load}$$

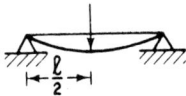

$$k = \frac{48EI}{l^3}$$

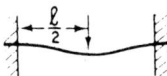

$$k = \frac{192EI}{l^3}$$

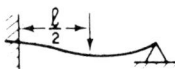

$$k = \frac{768EI}{7l^3}$$

Table of Spring Stiffness. (*Continued*)

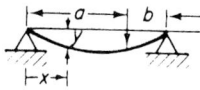

$$k = \frac{3EIl}{a^2b^2} \qquad\qquad y_x = \frac{Pbx}{6EIl}(l^2 - x^2 - b^2)$$

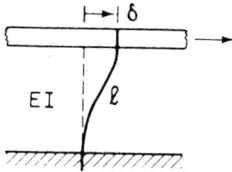

$$k = \frac{12EI}{l^3}$$

EI

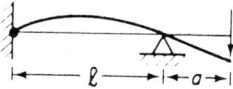

$$k = \frac{3EI}{(l + a)a^2}$$

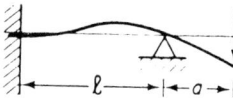

$$k = \frac{24EI}{a^2(3l + 8a)}$$

Chapter 2
Multi-degree of Freedom Systems

Single- and Multi-degree of Freedom Systems

Although the single degree of freedom system is a useful concept in studying basic theory, many practical cases involve more than one degree of freedom. A continuous structure such as a beam or plate on which the load is distributed over its area will in theory have an infinite number of degrees of freedom. With each degree of freedom is associated a mode of vibration and a natural frequency. Usually only the first few of the lowest frequencies are of importance. The mode of vibration is the shape of the deformed structure associated with one particular frequency. When free vibration takes place at one of the natural frequencies, the configuration of the vibration is referred to as the normal mode.

The mathematical complexity of analysis increases rapidly with the number of degrees of freedom. If a system has just two, then the motion can be expressed as two simultaneous differential equations and, in general, n degrees of freedom requires an array of n simultaneous equations. The solution of these is made much easier through the use of matrix methods in computer analysis. A number of proprietary packages are available to suit a variety of different types of computer, from the desktop personal computer to the the largest mainframe.

Coupled and Uncoupled Modes of Vibration

If excitation in a particular direction results in vibration in one or more modes, the modes are said to be coupled. If vibration in one direction results in vibration in only one mode, then the modes are decoupled. An example of coupled modes is shown in Figure 1, where a mass is supported by four springs. The modes of vibration might be considered as a vertical transition plus the various rotations of the mass about several centres as shown in the figure. The system, in theory at least, has five degrees of freedom (if one ignores lateral spring movements and assumes only vertical spring deflections), although not all may be significant. To some extent the choice of modes is arbitrary, so alternative modes might be the deflection of the left spring alone and deflection of the right spring alone. The resulting deformation of the structure under the influence of a particular disturbing force will be the same whatever choice is made. The choice will often be governed by the ease of computation. The test for coupling is whether the mode shapes appear in the same set of simultaneous equations. Those modes that appear in the same set are coupled; those that do not appear are uncoupled. It is desirable in any structural analysis that the system shall be reduced to one consisting only of uncoupled or normal modes.

There is one proviso that should be mentioned when considering coupled and uncoupled modes. Modes which are uncoupled in an undamped system may become coupled when damping is considered.

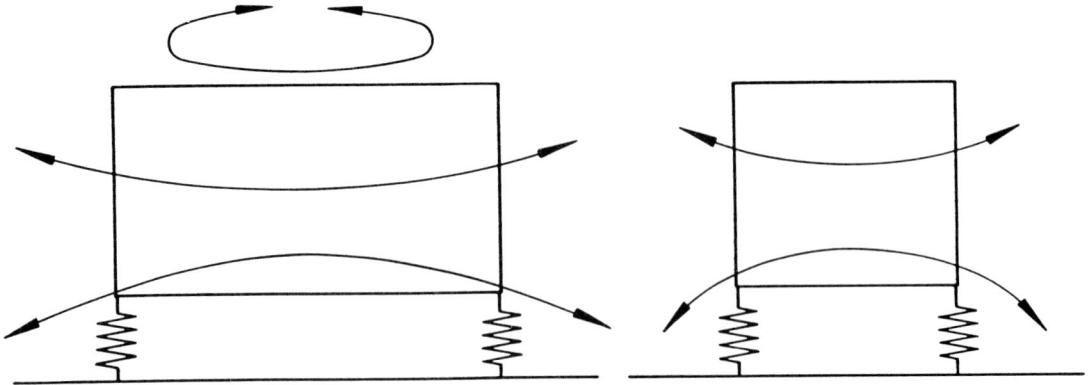

Figure 1 Modes of vibration of a practical system having resilient supports at each of four corners

Modes of vibration in continuous systems

A number of simple components can be analysed theoretically. These include beams, plates of various shapes, and shafts.

Table 1 presents the theoretical solutions for the natural frequencies of a selection of uniform beams. Figure 2 presents similar material in diagrammatic form. Note that, in this figure, A is equivalent to K_n. Table 2 is for a selection of bars, springs and strings. Table 3 is for a selection of plates. Square section plates are also shown diagrammatically in Figure 3, which also shows the nodal lines, or the lines along which the vibration is zero. Being able to identify the nodal lines, which can often be done very easily by sprinkling the surface of the horizontally vibrating plate with a powder and observing the position where the powder collects, is often a useful guide to the mode.

Vibration analysis of complex structures

Until powerful computers came into widespread use some years ago, much effort was spent in the search for analytical solutions to the problems of finding frequencies and modal shapes in

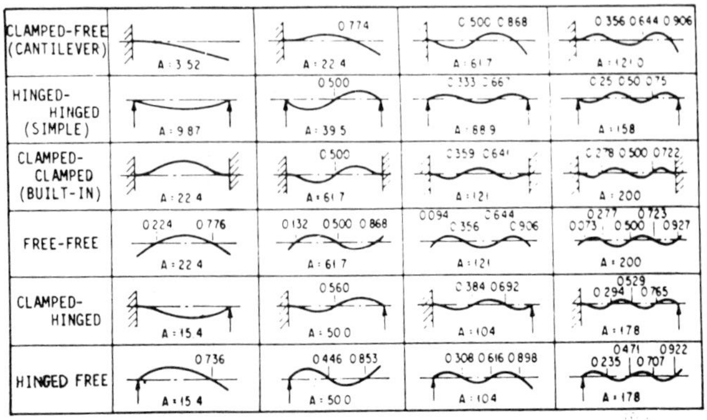

Figure 2 Boundary conditions and mode shapes for single uniform beams. (B. & K.)

TABLE 1 – NATURAL FREQUENCIES OF CONTINUOUS UNIFORM BEAMS

Support	Load	Natural frequencies							
Uniform beam; both ends simply supported	Centre load W, beam weight negligible	$f_1 = \dfrac{6.93}{2\pi}\sqrt{\dfrac{ELg}{Wl^3}}$							
	Uniform load w per unit length including beam weight	$f_n = \dfrac{K_n}{2\pi}\sqrt{\dfrac{ELg}{\omega l^4}}$	**Mode**	**K_n**	**Nodal position/l**				
			1	9.87	0.00	1.00			
			2	39.5	0.00	0.50	1.00		
			3	88.8	0.00	0.33	0.67	1.00	
			4	15.8	0.00	0.25	0.50	0.75	1.00
Uniform beam; both ends fixed	Centre load W, beam weight negligible	$f_1 = \dfrac{13.86}{2\pi}\sqrt{\dfrac{ELg}{Wl^3}}$							
	Uniform load w per unit length including beam weight	$f_n = \dfrac{K_n}{2\pi}\sqrt{\dfrac{ELg}{\omega l^4}}$	**Mode**	**K_n**	**Nodal position/l**				
			1	22.4	0.00	1.00			
			2	61.7	0.00	0.50	1.00		
			3	121	0.00	0.36	0.64	1.00	
			4	200	0.00	0.28	0.50	0.72	1.00
Uniform beam; left end fixed, right end free (cantilever)	Centre load W, beam weight negligible	$f_1 = \dfrac{1.732}{2\pi}\sqrt{\dfrac{ELg}{Wl^3}}$							
	Uniform load w per unit length including beam weight	$f_n = \dfrac{K_n}{2\pi}\sqrt{\dfrac{ELg}{\omega l^4}}$	**Mode**	**K_n**	**Nodal position/l**				
			1	3.52	0.00				
			2	22.0	0.00	0.783			
			3	61.7	0.00	0.504	0.868		
			4	121	0.00	0.358	0.644	0.903	
Uniform beam; both ends free	Uniform load W, per unit length including beam weight	$f_n = \dfrac{K_n}{2\pi}\sqrt{\dfrac{ELg}{Wl^4}}$	**Mode**	**K_n**	**Nodal position/l**				
			1	22.4	0.224	0.776			
			2	61.7	0.132	0.500	0.868		
			3	121	0.095	0.356	0.644	0.905	
			4	200	0.074	0.277	0.500	0.723	0.926
Uniform beam; left end fixed, right end hinged	Uniform load w per unit length including beam weight	$f_n = \dfrac{K_n}{2\pi}\sqrt{\dfrac{ELg}{\omega l^4}}$	**Mode**	**K_n**	**Nodal position/l**				
			1	15.4	0.0	1.000			
			2	50.0	0.0	0.557	1.000		
			3	104	0.0	0.386	0.692	1.000	
			4	178	0.0	0.295	0.529	0.765	1.000
Uniform beam; left end hinged, right end free	Uniform load W, per unit length including beam weight	$f_n = \dfrac{K_n}{2\pi}\sqrt{\dfrac{ELg}{Wl^4}}$	**Mode**	**K_n**	**Nodal position/l**				
			1	15.4	0.0	0.736			
			2	50.0	0.0	0.446	0.853		
			3	104	0.0	0.308	0.617	0.898	
			4	178	0.0	0.235	0.471	0.707	0.922

f = natural frequency Hz
K = constant for the mode of vibration
E = modulus of elasticity N/m^2
I = second moment of area m^4

TABLE 2 – NATURAL FREQUENCIES OF BARS, SPRINGS AND STRINGS

Support	Load	Natural frequencies
Uniform bar or spring vibrating along its longitudinal axis; upper end fixed, lower end free	Weight W at lower end, bar weight negligible	$f_1 = \dfrac{1}{2\pi}\sqrt{\dfrac{kg}{W}}$ for a spring $f_1 = \dfrac{1}{2\pi}\sqrt{\dfrac{AEg}{Wl}}$ for a bar
	Uniform load ω per unit length including bar weight	$f_n = \dfrac{K_n}{2\pi}\sqrt{\dfrac{AEg}{\omega l^2}}$ where $K_1 = 1.57$ $K_2 = 1.71\ K_3 = 7.85$
	Uniform load ω per unit length plus a load W at the lower end	$f_1 = \dfrac{1}{2\pi}\sqrt{\dfrac{kg}{W+\omega l/3}}$ for a spring $f_1 = \dfrac{1}{2\pi}\sqrt{\dfrac{AEg}{Wl+\omega l^2/3}}$ for a bar
Uniform shaft or bar in torsional vibration; one end fixed, the other end free	Concentrated end mass of J mass moment of inertia, shaft weight negligible	$f_1 = \dfrac{1}{2\pi}\sqrt{\dfrac{GK}{J_s l}}$
	Uniform distribution of mass moment of inertia along shaft; J_s = total distributed mass moment of inertia	$f_n = \dfrac{K_n}{2\pi}\sqrt{\dfrac{GK}{J_s l}}$ where $K_1 = 1.57$ $K_2 = 4.71\ K_3 = 7.85$
	Uniformly distributed inertia plus a concentrated end mass	$f_1 = \dfrac{1}{2\pi}\sqrt{\dfrac{GK}{(J + J_s/3)l}}$
String vibrating laterally under a tension T with both ends fixed	Uniform load ω per unit length including own weight	$f = \dfrac{K_n}{2\pi}\sqrt{\dfrac{Tg}{\omega l^2}}$ where $K_1 = \pi\ K_2 = 2\pi$ $K_3 = 3\pi$

G = shear modulus of elasticity N/m^2 k = spring constant N/m A = cross sectional area of a bar m^2

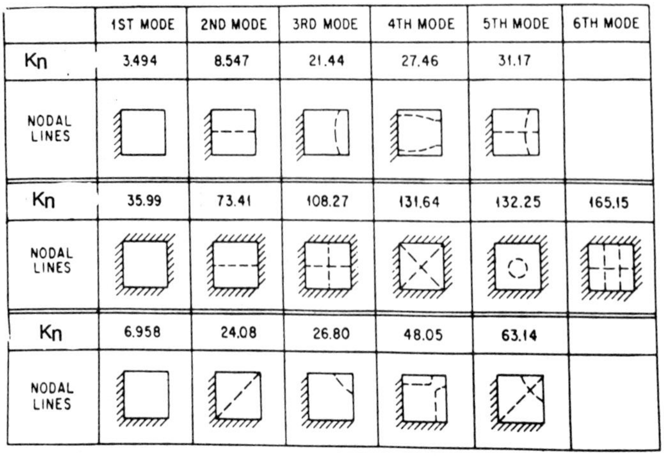

	1ST MODE	2ND MODE	3RD MODE	4TH MODE	5TH MODE	6TH MODE
Kn	3.494	8.547	21.44	27.46	31.17	
NODAL LINES						
Kn	35.99	73.41	108.27	131.64	132.25	165.15
NODAL LINES						
Kn	6.958	24.08	26.80	48.05	63.14	
NODAL LINES						

Figure 3 Nodal line configurations for square plates. (B. & K.)

TABLE 3 – NATURAL FREQUENCIES OF PLATES

Plate	Load	Natural Frequencies
Circular flat plate of uniform thickness t and radius r, edge fixed	Uniform load ω per unit area including own weight	$f = \dfrac{K_n}{2\pi}\sqrt{\dfrac{Dg}{\omega r^4}}$ where $K_1 = 10.2$ fundamental; $K_2 = 21.3$ one nodal diameter; $K_3 = 34.9$ two nodal diameters; $K_4 = 39.8$ one nodal circle
Circular flat plate of uniform thickness t and radius r, edge simply supported	Uniform load ω per unit area including own weight; $\nu = 0.3$	$f = \dfrac{K_n}{2\pi}\sqrt{\dfrac{Dg}{\omega r^4}}$ where $K_1 = 4.99$ fundamental; $K_2 = 13.9$ one nodal diameter; $K_3 = 25.7$ two nodal diameters; $K_4 = 29.8$ one nodal circle
Circular flat plate of uniform thickness t and radius r, edge free	Uniform load ω per unit area including own weight; $\nu = 0.33$	$f = \dfrac{K_n}{2\pi}\sqrt{\dfrac{Dg}{\omega r^4}}$ where $K_1 = 5.25$ two nodal diameter; $K_2 = 9.08$ one nodal diameter; $K_3 = 12.2$ three nodal diamters; $K_4 = 20.5$ one nodal diameter and one nodal circle
Elliptical flat plate of major radius a, minor radius b, and thickness t; edge fixed	Uniform load ω per unit area including own weight	$f = \dfrac{K_n}{2\pi}\sqrt{\dfrac{Dg}{\omega a^4}}$ where K_1 is tabulated for various ratios of $\dfrac{a}{b}$ a/b: 1.0 / 1.1 / 1.2 / 1.5 / 2.0 / 3.0 K_1: 10.2 / 11.3 / 12.6 / 17.0 / 27.8 / 57.0
Rectangular flat plate with short edge a, long edge b, and thickness t; all edges simply supported	Uniform load ω per unit area including own weight	$f = \dfrac{K_n}{2\pi}\sqrt{\dfrac{Dg}{\omega a^4}}$ where K_1 is tabulated for various ratios of a/b a/b: 1 / 0.9 / 0.8 / 0.6 / 0.4 / 0.2 / 0 K_1: 36.0 / 32.7 / 29.9 / 25.9 / 23.6 / 22.6 / 22.4
Rectangular flat plate with short edge a, long edge b, and thickness t; all edges simply supported	Uniform load ω per unit area including own wieght	$f = \dfrac{K_n}{2\pi}\sqrt{\dfrac{Dg}{\omega a^4}}$ where $K_n = \pi^2\left[m_q^2 + \left(\dfrac{a}{b}\right)^2 m_n^2\right]$ ($m_q = 1,\, m_n = 1$), ($m_q = 1,\, m_n = 2$), ($m_q = 2,\, m_n = 1$), ($m_q = 1,\, m_n = 3$) See table below
Rectangular flat plate with two edges a fixed, one edge b fixed, and one edge b simply supported	Uniform load ω per unit area including own weight	$f = \dfrac{K_n}{2\pi}\sqrt{\dfrac{Dg}{\omega a^4}}$ where K_1 is tabulated for various ratios of $\dfrac{a}{b}$ a/b: 3.0 / 2.0 / 1.6 / 1.2 / 1.0 / 0.8 / 0.6 / 0.4 / 0.2 / 0 K_1: 213 / 99 / 67 / 42.4 / 33.1 / 25.9 / 20.8 / 17.8 / 16.2 / 15.8

Table for the rectangular plate (all edges simply supported):

a/b	1.0	0.8	0.6	0.4	0.2	0.0
K_1	19.7	16.2	13.4	11.5	10.3	9.87
K_2	49.3	35.1	24.1	16.2	11.5	
K_3	49.3	45.8	41.9	24.1	13.4	

$$D = \frac{E t^3}{12(1-\nu^2)}$$

where t = plate thickness, and ν = Poisson's ratio

structures. It is now much more more common for these problems to be solved through finite element analysis, which consists in approximating a given structure by an assembly of simpler structural elements, each of which has a known behaviour when subject to external forces. The behaviour of the complete structure can be readily predicted. The mathematical procedure is straightforward in a conceptual sense, although arithmetically it would be very tedious in the absence of high speed computing methods. Desktop personal computers are capable of tackling a worthwhile range of problems. To handle most commercially available packages, a PC with a 80286 or 80386 Intel chip at least would be required. If one wishes to write one's own program using FORTRAN, a PC with a hard disk, would probably be adequate. Usually a great deal of development has gone into the preparation of the three-dimensional elements, but simple two-dimensional elements are readily generated.

There are a number of finite element systems on the market, which are suitable for solution of the most complex of problems, and one of these may be the best way of tackling a specific problem. But an alternative, economical answer could well be the writing of one's own computer program or adapting one already published. Usually proprietary systems are designed to do much more than tackle a particular limited range of structures and loading regimes, and may on that account be larger and more expensive than absolutely necessary. Most published program listings are in a high level language, usually FORTRAN but sometimes BASIC. These languages are now quite old and for many purposes have been replaced by more efficient ones, but despite that, they are likely to remain the popular choice of the practising engineer. Some of the commercial packages will allow a limited amount of programming by the user. Usually a consultancy service is provided by the suppliers of the commercial software packages.

A comprehensive finite element package, such as that provided by PAFEC, provides the following facilities:

   natural frequencies and mode shape calculations,

   determination of the response of a structure to sinusoidal, transient and seismic loadings,

   random vibration response,

   excitation by specified displacements, velocities and accelerations,

   modal damping and use of viscous damping (sometimes other forms of damping as well),

   relation between vibration and acoustic excitation.

These features just cover the vibration aspects of the analysis, but the packages are, of course, much more extensive than that. They were developed initially as linear structural analysis programs but have been expanded into problems of buckling, thermal stresses, creep, plastic analysis and orthoptropic materials.

The kind of elements that can be used are shown in Figure 4.

A typical example of a vibration analysis is the determination of free vibrations of a turbine blade used in a power plant, Figure 5. The finite element mesh with the degree of refinement shown (24 × 4 brick elements with 15 and 20 nodes) was proved to give the best results. The lowest five natural frequencies can be calculated. Other combinations (24 × 2 and 12 × 4 elements) gave inferior results when compared with the experimental results. This demonstrates that an ability to choose the number and type of finite elements is an important skill for efficient calculations. This analysis was performed on BERSAFE.

Element families

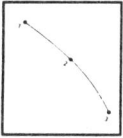

One dimensional elements

Beam elements may be straight, curved, with and without shear deformation, with and without bending, with or without offsets. There are curved three noded beams (which can be offset) compatible with semi-Loof and facet shells.

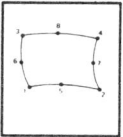

Two dimensional elements

Two dimensional elements are generally based upon isoparametric methods. They can all be used for plane stress, plane strain and axisymmetric analysis. They may be triangles or quadrilaterals with numbers of nodes varying from 3 to 17. Two dimensional isoparametric elements are also available for steady state and transient thermal analysis.

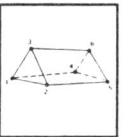

Three dimensional elements

Like the two dimensional elements, the isoparametric method is used for this family of curved elements. shapes may be brick, wedge or tetrahedron. Numbers of nodes can vary from 4 to 32. Similar elements exist for temperature calculations.

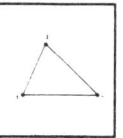

Facet shell elements

There are flat elements, both triangular and quadrilateral for both thick (shear deflection included) and thin (no shear deflection) plates in which both membrane and bending actions are important.

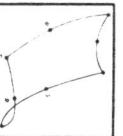

General shell elements

For thin general shell problems, the triangular and quadrilateral semi-Loof elements are used. For thick general shells the Ahmad formulations are used.

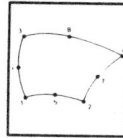

Temperature calculation elements

These two and three dimensional elements, as mentioned earlier, can be used for calculating both steady state and transient temperature distributions.

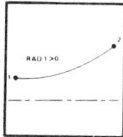

Axisymmetric shells and solids

For thin axisymmetric shells there is a special element which appears to be a one dimensional curved element, but on consideration of the axisymmetry becomes a thin axisymmetric shell with meridional curvature. Each element may have 0, 1 or 2 mid-side nodes.

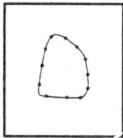

Boundary elements

For two dimensional regions there is a three noded line element whereas for three dimensional regions there are curvilinear triangular and quadrilateral elements.

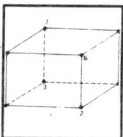

Acoustic elements

There are three dimensional and axisymmetric acoustic elements for finite fluid regions with hard or flexible boundaries. The boundary element method can be used for infinite regions of fluid. The acoustic elements can be coupled with structural elements.

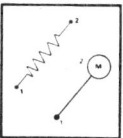

Miscellaneous elements

There are a number of other elements which fit into no clear family. These include:

lumped mass element
spring element
piezoelectric element families
user defined element – the element matrices are defined numerically.

Figure 4 Finite elements for general structural analysis

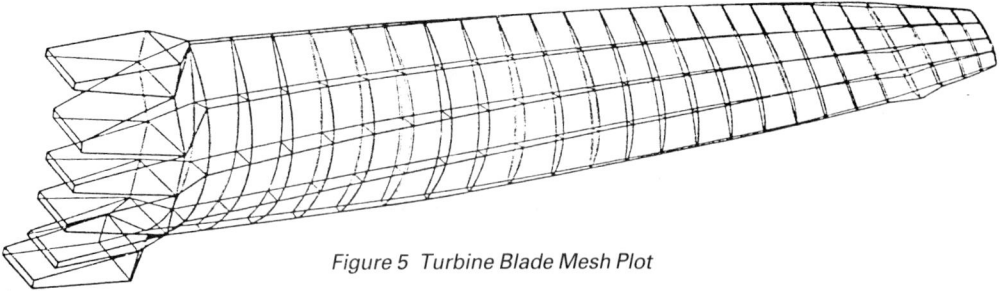

Figure 5 Turbine Blade Mesh Plot

Chapter 3
Torsional Vibrations

Torsional vibrations occur in rotating shaft systems and are associated with the torsional stiffness of the shafts and the rotational inertia of the wheels, rotors and discs that make up the power train. These systems are excited by various mechanisms. The most common is probably that produced by torque from the engine or other source of power. In any internal combustion engine there are fluctuations, in output torque produced by the gas pressure throughout the cycle. The larger the number of cylinders, the smaller the fluctuations but in all engines there is a varying torque superimposed on the uniform output torque. A flywheel is usually introduced between engine and drive to smooth out the variations in speed, which it can readily do, but the added inertia of the flywheel can change the vibration characteristics of the system, usually by lowering the natural frequency, which may cause further problems. An analysis of the complete system is desirable to ensure that the engine speed does not coincide with a resonance of the power train. Even rotary piston machines such as the Wankel are not free from cyclic torque variations, although their magnitude is small.

Excitation can also be caused by variations in torque of the output device. A reciprocating compressor, for example, experiences the same kind of cyclic pressure changes as an engine. A further possible source of excitation is caused by non-linear characteristics in the drive train. Vibration can be introduced by transmission error, by slackness in a gear drive as the gears take up the torque, by variations in tooth form or roughness of surface finish.

A characteristic of torsional vibration which makes it different from linear or spring-mass vibration is that the vibrational displacements are superimposed on steady state movement (the rotation of the shafting). While the presence of the steady state does not affect the analysis, it must be known and taken into account because usually the excitation is a harmonic of the rotational speed.

A further difficulty occurs in the measurement of torsional vibration. One cannot simply place an accelerometer on the part of the shaft being measured, because the centripetal acceleration would swamp the readings. Ways of dealing with the measurement of torsional vibrations will be considered later.

Mathematically, there is a direct relationship between torsional and spring-mass vibrations, as the following table shows:

	Spring-mass system	**Torsional system**
Mass or inertia	m (kg)	J (kg m^2)
Spring	k (N/m)	K (N m/rad)
Damping	c (N s/m)	C (N m s/rad)
Acceleration	a (m/s^2)	(rad/s^2)
Velocity	v (m/s)	(rad/s)
Displacement	y (m)	(rad)

In a general case a torsional system can be considered as a series of shafts possessing stiffness and rotors possessing inertia. In practice, shafts have inertia and rotors have stiffness, but for ease of analysis it is usually satisfactory to break down (in a mathematical sense) each component into its equivalent inertia and stiffness. The analysis of rotating systems is quite straightforward, because the components are connected together singly (i.e. inertia followed by stiffness followed by inertia etc.). Modelling is easily done, made only a little more complicated by the presence of gearing and split shafting. Two simple configurations can be discussed as an analogue of the single degree of freedom spring–mass. These are shown in Figures 1 and 2.

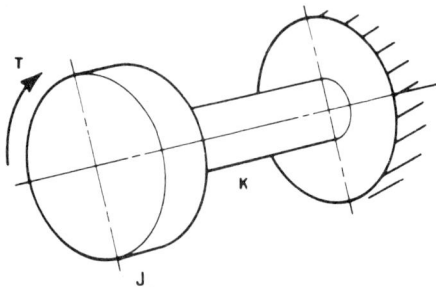

Figure 1 Single-degree-of-freedom system

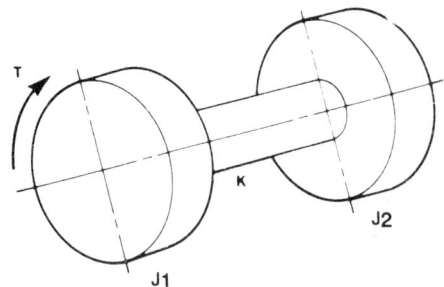

Figure 2 Two-inertia, one-spring system

In Figure 1, the natural angular velocity is $\sqrt{K/J}$

In Figure 2 where the two rotors can rotate independently, the natural angular velocity is $\sqrt{K(J_1 + J_2)/(J_1 J_2)}$. Although the rotors rotate independently, there is only one natural frequency, so only one degree of freedom.

More complex rotating systems are best solved by means of a computer for which a variety of programs are commercially available.

The torsional stiffness of shafts can be assessed from the following table

Section	**Stiffness Factor F**	**Stress Factor S**	
Circular inner diameter 2r$_1$ outer diameter 2r$_2$	$\dfrac{(r_2 4 - r_1 4)}{2}$	$\dfrac{2}{r_2^3}$	at boundary
Solid elliptical major diameter = 2a minor diameter = 2b	$\dfrac{a^3 b^3}{(a^2 + b^2)}$	$\dfrac{2}{a b^2}$	at end of minor axis
Solid square side 2a	2.25 a^4	$\dfrac{0.601}{a^3}$	at midpoint of each side
Solid triangular side a	0.0217 a^4	$\dfrac{20}{a^3}$	at midpoint of each side
Hexagon across flats 2a	1.728 a^4	$\dfrac{0.675}{a^3}$	at midpoint of each side

The actual stiffness is

$$k = \frac{FG}{L}$$

and the maximum shear stress is

$$f = TS$$

G is the torsion modulus of the material, L is the length of the shaft and T is the applied torque.

For more complex sections such as shafts with splines or keyways, refer to a standard text such as *Roark's Formulas for Stress and Strain*.

Holzer Analysis of Torsional Systems

This is a technique for determining the natural frequencies and mode shapes of torsional systems, which is particularly suited for drive-trains with an engine at one end and some driven device at the the the other, illustrated in Figure 3. It can also be used for any lumped parameter linear system.

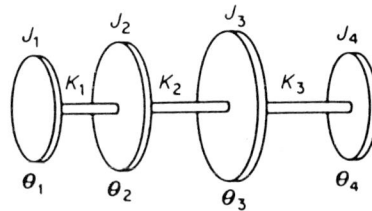

Figure 3 Idealised torsional vibration problem

The technique consists in guessing a frequency and applying unit amplitude at one end, and progressively calculating the torques and angular displacements along the system until the far end is reached. The object is to satisfy a specific boundary condition at the far end e.g. zero torque at a free end or zero angular displacement at a fixed end. If the condition is not satisfied a new frequency is chosen until it is. The method is quite straightforward and can be used successfully where manual computation only is available, but it is made particularly easy with a desktop computer.

By careful choice of the initial frequency, the method converges well and all the natural frequencies and mode shapes can be calculated.

In Figure 3, the inertia torque of the first disk is

$$J_1 \ddot{\theta}_1 = J_1 \omega^2 \theta_1 = J_1 \omega^2 1$$

(θ is assumed $= 1$, initially)

This torque acts on the first shaft, and twists it by an amount equal to the inertia torque divided by the stiffness.

$$\frac{J_1 \omega^2}{K_1} = \theta_1 - \theta_2 = 1 - \theta_2$$

$$\theta_2 = 1 - \frac{J_1 \omega^2}{K_1}$$

With θ_2 known, the inertia torque on the second disc can be calculated and so on progressively through the system, until the end disc is reached.

The resulting torque at the end is given by

$$\sum_{1}^{k} J_n \omega^2 \theta_n$$

This should equal zero for a free end. If it does not, a new guess has to be made. Usually, if a computer is used, an accurate determination of the natural frequency requires only about three iterations. The programming for a Holzer calculation is quite straightforward. A further quantity S is required from the Holzer calculation at each frequency. This is used below in calculation of the actual amplitudes.

$$\sum_{1}^{k} J_n \omega \theta_n^{\ 2}$$

Drive Train Analysis

All the natural frequencies of interest can be calculated using the Holzer technique. When examining engine driven torsional systems (which is the usual type of problem to be solved), the search has to be carried out to include all the modes that can be excited by the torque excitations from the engine, which means carrying the calculations up to a frequency corresponding to 15 × the maximum continuous speed of the engine. This is equivalent to a 12th order torque excitation at 125% of the maximum speed. The range to be covered is from 10% to 125%. There may be several possible frequencies within this range, and all have to be examined, although some may be more significant than others.

Each mode of vibration with a frequency, f Hz, is related to a critical speed, Nc rev/min, and the order of the excitation, m, by the relationship:

$$N_c = \frac{60f}{m}$$

A natural frequency at a low order close to an engine operating speed is clearly of concern and may point to a redesign or the use of a vibration damper, which will be dealt with later. Dampers are however not to be recommended if the critical frequency is in a range close to the maximum engine speed, because of the problem of dissipation of heat. Flexible couplings placed between engine and drive train can be used to modify the torsional characteristics as well as to solve alignment problems.

The Holzer analysis gives the mode shapes relating to a modal amplitude of 1 rad at the engine. The relative shaft stresses can also be calculated on the basis of this amplitude, but at this stage the actual values are unknown. If the amplitude at some point can be measured, the actual amplitudes and stresses can be determined throughout the system, but if that is not possible, the magnitude of the excitation torques and the damping components have to be estimated.

Vibration Amplitudes

The actual torsional component of the engine drive, T_m, can be found from the use of curves such as in Figure 4. These curves are obtained from measurements on a wide variety of installations and are adequate for initial design purposes, but as in all problems of vibration, the final torsional system needs to be checked experimentally, and as far as possible the harmonic components measured. The curves are convenient for calculation purposes, in that the relationship between the torsional component and the mean indicated pressure is a linear one, and so can readily be represented by a data block in a computer program.

Figure 4, gives the values of T_m in N/mm^2, for orders between 0.5 and 6.0. Orders up to 12 can be obtained by reference to the data contained in Lloyd's Register of *Shipping Rules and Regulations for the Classification of Ships*. In addition to providing a comprehensive set of tables for determining the alternating torque produced in gas engines, the Rules also give guidance in assessing the permissible stresses in shafting. The Rules have a much wider application than for marine engine drives.

The actual harmonic torque produced by the engine must take into account the contributions from all the engine cylinders, their firing order and their angular disposition (in-line or vee). This requires the calculation of the phase vector sum of the contributions from all the cylinders. A vector diagram has to be drawn for each order, in which one revolution equals the time for one vibration, and each crank is shown in its firing position by a vector at an angle corresponding to the crank angle multiplied by the order of vibration, and with a length derived fron the relative amplitudes of the modal diagram. Different firing orders will produce different phase vector sums, so that in some circumstances, changing the firing order will result in a reduction in the amplitude of vibration to such an extent that the problem is solved.

Finally the vibration amplitude at the engine (station 1) in each mode is found from:

$\theta_1 = \pm M\theta_0$, where M is the dynamic magnifier and

$$\theta_0 = \frac{T_m A R \vec{\Sigma} \theta}{\omega^2 S} \quad \text{where}$$

A = the area of each piston (mm^2),

R = the crank radius (m)

$\vec{\Sigma}$ = phase vector sum of all the engine cylinders

S = summation of terms ($J\theta^2$) in the Holzer calculation (see above)

ω = natural phase velocity of the mode (rad/s)

The amplitudes and stresses throughout the complete train can be found by reference to the modal shape.

The dynamic magnifier, M, is introduced to take account of the damping in the system. In many situations the only effective damping comes from the engine, but in others it can come from a deliberately inserted vibration damper, from a flexible coupling, or from a compressor, a generator or some other equipment.

M for an engine is expressed as

$$M_E = \frac{3.8}{(\theta_0)^{0.25}}$$

In practice the value of M_E is limited to a maximum value of 50.

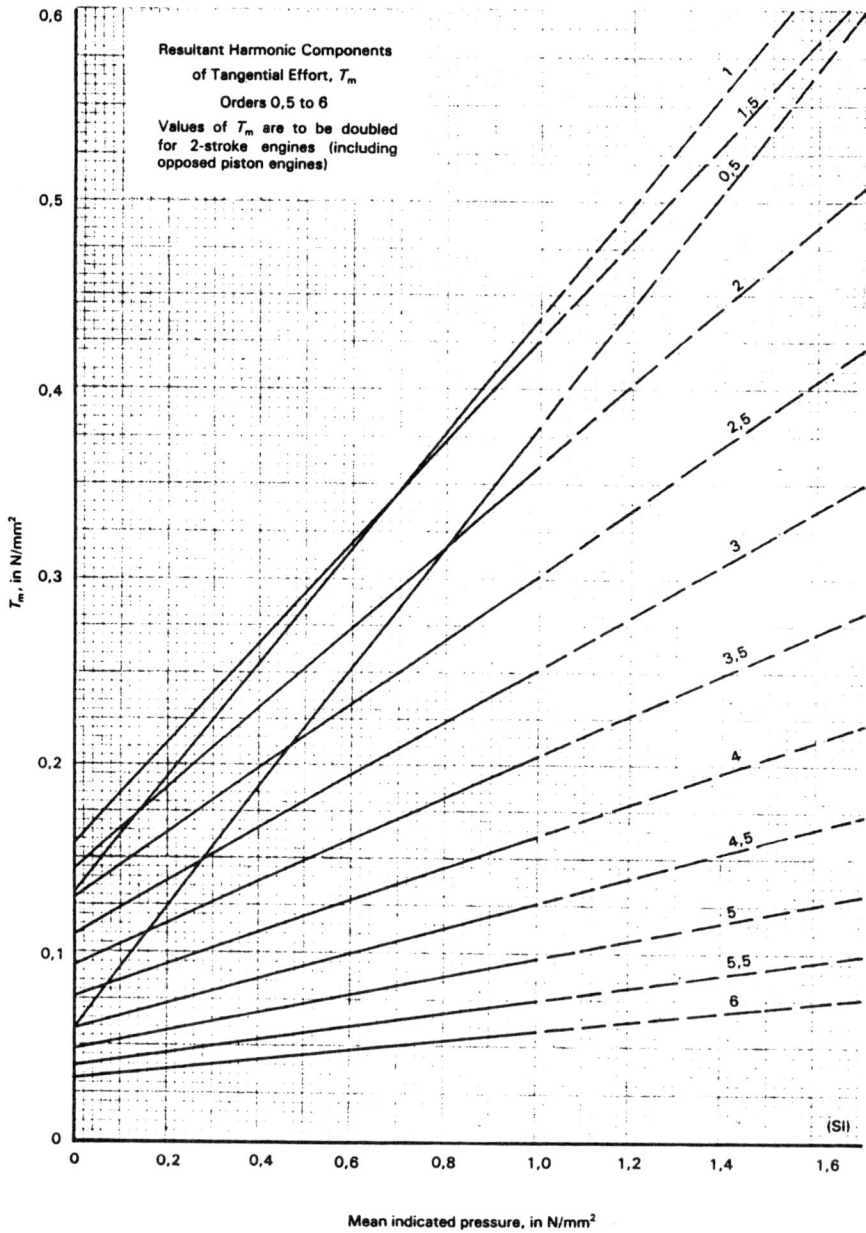

Figure 4

For other types of damping, reference should be made to the manufacturer.

When combining several sources of dampingk

$$M = [(1/M_E)^2 + (1/M_1)^2 + \ldots \ldots (1/M_n)^2]^{-0.5}$$

Following this procedure it is possible to estimate the vibratory stresses and displacements through the system. If desired, the stresses, torques and displacements can be added to those produced by the steady driving torque from the engine, to meet a particular design condition. Refer to LLoyd's *Rules* for guidance on ship's propulsion units and the stress conditions to be met in different kinds of shafting.

Another useful piece of information can be gained from this analysis: in the case of drive systems incorporating gears or splines, in which it is desirable that the contact forces shall always be kept positive, so that there is no possibility of gear hammer caused by the backlash in the drive, an examination of the mode shape will indicate if the position of the gears coincides with a node, or is so close to it that the net torque becomes zero at that point.

At critical speeds near the maximum, the vibratory torque should not exceed one-third of the full transmission torque.

In some cases, a high level of vibration can occur at a rotational speed different from one of the critical ones. This can happen when two or more critical speeds are close together or when the maximum continuous operating speed corresponds with the flank of a powerful critical. In this event the two off-peak criticals are added together.

Chapter 4
Analysis Techniques

Random Vibrations

Random vibrations are met with frequently: typical examples are the motion of the suspension of an automobile driving over a rough road, an aircraft in turbulent air, a power tool cutting material and machinery noise in general. The main characteristic of random vibrations is their unpredictability, i.e. they are non-deterministic so that it is not possible to say what the vibration amplitude will be at a given moment in time. To obtain a complete description of the vibration, an infinitely long record would be necessary, but in practice, a finite record has to be used. Even so, analysis can be quite difficult, and methods and concepts that have been developed for handling problems in statistical mechanics have to be used. These include amplitude probability distributions in terms of probability densities and continuous vibration frequency spectra in terms of power spectral density. Practical examples that can be dealt with are of a stationary random nature, which are defined as random vibrations whose statistical characteristics do not change with time.

The theory behind these concepts can be quite complex and difficult to use by a practicing engineer; however in order to perform any sort of analysis on systems that are characterised by this kind of vibration, it is necessary to acquire a basic understanding.

Dynamic Analysis of Vibration

Two forms of dynamic analysis can be used to obtain data from which inferences can be made concerning the properties of the system or structure under investigation.

Time domain analysis, in a basic form, consists of examining the signal amplitudes plotted on a time scale. This can be an oscilloscope record or a paper trace for example. The equipment used for this analysis includes a variety of capture and display instruments and correlation circuits. These tools provide a means of estimating the reponse characteristics of a system, particularly when studying transients. When signals are of a random or apparently random nature, they are often very difficult to understand, appearing as a trace without any apparent form or coherence. To make analysis easier, it has been found that analysis in the frequency domain can give much more information to the investigator. A signal in the frequency domain will enable him to determine the frequencies that are relevant to an understanding of the behaviour of the structure.

There is a difference between mathematicians and practical engineers in the way they think of vibration analysis. The mathematician or theoretician uses the concepts of Fourier analysis and Fourier transforms, wheras the engineer prefers to think of vibration measurements taken with filters tuned to different frequencies. The emphasis here is on the practical approach, but some theoretical principles need to be understood also.

Probability Density

An important basic concept is the probability density, defined as the probability of finding a specific amplitude level within a certain amplitude interval, divided by the size of the interval. Figure 1 shows part of a random vibration signal to illustrate this.

$$p(x) = \text{Lim } (\Delta x \to 0) \ \frac{P(x) - P(x + \Delta x)}{\Delta x}$$

p(x) is the probability density, P(x) is the probability that any instantaneous amplitude level exceeds the level x, and P(x + Δ x) is the probability that any instantaneous amplitude level exceeds the value x + Δ x.

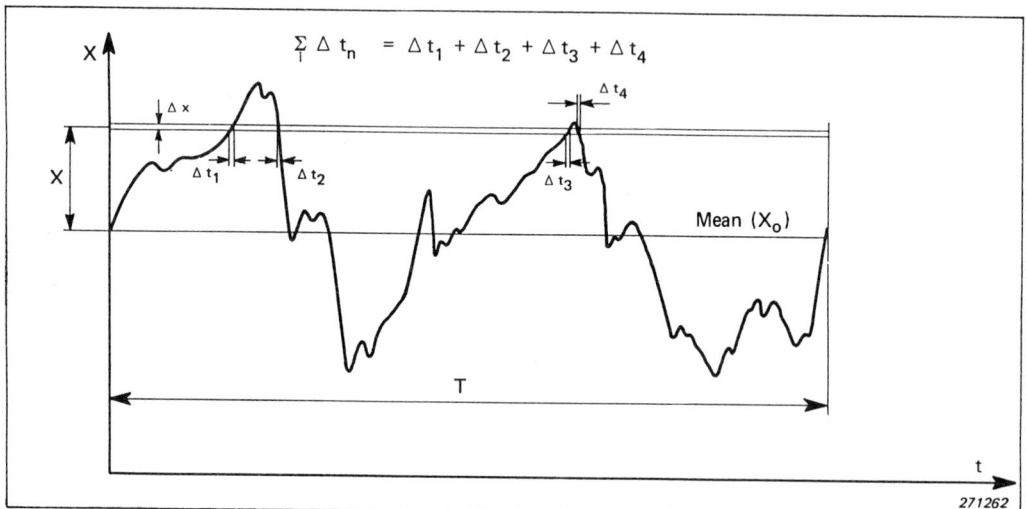

Figure 1 The concept of Probability Density

In principle it is possible to determine p(x) for all values of x that occur in the vibration signal, and then to plot its value. This is known as the probability density curve. The advantage of this curve is that it is possible to integrate the curve between two values of x, say x_1 and x_2 and so determine the probabilty of occurrence of amplitude values within the range x_1 to x_2. Following classical probability theory, the integral of the probability of finding an amplitude between zero and the maximum must be 1. For distributions which are symmetrical about the mean, statisticians use the Gaussian (or normal) curve as in Figure 2 for a variety of statistical processes and this is still the most commonly used curve to analyse vibration data, since it is possible analytically to determine the response of a system to a random input if its distribution is Gaussian. For non-symmetrical distributions the Rayleigh distribution may be used. If the process is not Gaussian, it is much more difficult to find an analytical solution to most problems.

Autocorrelation Function

Probability density is useful in assessing the distribution of amplitudes in a specific vibration regime, as for example when assessing the stress fatigue of a component or a human exposed to the vibration. Its use however is limited because it gives little information about the time history

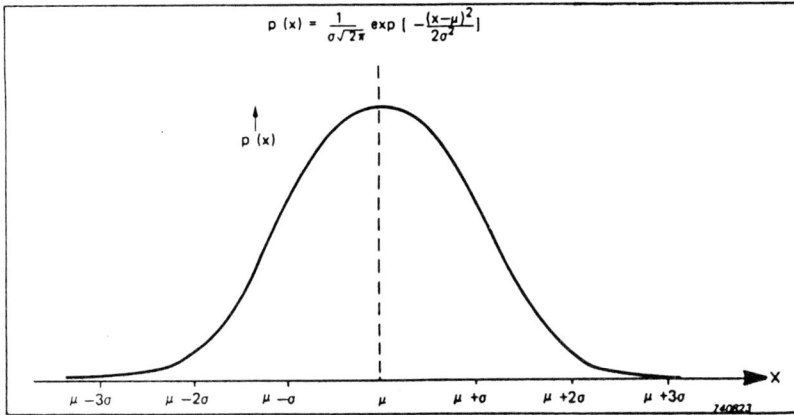

$$p(x) = \frac{1}{\sigma\sqrt{2\pi}} \exp\left[-\frac{(x-\mu)^2}{2\sigma^2}\right]$$

Figure 2 Gaussian distribution

of the vibration and how the amplitude at a particular time depends on the preceding instantaneous vibration magnitudes. A function which can be used to this purpose is the autocorrelation function. It is a measure of how a vibration signal correlates with the time displaced function of itself. It is defined as follows:

$$R_{xx}(\tau) = \text{Lim } (T\to\infty)\frac{1}{T}\int_{-T/2}^{T/2} f_x(t)\,f_x(t+\tau)\,dt$$

$R_{xx}(\tau)$ is the autocorrelation function of the magnitude of the vibratory function $f_x(t)$, and $f_x(t+\tau)$ is the magnitude of that function at a time τ later. The vibratory function can be amplitude or the acceleration or any other chosen function. If $\tau = 0$ the autocorrelation function is equal to the mean square value (its maximum value). The autocorrelation function can be normalised to a maximum value of 1 by dividing by the mean square value.

If the signal were completely random, the autocorrelation function would be a line centred on $\tau = 0$. In practice the function is never quite a line of zero width, but is a symmetrical curve with a width which depends upon the frequency range of the vibration. Figure 3 demonstrates the possible forms of the function, all for a random process. The autocorrelation function of a periodic wave, a sine wave for example is itself a wave of the same periodicity, since shifting of a signal by one period brings it into coincidence again.

The autocorrelation function can be used for the detection of periodicity which may be buried in what seems to be a random signal. It can also detect echoes, their strength and time separation.

Cross correlation function

If it is required to relate two measurements made at two different points in a system, the cross correlation function can be used. It may not be apparent by examination what relationship, if any, there is between two signals, so one way of quantifying any relationship is through this function defined as:

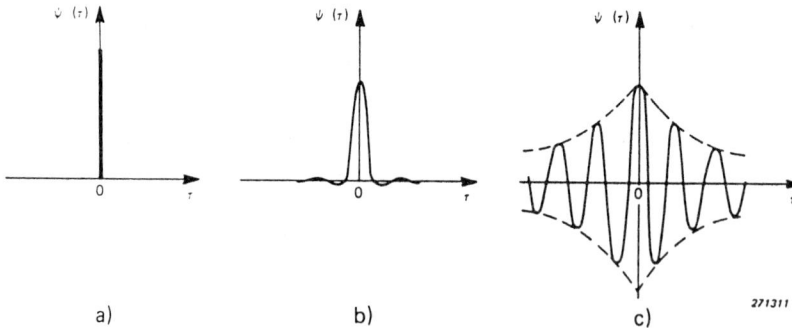

Figure 3 *Example of autocorrelation functions*
a) Autocorrelation function for an ideal stationary random process containing frequencies from 0 to ∞ (constant spectral density)
b) Autocorrelation function for a 'practical' wide band stationary random process
c) Autocorrelation function for a narrow band stationary random process

$$R_{xx}(\tau) = \text{Lim}\,(T \rightarrow \infty)\,\frac{1}{T}\int_{-T/2}^{T/2} f_x(t)\,f_y(t + \tau)\,dt$$

R_{xy} is the cross correlation function of the two functions f_x and f_y, where f_x is the magnitude of the signal at point x, and f_y is the magnitude of the signal at point y, measured at a time τ later. The cross correlation function can be useful in determining how a signal measured at one point affects the response at another and with what time delay. It can also detect the existence of a signal buried in extraneous noise. Figure 4 shows a typical cross correlation function. At $\tau = 0$ there is no correlation but at $\tau = \tau_0$ there is maximum correlation.

Fourier Transforms

The concept of Fourier analysis was discussed in the chapter on Linear Vibration Theory. The concepts introduced there were applicable only to periodic motion with a finite time period. Fourier analysis is such a powerful tool that its application has been extended to the analysis of

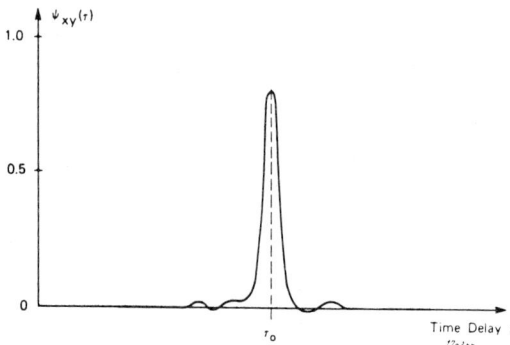

Figure 4 *Sketch indicating the cross-correlation function for a hypothetical frequency independent random process*

random vibrations. An outline of the theory is included here but for a deeper treatment refer to a specialist text. Random vibrations do not possess the periodicity essential for classical Fourier analysis, but for the stationary statistical processes that are considered here, a comparable treatment is possible.

If the coefficients of the Fourier series are plotted on a frequency axis, the result is a series of discrete lines forming what can be called a Fourier spectrum. Usually, to define a function, two spectra are needed: these can be either the ak and the bk coefficients, but more usually the ak and the k terms. It is in the latter form that it is usually presented.

A random vibration as described above can be looked upon as having a periodicity of an infinitely large time, and the Fourier spectrum would then not have discrete lines but be a continuous curve. The presence of a discrete line in a spectrum indicates the presence of a periodic function hidden in the signal. When dealing with a stationary random signal, a short length of the signal is used to represent its characteristics, the assumption being that the frequency domain analysis of the short length is a good approximation to that of an infinitely long length.

Fast Fourier Transforms

The process of converting a signal in the time domain into its equivalent spectrum in the frequency domain is called transformation. The calculation involved in this transformation is done internally in an analyser, using a sophisticated algorithm known as a Fast Fourier Transform (FFT). Although it is fast, it still takes time for the analysis to be done. In performing the analysis, a window of the data in the time domain has to be processed to produce a spectrum. The time taken depends upon the maximum frequency of interest: as the frequency of interest goes up, the length of the time window comes down.

The consequence of this is that, as the duration of the time window comes down, it eventually becomes shorter than the calculation time for the transform. It is not then possible to take contiguous signals and transform them in real time. The real time speed of the FFT analyser is when this point is reached.

FFT analysis is of a constant bandwith across the frequency range. The spectrum lines are separated by equal frequencies, which may give too coarse a definition at low frequency and too fine a definition at high frequency.

Time Windows

In Fourier analysis, time and frequency are two ways of observing a signal. By changing the nature of a signal in the time domain, the nature of the signal in the frequency domain is also changed. This is done in an analyser by applying a weighting function or time window to the signal. In the case of continuous signals the time window slices up the signal, but in the case of transient data, the windows edit the record so that the analyser works on the transient and not the portion of the signal that contains noise before and after the transient. See Figure 5.

The type of weighting or filtering in the frequency domain is determined by the shape of the window through which the analyser sees the data in the time domain. There are many types of window, the choice of which depends on the type of signal and the application.

An analyser multiplies the window and the signal together in the time domain. In quantifying the effects of a filter the following definitions are used. See Figure 6.

Effective Noise Bandwith is the width of an ideal filter with the same transmission level and which transmits the same power from a white noise source.

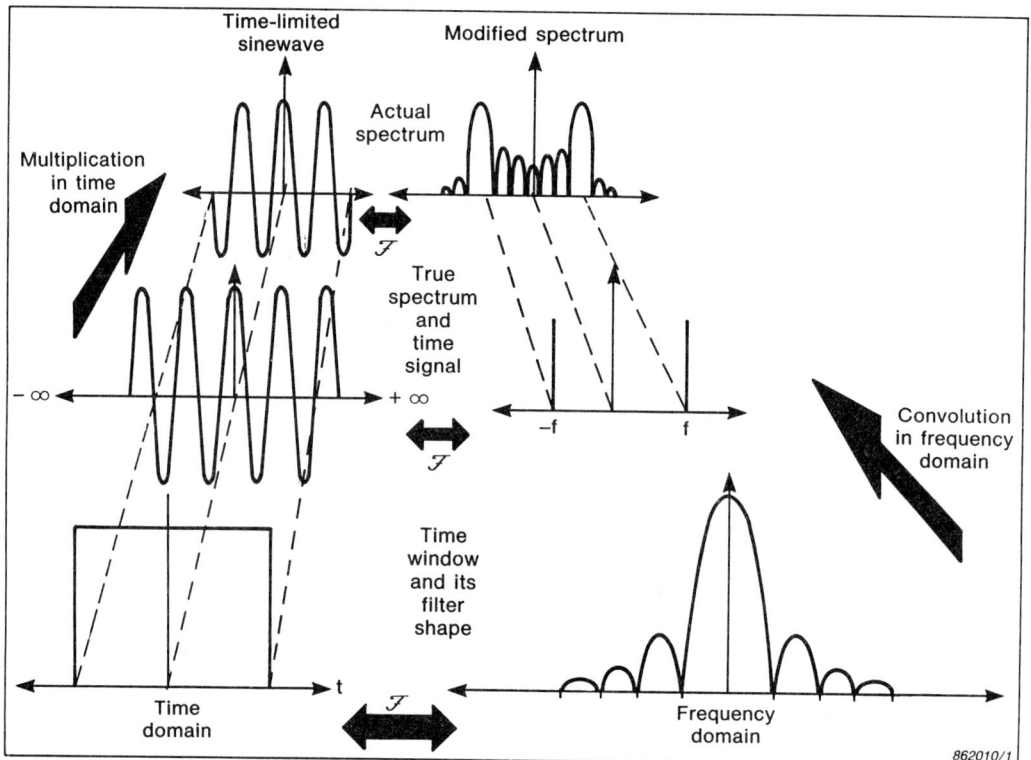

Figure 5 FFT Analyses over time signals through windows and so change the frequency spectrum

3 dB Bandwith is the distance in Hertz between the half power (−3dB) points on the amplitude axis.

Selectivity is a measure of how well a filter separates components of different levels. The basic measure of selectivity is the ratio of the −3 dB bandwith to the −60 dB bandwith known as the Shape Factor.

There are several types of filter in common use, which can be implemented in proprietary instruments.

The rectangular or flat window as in Figure 7 is defined simply as having a unit value for the record length and zero elsewhere. There are discontinuities at the start and end of the time T which cause leakage of the energy from the main frequency of the sine wave into nearby frequencies. As can be seen from the transform in the frequency domain, the main lobe has a width equal to twice the line spacing. The first lobe is attenuated 13 dB relative to the main lobe and the fall-off rate is 20 dB per decade. The selectivity is very poor and is not recommended for use on continuous signals. If the frequency of the sine wave coincides with the edges of the window, the spectrum is sampled in the centre of the main lobe and the amplitude errors are zero. If however the frequency lies between two lines on the frequency axis, the amplitude can be underestimated by 3.9 dB.

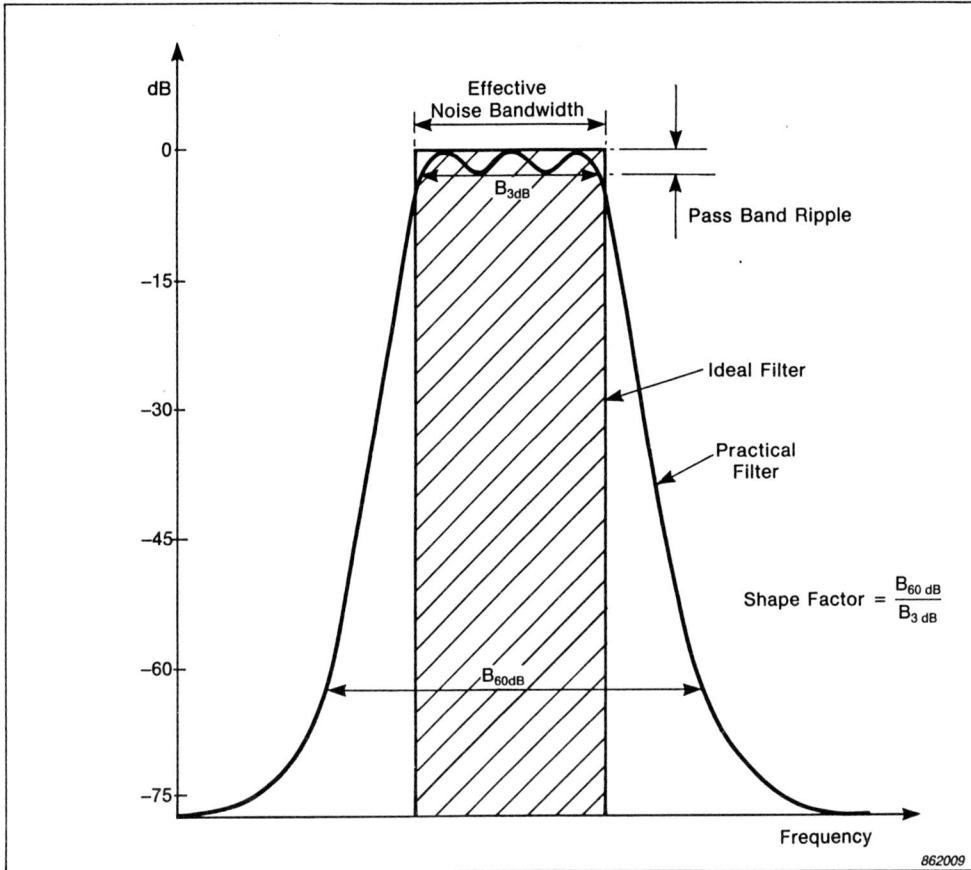

Figure 6 Filter parameters

The Hanning window avoids some of the difficulties in the rectangular window, Figure 8. It is defined as

$$w(t) = 1 - \cos 2\, t \text{ for } 0 \leq t < T$$
$$w(t) = 0 \text{ elsewhere}$$

As can be seen, in the frequency domain the fall-off rate is more rapid, with a rate of 60 dB per Octave.

There is a variety of other windows in use, but the two most common ones are the rectangular for transient analysis and the Hanning for continuous signals. Some FFT analysers offer the user a choice of windows and allow him to create his own for special purposes.

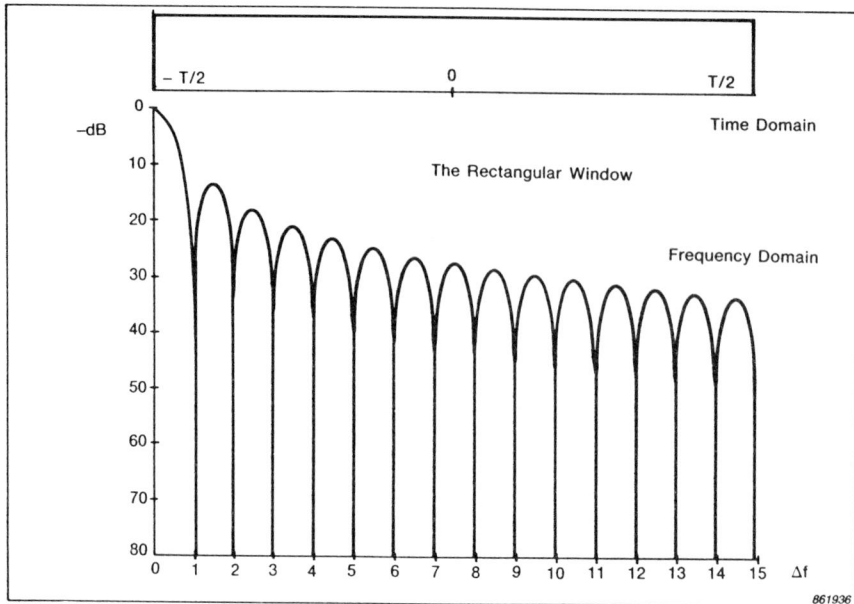

Figure 7 Frequency and time characteristics of the rectangular windows

Real Time Analyser

An alternative form of analysis in the frequency domain is achieved through the use of filter circuits. It will be recalled that a sound level meter employs a set of filters in octave or one-third octave bands. A sound level meter can be used to measure vibration levels when supplied by a accelerometer and a suitable amplifier. A real time analyser (RTA) is similar in function to a sound level meter in that the vibration signal is passed through a filter set; the analysis is of constant percentage bandwith (CPB), where the bandwith increases with centre frequency. The filters are arranged in parallel and the implementation is digital rather than electrical. The analysis is performed in real time so it does not suffer from the inherent drawbacks of FFT.

The fineness of the analysis of which a RTA is capable depends on the number of parallel filter circuits. Most RTAs are capable of octave and one-third octave analysis, but some can give as small as one-tenth or one-twenty fourth octave analysis.

Each type of analysis (FFT and RTA) has its advantages and range of applications. For mechanical vibrations of structures, machinery and systems, the FFT is preferred but for acoustic excitation, power tool vibrations, vibrations in the environment, or for human response analysis, the RTA is more frequently used. Some instruments are sufficiently versatile that they can perform either kind of analysis at the user's choice.

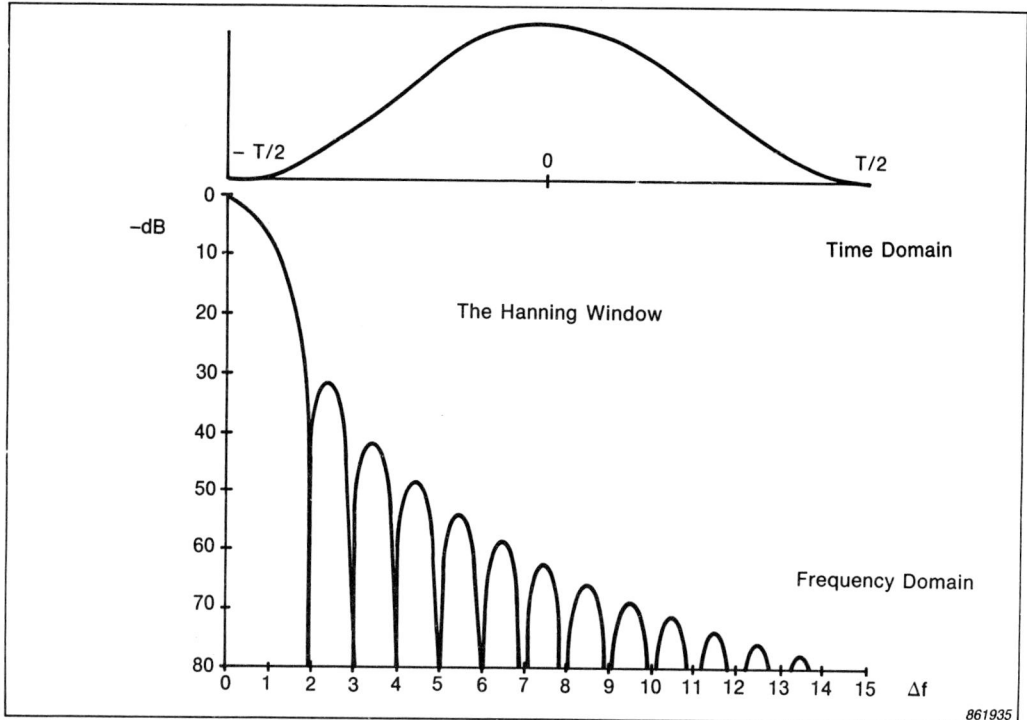

Figure 8 Frequency and time characteristic of the Hanning windows

Chapter 5
Human Response to Vibrations

One can distinguish two areas where standards have been developed to assess the effect of vibration applied to the human body: vibration to the hand and vibration to the whole body. The former can give rise to diseases, in particular vibration white finger; the latter can cause a variety of health problems ranging from motion sickness to tissue damage. It should be understood that the relationship between exposure and disease is a complex one still being studied, and many questions in this area remain unanswered. A number of standards incorporating the latest state of knowledge have been introduced over the last few years with the object of giving guidance to those required to assess the importance of human vibration exposure. For the hand–arm system, refer to BS 6842 (ISO 5349) and for whole body vibration, BS 6841 (ISO 2631). For vibration in buildings refer to BS 6472 and BS 6611.

Hand-Arm Effects

Very high levels of vibration can be transmitted to the hands of industrial workers who use vibrating hand tools such as chain saws, rock drills, road breakers, chipping hammers and hand grinders. Not only hand tools but fixed equipment such as pedestal grinders and swaging machines have been implicated in the incidence of vibration disease. At the lowest level, vibration can cause discomfort and consequently impaired performance in carrying out a prescribed task. Habitual use of a vibrating tool has been connected with various diseases affecting the blood vessels, nerves, muscles and connective tissues of the hand and arm, and there is sufficient epidemiological evidence available to be certain of a causal link between vibration and disease.

A medically observed syndrome, first described by Dr Maurice Raynaud in 1862 and subsequently called Raynaud's Disease, is characterised in the initial stages by blue or white fingers (sometimes toes also) brought about by exposure to cold. It was found to be more common in females. A proportion of sufferers went on to suffer from atrophy of the skin, ulcerations and gangrene. Raynaud was not concerned with industry or the effect of vibration. He was describing a naturally occurring disease for which he had no explanation; however those same symptoms are to be found in users of vibrating equipment, so his name and the syndrome he described has been adopted by workers in the field, the implication being that there is a common physiological basis for the disease. When the symptoms of the disease are caused by vibration, it is often referred to as Raynaud's Disease of Occupational Origin, or Vibration White Finger (VWF).

Primary Raynaud's Disease or constitutional white finger has often a hereditary element and occurs early in life, mostly before the age of 30. Secondary Raynaud's Phenomenon is caused by many diseases such as arterial disease, neurovascular disease and trauma, all associated with changes in blood flow.

It is now generally believed that vibration causes damage to the blood vessels which are thus made incapable of circulating blood to the extremities. There is no known cure, and the symptoms are irreversible, so it is clearly important to establish the levels of vibration and the duration of exposure that are statistically likely to cause the disease and to take precautions so that the troubles are stopped before they start. The problem is a complex one, and despite much research there is a shortage of quantitative data concerning occupational health. It is only in recent years that standardised methods of measurement of vibration levels have been developed. The study of the disease is further complicated by the time taken between first exposure and the clinically observed onset of the disease.

The severity of the symptoms and the time taken for them to develop are influenced by:

the magnitude and frequency spectrum of the vibration

the exposure pattern, i.e. the length of time the tool is used during the day and the nature and frequency of the rest periods

the cumulative exposure to date

the direction of the vibration

the posture and grip of the operator when holding the tool

the skill of the operator

the care with which the tools are maintained

the general health of the operator and his use of drugs which may affect his circulation; smoking and alcohol may be influencing factors

environmental conditions, such as the coldness of the workshop or (when working outside) the climate; noise may also be a contributory factor

The relative importance of each of these factors is not fully known, but it clearly makes sense for an employer to take obvious steps to maintain the health of his workers. Some of these may stem from an obligation imposed on him by national legislation, such as the UK Health and Safety at Work Act 1974. In view of the need for much more epidemiological information to enlarge the knowledge about the disease and develop a precise set of recommendations for designers of equipment and employers, research workers are advised to note as many of the above factors as they think reasonable.

The following are suggested as precautions which a careful employer should consider:

Ensure that those tools which are purchased should as far as possible be ones which have been designed for minimum vibration. It would be unwise however to choose tools whose sole merit is their reduced vibration level; unless the design has been carefully done, they may be heavier, bulkier or less easy to use. Some pneumatic tools may discharge a cold exhaust over the hands of the operator; they should be avoided. The wearing of gloves may help the wearer to keep warm but they may make it less easy to control the tool.

Since cold is known to be a predisposing factor, workshops should be kept warm, and work should not start until a comfortable temperature has been reached. If the employee has become cold in travelling to work, he should be allowed to get warm before starting.

Ensure that the work pattern allows for regular rest periods, and try to rotate the work schedules so that no one is expected to spend his working life using vibrating tools.

Ensure that tools are properly maintained, and where cushioned hand grips are supplied that they are kept in good condition. Operators should be encouraged to report any occasions where they notice numbing or blanching of their fingers.

Medical Conditions Other Than VWF

A number of other medical conditions have been associated with vibration applied to the hand. These include carpal tunnel syndrome, osteo-articular injuries affecting the elbow and wrist joints, and callus formation. There may be more remote disorders: pain in the collar bone, muscular pain, digestive disorders, vertigo and insomnia; the effort needed to control percussive tools necessitates powerful muscle contractions which promote the transmission of pain to the whole organism.

In the present state of knowledge it is not possible to identify vibration regimes specifically responsible for these other conditions, so the assumption has to be made that precautions taken against VWF are effective in these instances also.

Medical Preventive Measures

In some work situations where the workers are known to be subject to hand-arm vibration, medical screening may be necessary. These will include road works where concrete breakers are in constant use, mines where hand-held rock drills are used, fettling shops, forestry work with chain saws, swaging shops and grinding shops for dressing castings.

The general principles of medical screening will follow those for hearing protection. Before employment in a high vibration area, there should be an initial screening which would include asking questions about previous vibration exposure and relevant medical conditions. Factors which should be explored will include the presence of Primary and Secondary Raynaud's, disorders of the peripheral nervous system, injuries or deformities of the hands and arms, current drug treatment and observed incidence of numbness and tingling. Smoking history should be recorded.

If any of the factors mentioned above give cause for concern, the prudent course would be to exclude the worker from exposure.

Further regular screening examinations should be carried out as long as the worker is exposed to vibration.

Anti-Vibration Gloves

There are several suppliers of anti-vibration gloves incorporating a polyurethane elastomer layer on the palm and fingers (either Sorbothane, a product of the Leyland and Birmingham Rubber Company, or Viscolas from the Chattanooga Corporation). A reduction in vibration intensity is claimed by the manufacturers of these gloves and tests would appear to substantiate these claims, although it is not easy to devise an objective test which is capable of measuring the properties of a glove. The effect is largely limited to frequencies above 500 Hz; in some tests an increase in vibration magnitude at lower frequencies has been noted. In the UK, gloves are available from Guardsman and Hypasafe.

The user of a chipping hammer or similar tool normally holds the handle of the tool in his right hand (for a right handed person) and guides the chisel with his left. It is the left hand which is subject to high frequency shock waves and so is the one most likely to benefit from the protection afforded by a glove. Some pneumatic tools discharge a freezing air stream from their exhaust ports so the wearing of gloves may help to preventing chilling of the hands. Care should be taken that the wearing of gloves does not interfere with the proper operation of the tool and does not introduce an extra risk of injury through trapping in moving machinery or lack of sensitivity and control.

The dynamic characteristics of the hand-arm-glove system depends on the coupling between the hand and the tool, which in turn depends on the grip. If the wearing of a glove means that the user has to increase the grip force, the gloves may do more harm than good, so the gloves must be comfortable and supple to be effective. Some tools (chain saws, for example) incorporate heated handles, which have the same warming effect as the wearing of gloves.

Depending upon the frequency of the vibration, gloves may have an advantageous effect on vibration. The subjective effect on the tool operator is probably the best guide to acceptibility.

Vibration Measurements

The primary quantity to be measured is the acceleration of the tool handle. Note that in what follows, it is assumed that the vibration is generated at the handles of a vibrating tool, but exactly the same principles apply if the vibration comes from holding a work piece against a fixed tool.

Acceleration is measured by an accelerometer mounted on the handle of the tool and is expressed in m/s^2 rms. The measured spectrum on a tool is rarely as simple as a pure sine wave. It usually has a very complex character. It is believed that the most damaging frequency range of vibration is from 8 Hz to 1 kHz, so measurements need to be taken at least between those two frequencies. The nature of the vibration, particularly on a percussive tool, is a combination of the motion of the handles caused by the reaction of the fluid pressure on the body and a reflective shock as the tool bit impacts the work surface. This reflective shock can produce an acceleration as high as 100 000 g. Much of the difficulty experienced in measuring vibrations is brought about by the need accurately to determine levels of about $1000 m/s^2$ in the range up to 1 kHz, while there are shock waves present in the vibration signature with many thousand times greater levels. Techniques for dealing with this are discussed later, but it should be noted at this stage that measurement is not as easy as it might seem at first sight.

Some authorities prefer some other measurement parameter to rms, for example rmq (root mean quad), which gives a better measure of discomfort. At present the general preference is for rms which is well understood and for which instruments are readily available. Root mean quad values are applicable to those vibration signals characterised by high crest factors (exceeding 6); rmq is analogous to rms and is defined by:

$$rmq = \left[\frac{1}{T} \int_0^T a^4(t) \, dt \right]^{1/4}$$

Vibration Levels

Vibration can in general occur in the three orthogonal axes as shown in Figure 1. Vibration axes related to the tool are known as basicentric and when related to the body are anatomical.

Usually the vibration levels in one axis will predominate, and it may be adequate to measure just in that direction but in other cases the magnitude of the vibration in all three orthogonal directions needs to be recorded, although for assessing damage, the Standards require the maximum to be used. Results can be presented in the form of a frequency weighted acceleration, or in octave, 1/3rd octave or narrow bands. The weighted value is a useful concept for quick assessment purposes, in a similar way to A-weighting in noise measurement, but for research purposes the full frequency spectrum may be more useful. The frequency weighting network required to

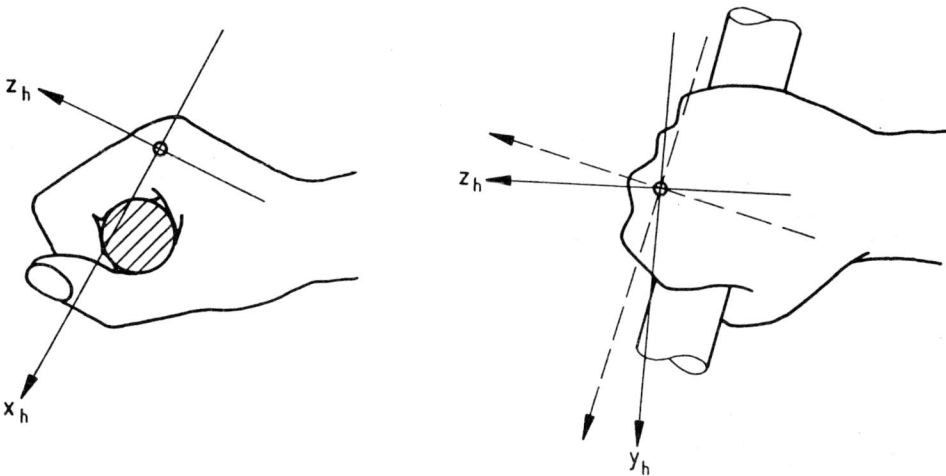

(a) 'Handgrip' position (showing a standardized grip on a cylindrical bar of radius 2 cm)

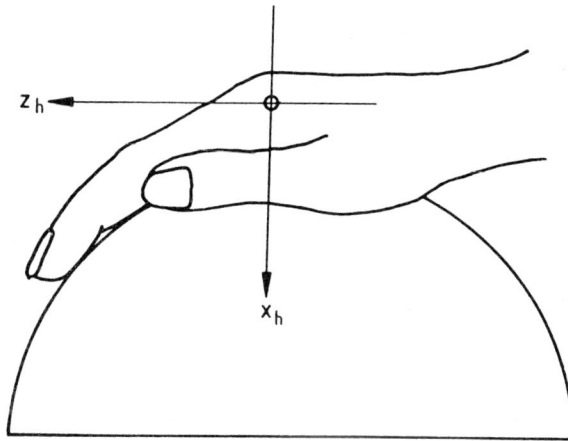

(b) 'Flat palm' position (showing a hand pressed onto ball of radius 5 cm)

Figure 1 Hand co-ordinate system
The solid lines are for an anatomical co-ordinate system.
The dashed lines are for the basicentric

give the weighted value is given in BS 6842. When vibrations are measured in 1/3rd or octave bands, it is possible to convert the readings into the weighted measurement through the use of Table 1 which gives the appropriate multiplying factors at each frequency.

The frequency weighted acceleration $a_{h,w}$ can be found from

$$a_{h,w} = \left[\sum_{i=1}^{i=n} k_i \, a_{h,i} \right]^{1/2}$$

TABLE 1 – FACTORS TO PRODUCE WEIGHTED VIBRATION VALUE

Frequency (Hz)	Factor K (1/3rd octave)	Factor K (octave)
6.3	1.0	
8.0	1.0	1.0
10.0	1.0	
12.5	1.0	
16	1.0	1.0
20	0.8	
25	0.63	
31.5	0.5	0.05
40	0.4	
50	0.3	
63	0.25	0.25
80	0.2	
100	0.16	
125	0.125	0.125
160	0.1	
200	0.08	
250	0.063	0.063
315	0.05	
400	0.04	
500	0.03	0.03
630	0.025	
800	0.02	
1000	0.016	0.016
1250	0.0125	

where K_i is the i'th factor taken from Table 1 (according to whether octave or 1/3 octave band data are used) and $a_{h,i}$ is the acceleration measured in the i'th octave or 1/3 octave. n is the number of bands being used.

The assessment of the damaging effect of vibration is based on a daily exposure of 8 hours. If the exposure is different from 8 hours, then a correction has to be made:

$$A_8 = (T/8)^{1/2} A_T$$

where A_8 is the frequency weighted 8 hour energy equivalent and A_T is the frequency weight energy equivalent for a period of T hours.

The magnitude of vibration can be expressed as an acceleration level using the decibel notation and a reference level of 10^{-6} m/s^2 as follows:

$$Ln = 20 \log (a/a_0)$$

where a is the acceleration in m/s^2 and a_0 is the reference level. Where the acceleration used is the weighted value $a_{h,w}$, the frequency weighted acceleration level is then:

$$L_{h,w} = 20 \log (a_{h,w}/a_0)$$

The Effect of Regular Rest Periods

The analysis above takes no account of the pattern of tool use or other exposure. It seems intuitively likely from general principles that the risk of contracting vibration diseases will be greater if all the exposure in a working day is crowded into one continuous period rather than spread out during the day with periods of rest interspersed. There is no epidemiological evidence to support such a hypothesis, however most workers feel the need to take regular periods of rest and this should be encouraged.

The probability of a worker developing symptoms of VWF depends on a number of factors, the most important as discussed above being the magnitude of the weighted acceleration and the exposure time.

Table 2 reproduces the best available evidence on the likelihood of symptoms developing.

TABLE 2 – MAGNITUDE OF $A_{h,w}$ (m/s²) WHICH MAY BE EXPECTED TO PRODUCE FINGER BLANCHING IN 10% OF PERSONS EXPOSED

Daily Exposure	Lifetime Exposure (Years)					
	1/2	1	2	4	8	16
8 hour	45	22	11	6	3	1
4 hour	64	32	16	8	4	2
2 hour	90	45	22	11	6	3
1 hour	128	64	32	16	8	4
30 minutes	180	90	45	22	11	6
15 minutes	256	128	64	32	16	8

These are approximate figures obtained by calculation from a limited amount of data. They can be extrapolated in any direction by multiplying by $2^{1/2}$ for a halving of daily exposure and by 2 for a halving of the lifetime exposure.

The equivalent table in terms of $L_{h,w}$ is

TABLE 3 – MAGNITUDE OF $L_{h,w}$ (dB re 10^{-6} m/s²) WHICH MAY BE EXPECTED TO PRODUCE FINGER BLANCHING IN 10% OF PERSONS EXPOSED

Daily Exposure	Lifetime Exposure (Years)					
	1/2	1	2	4	8	16
8 hour	153	147	141	135	129	123
4 hour	156	150	144	138	132	126
2 hour	159	153	147	141	135	129
1 hour	162	156	150	144	138	132
30 minutes	165	159	153	147	141	135
15 minutes	168	162	156	150	144	138

When determining damage risk criteria, it is recommended that these tables are used rather than exposure limits suggested in early versions of ISO 5349, or in BSI DD 43 (Draft for Development 43) now withdrawn. There have been many proposals and recommendations in recent years, issuing from Japan, Russia and Sweden among others, which exhibit considerable differences among themselves, reflecting the paucity of research on this subject. All these recommendations ought now to be submerged by the present proposals in BS 6842 (ISO 5349).

Whether they will remain the last word on this subject is doubtful, but they do represent a framework within which surveys can be done.

There are, as yet, few national recommendations as to the acceptable vibration magnitudes for specific tools, but it is expected that they will be produced eventually and based on ISO 5349. It is generally believed that if the vibration has a weighted value of less than 1 m/s^2, in use through the working day, symptoms will not usually occur, but at the present state of technology there are few vibrating tools that are able to reach such a small value. The placing of an onerous burden on employers by limiting the time for which tools may be used so that the weighted value does not exceed 1 m/s^2 is not justified by the present state of knowledge. Some national authorities have ventured to make recommendations on exposure limits although the evidence on which these are based is limited; Table 4 summarises some of these. It seems likely that a degree of consensus will be reached at a value of $a_{w,(4\ hours)} = 4$ m/s^2, equivalent to 10% of the exposed population expected to produce finger blanching after 8 years. In the UK, at the present time, it seems likely that an exposure based on an 8 hour day will be used to define an Action Level $a_{w,(8\ hours)} = 2.8$ m/s^2. The two values are, of course, equivalent.

Each National Standardisation Board (NSB) is expected eventually to specify a recommendation for the maximum value for $a_{h,w}$. Working within any such a prescribed value will not guarantee freedom from vibration injury. The relationship between exposure and damage risk is a statistical one, and any prescription has to be based on acceptable percentage risk. Figure 2 can be used to assess the risk for exposure percentiles other than 10%.

Methods of Measurement of Vibration Magnitude

For some tools, methods of measurement have been prescribed through standards or codes of practice. As might be expected, those tools which are known to have caused incidences of VWF in the past are the ones to which most attention has been given. The appropriate standards, where they exist, should give guidance on the position of the vibration transducers and the manner in which the tool is used.

Chain saws have been known for many years to cause the disease, primarily because forestry workers tend to use their saws over long periods of time and outdoors in cold weather. BS 6916 (ISO 7505) specifies the method of mounting the accelerometers and the type and size of log to be cut. This standard is clearly an attempt to regularise the measurement regime in order that the various tools can be compared; it does not necessarily identify the most severe way in which a particular tool may be used.

This problem of choosing a representative regime is one faced by all bodies concerned with setting standards for vibration measurement: the actual way that the tool is used can make a considerable difference to the vibration level. A skilled operator, through much experience, adopts a way of working which is a comfortable balance between vibration and efficiency. It may result in considerably less vibration exposure than that suffered by an unskilled operator.

Vibration test codes for hand-held grinders and chipping hammers have been developed (Pneurop Publication 6610 and 66160, see section on *Standards*). Once again, the tests are for comparative purposes only and the vibrations measured may not be typical of an ideal operation.

The whole issue of vibration applied to the body is one in which there remain many outstanding questions, in particular when trying to assess the relationship between exposure and injury. There is an undoubted connection and the careful employer will take steps to limit the exposure of his employees.

TABLE 4 – HAND-ARM VIBRATION: LIMITS AND GUIDE VALUES FOR ASSESSMENTS

Source	Assessment criteria and limits		Comments
	Daily exposure (hrs)	Limit a_{wh} (m/s^2)	
US (ACGIH) 1984	4 – 8 2 – 4 1 – 2 <1	4 6 8 12	
Netherlands NVVA 1987	8 4 2 1	2 3 4 6	Absolute limit regardless of duration = 10 m/s^2
Netherlands Ministry of Social Affairs and Employment Directorate General of Labour	Health Boundary Value (4 hours) Action level	1.5 3	Limit for the vector sum
UK BS 6842:1986 Appendix A	No symptoms in normal usage Symptoms likely after about 8 years in more than 10% of vibration exposed population 4 hrs/day	<1 < 4	For guidance only Refers to exposures which are regularly repeated on a daily basis
Germany VDI 2057: 1987	8	2.5	Guidance in the form of a curve. Lower limits for elderly and sick.
Australian Council of Trade Unions H&S Bulletin 1982	4	1	
USSR Health Standard 30414 – 84	8	2	8 hours basis – weighting similar to ISO. Implied Health Effect Boundary at 20% of this value
France AFNOR NFE – 90 – 402	1 Continuous exposure 2 Short duration exposure		Short duration limit 30 m/s^2
Poland Biuletyn Zoszyt 1986	8	1 at 20Hz 2 at 40Hz 4 at 80Hz 8 at 160Hz 16 at 320Hz 32 at 640Hz	
China IVSS Paper Vienna 1989	4	5	Proposed by researchers developing a hygiene standard
Denmark IVSS Paper Vienna 1989	Low risk 4 hr Max limit for Daily exposure 4 hr	<1 3.2	

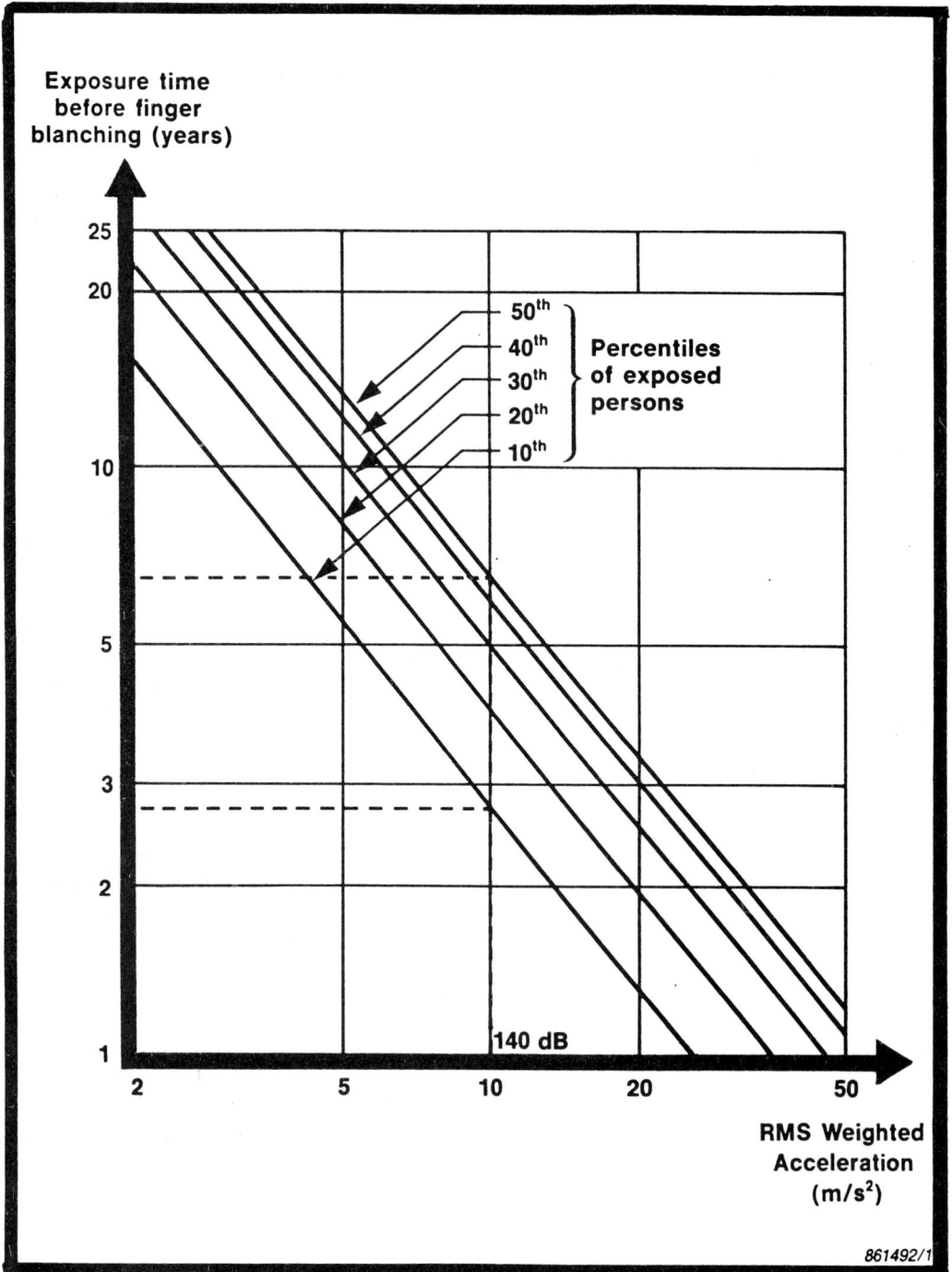

Figure 2 Risk of contracting Vibration White Finger

Chapter 6
Whole Body Vibration

Whole body vibration is that which is applied through a supporting surface – a seat or platform as shown in Figure 1, which also defines the basicentric coordinate system used in the standards. Exposure of the body to vibration or shock of this kind produces a complex distribution of oscillatory motions and forces within the body which can degrade health, impair activities, impair comfort and cause motion sickness.

One possible way to study the response to the human body to vibration is by treating it as a collection of springs, masses and dashpots (the bio-mechanical analysis). A number of attempts have been made to do this but the usefulness of such an approach is dubious.

Degraded health includes back ache and spinal damage resulting from exposure to seat vibration. Almost any part of the body can be damaged by vibration or shock, in some cases by a single event, in others by long term exposure. Exposure can disturb the central nervous system and can affect the circulatory and urological systems.

Vibration of the body may affect visual perception. Even small movements of the head (of the order of 1 mm) produces a similar movement of the retina of the eye and can disturb visual acuity. Sensations of touch and hearing may also be affected. There appears to be a complex relationship describing a combined exposure to noise and vibration. The results of disturbed perception may affect a person's ability to aim a sight or track a target.

It is commonly observed that vibration can disturb one's comfort. It is also apparent that some kinds of vibration, perhaps better described as oscillation, can produce an enhanced sense of well-being. One of the problems in this area is how to define the difference between them. Much the same problem occurs in trying to distinguish between an offensive noise and a pleasurable sound.

Low frequency vibration (less than about 0.5 Hz) can cause the motion sickness syndrome characterised by pallor, sweating, nausea and vomiting. Human reactions to vibrations in this range vary widely – some people can become accustomed to motion and after a time no longer experience sickness. It would clearly be of considerable advantage if the probability of vomiting could be predicted.

BS 6841 discusses the evaluation of vibration and shock in respect of the effects on:

    health
    human activities
    discomfort and perception
    incidence of motion sickness

These are indicated in detail in Table 1, with the frequency ranges and the predominant axes.

TABLE 1 – FREQUENCY WEIGHTINGS FOR THE VARIOUS EFFECTS ON BODY VIBRATION

Frequency Weightings (Hz)	Health	Hand control	Vision	Discomfort	Perception	Motion sickness
Wb (0.5 – 80)	z-seat	z-seat		z-seat x, y, z-feet z-stand x-prone	z-seat z-seat x-prone	
Wc (0.5 – 80)	x-back			x-back		
Wd (0.5 – 80)	x-seat y-seat	x-seat y-seat		x-seat y-seat x, y-stand x, y-prone y, z-back	x-seat y-seat x, y-stand x, y-prone	
We (0.5 – 80)				rx, ry, rz-seat		
Wf (0.1 – 0.5)						z-vertical
Wg (1.0 – 80)		z-seat	z-seat			

As with hand-arm vibration and noise generally, the sensitivity of the human body to vibration varies throughout the frequency range, which in the case of whole body is from 0.1 to 80 Hz. The weighting curves are different for the different axes of measurement. For motion sickness, the range is from 0.1 Hz to 0.5 Hz in a vertical direction; for other forms of vibration the range is 0.5 to 80 Hz. BS 6841 and ISO 2631, which overlap but cannot be considered equivalent, should be referred to for information on the various weighting curves. Table 1 identifies the six weighting curves that are applicable to the axes of Figure 1. The curves themselves are presented in outline form in Figure 2, but for detailed information refer to the standards. Note that for linear vibration the units used are m/s^2 and for rotational vibrations, rad/s^2. Many of the suggestions made in the standards are of a tentative nature and will be moderated as further knowledge is gained.

ISO 2631 defines three criteria which can be used to assess the significance of vibrations

preservation of working efficiency – the fatigue decreased proficiency boundary

preservation of health or safety – exposure limit

preservation of comfort – reduced comfort boundary

The fatigue decreased proficiency boundary is used to assess exposure limits for the tasks where fatigue is known to impair prformance. Exposure limit is the criterion which if exceeded is likely to impair a person's health. Reduced comfort boundary is for assessment of people exposed to vibration in vehicles. Figure 3 relates the weighted acceleration to the allowed exposure time for each of the criteria above. When vibration occurs in several axes simultaneously, the effective vibration can be calculated by taking a 'vector' sum:

$$a = [(1.4\,a_x)^2 + (1.4\,a_y)^2 + a_z^2]^{0.5}$$

Some instruments, utilising a triaxial accelerometer, are capable of calculating this value.

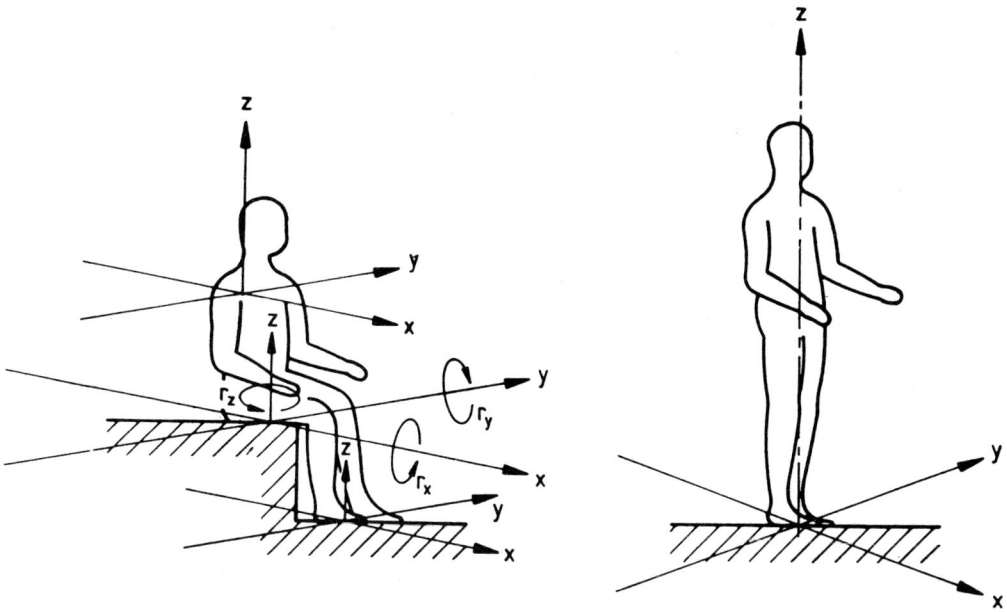

(a) Principal basicentric axes for a seated person

(b) Basicentric axes for a standing person

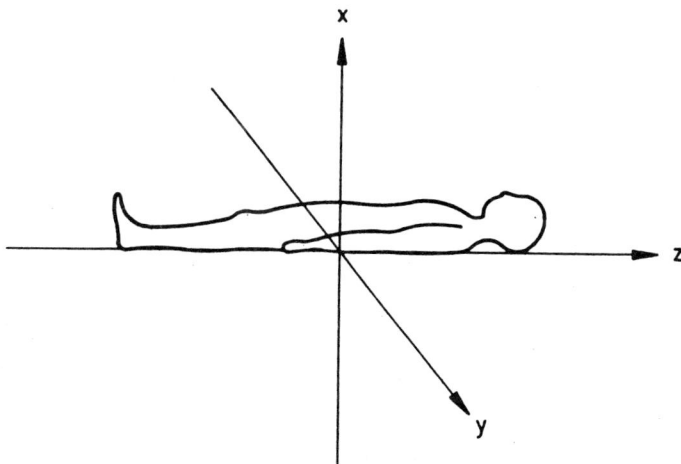

(c) Basicentric axes for a prone person

Figure 1

Whole-body vibrations should be measured in the directions of an orthogonal coordinate system having its origin at the location of the heart (see diagram). The longitudinal direction (head-to-toe) is called the z-direction. In this direction, the body is most sensitive to vibrations in the frequency range from 4 –8 Hz. Human response to vibrations in the x-direction (front-to-back) and the y-direction (side-to-side) do not differ and, in this lateral (transverse) plane, human response is greatest in the frequency range from 1–2 Hz.

Vibrations in the frequency range 0.1–O.63 Hz are considered to be responsible for causing discomfort or acute distress (commonly known as motive sickness), in people who are exposed to them. Individual human reactions to vibrations in this frequency range vary widely and are dependant not only on the vibration (motion) itself, but on factors such as vision, odours and age, which makes a study of this type of human vibration particularly complicated.

It is important to note that the co-ordinate system always follows the position of the body. This means, for example, that the z-direction for a person in a lying-down position will be horizontal.

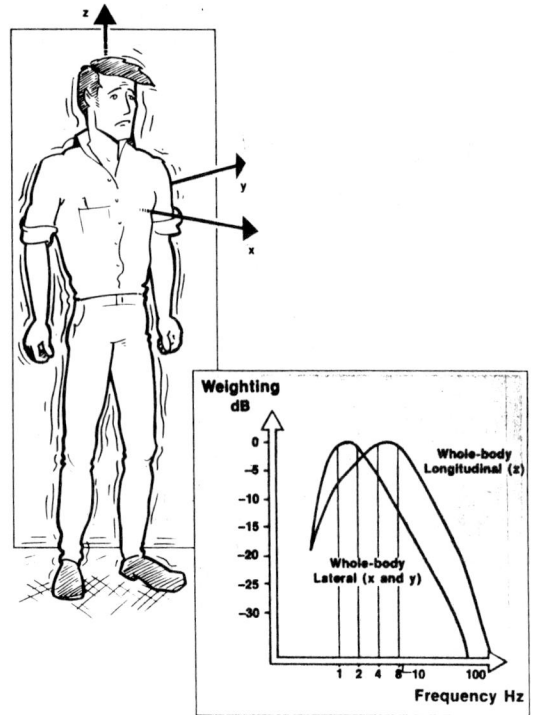

Figure 2 Whole-body weighting curves and reference axes

Effect of Vibration and Shock on Health

The main regime is that of a seated driver or operator, subject to a series of repeated shocks (W_b, W_c, W_d). Where the vibration is of constant magnitude, characterised by a crest factor of 6 or less, the rms of the vibration weighted according to the curves of Figure 3 can be used. For high crest factors (greater than 6), the concept of Vibration Dose Value (VDV) has been introduced, defined as

$$VDV = [\int_0^T a^4 (t) \, dt]^{1/4}$$

which has the units m/s$^{1.75}$

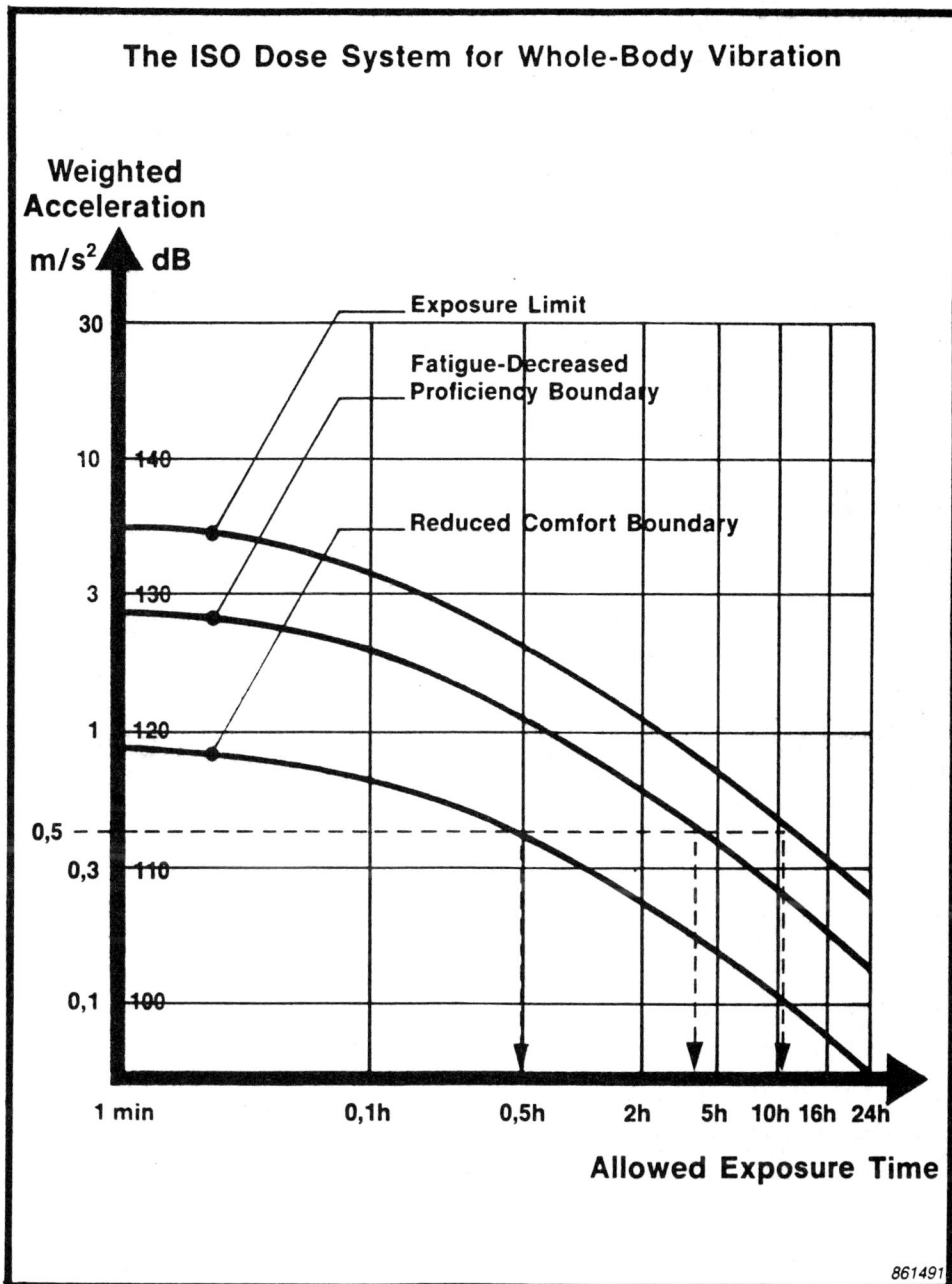

Figure 3

a(t) is the frequency weighted acceleration. The time T is the total period in the day during which the vibration may occur. It is believed that VDV is a better measure by which to assess biological damage to tissues than rms. To use this method the appropriate measuring equipment has to be available; this may present a problem.

When the crest factor is below 6 and the exposure is reasonably constant, the estimated vibration dose value (eVDV) can be determined from

$$eVDV = [(1.4\,a)^4\,b]^{1/4}$$

a is the rms acceleration in m/s^2; b is the duration in seconds; and the factor 1.4 is an empirical value determined from typical vibration environments with low crest factors. eVDV has the units of $m/s^{1.75}$. This method is likely to be much easier to use but is not as reliable as the one above.

Values of vibration dose values in the region of 15 $m/s^{1.75}$ are likely to cause severe discomfort. For periods of exposure in excess of 4 hours, this value is above the 'exposure limit' of ISO 2631.

For vibration regimes where the peak acceleration exceeds 1 g ($9.8\,m/s^2$), resulting in the person lifting off his seat or platform, the shock produced by impact may impose more severe damage.

Effect of Vibration and Shock on Human Activities

There are several possible effects in this category: hand manipulation and control, and vision (W_d, W_g)

Where hand or finger control is required to an accuracy of 5 mm rms or 2.5 N rms, the weighted acceleration should not exceed

0.5 m/s^2 rms. If less accuracy is acceptable, the weighted values can be increased in linear proportion.

For vision, where it is necessary to resolve detail subtending less than 2 minutes of arc, the weighted acceleration should not exceed 0.5 m/s^2. For every increase of $\sqrt{2}$ in the resolution, the vibration can be doubled. This criterion is based on an error in reading of less than 5% in good lighting using healthy subjects with good vision. Any other circumstances might require lower vibration magnitudes.

Effect of Vibration on Discomfort and Perception

This is concerned with the effect on comfort of people exposed to vibration and shocks in travel, work or leisure (W_b, W_c, W_d, W_e). Another aspect is the procedure for predicting the perception thresholds for vibration.

As well as vibration along the main axes, comfort can also be affected by rotational vibration of the seat, and by linear vibration of the backrest and feet. Different multiplying factors can be applied to assess the relative significance of vibration in these directions; the recommended factors are seen in Tables 2 and 3.

For vibration thresholds, the factors are unity throughout. The median perception threshold is 0.015 m/s^2, but there is considerable variation between individuals.

Effects of Vibration on Motion Sickness

In assessing motion sickness, a motion sickness dose value is defined as:

$$MSDV = [\int_0^T a^2\,(t)\,dt]^{1/2}$$

**TABLE 2 – MULTIPLYING FACTORS FOR PREDICTING
COMFORT FOR SEATED PERSONS**

Axis	Weighting Curve	Factor
Seat		
x	W_d	1
y	W_d	1
z	W_b	1
rx	W_e	0.63 ⎫ rotational
ry	W_e	0.4 ⎬ vibration in
rz	W_e	0.2 ⎭ rad/s²
Seat back		
x	W_c	0.8
y	W_d	0.5
z	W_d	0.4
Feet of seated persons		
x	W_b	0.25
y	W_b	0.25
z	W_b	0.4
Standing		
x	W_d	1
y	W_d	1
z	W_b	1
Prone		
vertical	W_b	1
horizontal	W_d	1

TABLE 3 – REACTIONS TO VIBRATION MAGNITUDE

Level (rms m/s²)	Reaction
< 0.315	not uncomfortable
0.315 – 0.63	a little uncomfortable
0.5 – 1.0	fairly uncomfortable
0.8 – 1.6	uncomfortable
1.25 – 2.5	very uncomfortable
> 2	extremely uncomfortable

which has the units m/s$^{1.5}$, and can be seen equal to the rms value of the acceleration a, integrated over a time T, multiplied by $T^{1/2}$.

Alternatively, if the motion is continuous and of constant magnitude:

$$\text{MSDV} = (a^2\, t_0)^{1/2}$$

where t_0 is the exposure period.

It is found that, if effect of motion sickness is taken as the likelihood of a person vomiting, the percentage of such people in a healthy population comprising males and females exposed to motion is 1/3 MSDV. The empirical value of 1/3 is taken from experimental studies and data from ship's passengers.

Evaluation of Human Exposure to Vibration in Buildings

This is another aspect of whole body vibration, for which different criteria apply. The weighting curves in the range 1 to 80 Hz, as might be expected, are similar in shape to the ISO curves discussed above. The interpretation is a somewhat different. Building vibrations may be classified as impulsive, intermittent or continuous. Impulsive is a rapid build up to a peak, followed by decay and may involve several cycles or can consist of a sudden application of several cycles lasting less than 2 seconds; intermittent is a string of vibration incidents separated by much lower magnitudes; continuous vibration is uninterrupted for a period of 16 hours.

The appropriate basic exposure curves are shown in Figure 4. To the basic curves one can apply multiplying factors varying from 1 to 128 according to the circumstances as given in Table 4.

TABLE 4 – MULTIPLYING FACTORS TO DETERMINE SATISFACTORY MAGNITUDES OF BUILDING VIBRATION

Place	Time of day	Multiplying factors for vibration type	
		Continuous	Intermittent and impulsive
Critical areas hospitals,	Day	1	1
laboratories etc.	Night	1	1
Residential	Day	2 to 4	60 to 90
	Night	1.4	20
Office	Day	4	128
	Night	4	128
Workshops	Day	8	128
	Night	8	128

Note: There are trade-offs between the number of events per day their magnitude and durations, and it may be difficult in any given case to select the proper factor to use. More detailed information is given in BS 6472.

A separate standard applies to horizontal building vibration (perhaps better described as repeated motion) in the frequency range 0.063 to 1 Hz. Three types of structure are considered: general purpose buildings, offshore structures and buildings for special purposes. Three categories of human response are considered: basic threshold effects; intrusion, alarm and fear; and interference with activities. As might be expected, the motion of offshore structures is generally greater than for onshore buildings and the people stationed there are expected to carry out work in a more arduous environment. Offshore, satisfactory vibration levels are six times as great as for buildings on land.

General Conclusions on Whole Body Vibration

The various standards and recommendations in this area are not always consistent, reflecting the developing knowledge about the subject. There is rather less consistency in whole body than in hand-arm vibration, probably because the results of an excessive vibration applied to the hand has been shown to cause disease, whereas vibration applied to the whole body will cause a combination of injury and subjective effects. In a developing subject, more research and consistent reporting of results are called for and many of the standards are directed in part towards that end.

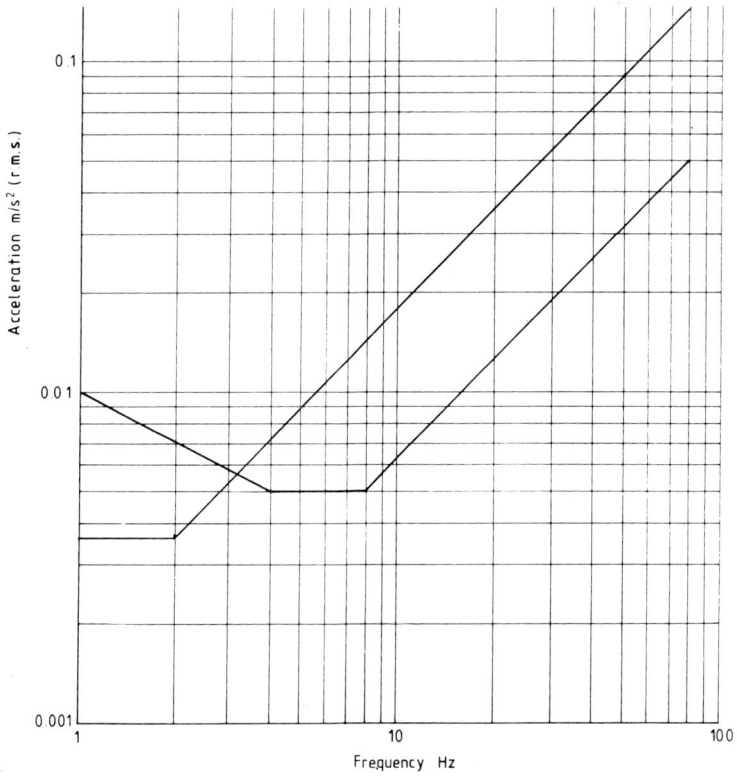

Figure 4 *Basic exposure curves for estimating the effect of Human exposure to building vibration. The upper curve is for x and y axis, the lower curve for the z-axis.*

SECTION 3

Chapter 1
Measurement Techniques

Noise measurement

The first step in noise measurement is to choose the sound metering equipment that best satisfies the requirements of the test. In most cases, where a particular test code is in use, the quality of the hardware is specified by reference to an appropriate standards authority. It is important that the instructions in the test code are followed implicitly, which will include the type of meter to be used; otherwise the test will be vitiated. Laboratories that are accredited for Type testing, for example to meet the requirements of the EEC legislation can only obtain such approval through strict adherence to the specified procedure. Thus in UK, the National Measurement Accreditation Service (NAMAS) operated by the National Physical Laboratory is the appropriate approvals body; similar bodies exist throughout the rest of Europe.

Ordinary commercial laboratories may not wish to meet the highest standards, but ultimately if the work they do and the developments they undertake have eventually to be submitted for approval, the best available equipment must be used, and proper calibration regimes must be adopted to ensure consistency.

In most laboratories, equipment has to be multifunctional, so for example if both noise and vibration have to be measured, a meter capable of serving a dual purpose should be used. The recording and analysis part of the system should be considered in light of all the requirements to be met.

Sound Pressure Measurements

A microphone, by the nature of its design responds to pressure changes only. Pressure is non-directional so a sound meter connected to a microphone cannot give any clear information about the orientation of the sound source. In this it is similar to the human ear (it takes two ears to give a limited amount of directional information). The primary use of a sound level meter is to give information about the effect that the sound can have on a person in the same position as the microphone. A secondary use is for determining the effect of pressure changes on a structure.

A microphone responding to pressure fluctuations is not able to determine with any degree of accuracy the directional quality of the sound. Some commercially available directional microphones, used for sound recording purposes, have a good front to back discrimination, but otherwise have rather poor directional qualities. The best way of determining the directional properties of a sound source is through the use of an intensity probe as discussed below. In measuring sound for the purpose of assessing human exposure, a simple microphone is a reasonable analogue for the ear, and the frequency weighting in a sound meter can be considered as equivalent to the signal processing that goes on in the brain. In evaluating human exposure, a sound meter with an attached microphone is perfectly adequate.

Certain simple precautions need to be taken. Measurements are commonly made out of doors with a hand held microphone pointed approximately at the source. This is assumed to be a free field condition (i.e. the sound pressure is not affected by the presence of reflections), but may not be so. There may be local reflections from quite small objects in the vicinity.

Even the body of the operator can reflect or absorb some of the sound unless the microphone is held at arm's length; wind noise can be present (many test codes set a limit to the wind speed during the time that testing is permitted); the microphone may be dirty; the meter may need to be calibrated; the battery charge may be too low.

Outdoor measurements may be affected by the wind developing a turbulent air stream around the microphone. This can be avoided by the fitting of a windshield, provided it is one approved by the makers. It is worthwhile checking on the effect of the wind by turning the microphone through 90° or 180°.

Sound pressure readings should normally be taken in the far-field rather than the near-field. It will be recalled that far-field conditions are normally to be expected at a distance from the source greater than about three or four times the characteristic length of the radiating object. Alternatively in the far field, the sound pressure level should decay at 6dB for a doubling of the distance from the source. It should not be assumed that, even in the far-field, the level is the same either in volume or in frequency content at equal distances from the centre of the source. Some machines may radiate differently in different directions; for example, an engine exhaust will increase the sound level in the direction in which it is pointed.

If a machine has a number of operating modes, for example off- and on-load, sound pressure levels should be checked at them all. Some machines exhibit surprisingly large variations in level from day to day, depending on environmental conditions and on other random circumstances such as the level of maintenance, lubrication, type of fuel, and voltage variations in electrical machines all of which are outside the control of the operator.

Near-field measurements close to a source may need to be taken if an operator is required to work there, but the readings will only be correct for that position.

Average Noise Levels in Offices

Usually in offices, the sound sources are distributed throughout the working area, and although there may be several machines such as computer printers, typewriters and shredders which are themselves particularly noisy, the usual technique for establishing typical sound pressure levels is to take the measuring instruments around the room at fairly evenly spaced locations and at several times during the day. Figure 1 shows recommended measurements for a typing pool. An average SPL is likely to give a fairly accurate idea of the exposure of a worker. Reflections off the walls ceilings and floors normally ensures an even distibution, except for particularly noisy items. If any such are identified by a few spot measurements, that will indicate where the sound reduction treatment is best applied and will also indicate the level of reduction required to bring it down to the the background level.

Sound Pressure Level Measurement for Machines

Typical of the standards introduced for assessing the sound pressure level of a machine is the Pneurop/CAGI Standard (Pneurop is the European Committee of Manufacturers of Compressors, Vacuum Pumps and Pneumatic Tools, and CAGI is the American Compressed Air and Gas Institute). This is a jointly developed technique, once widely used although it is now being superseded for most purposes by more modern methods based on sound power.

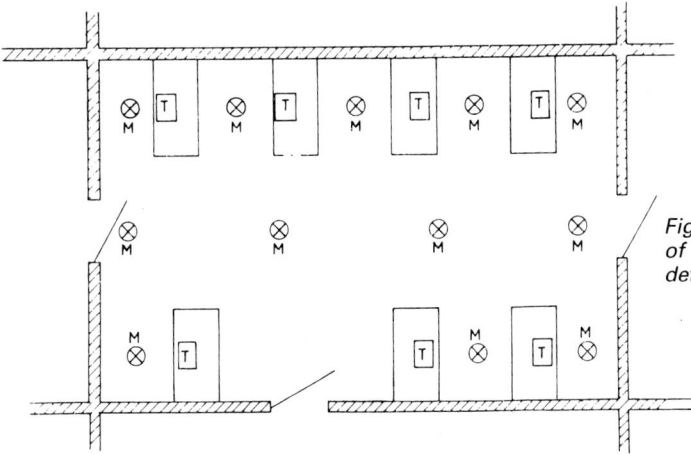

Figure 1 Example of the distribution of measurement positions used to determine the average noise level in a typical typing pool

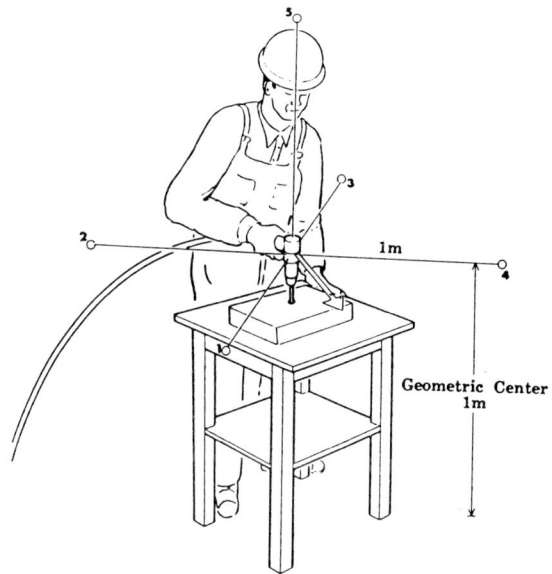

Figure 2 Sound Pressure measurement for hand tools

Two measurement procedures are defined – the first suitable for small, hand operated tools, and the second for large items of equipment such as compressors and track-mounted drills. For small tools the measurement distance is 1 m from the source; five measuring stations are defined as in Figure 2. For large equipment, measurements are taken at 1 m and 7 m from the side of the machine (as well as in the direction of maximum source) as in Figure 3.

Note that the distance is from the *side* of the machine not from its centre. The justification for this is that, as far as the operator or the passer by is concerned, what matters most is his proximity to the machine not to some arbitrary point at its geometric centre; 1 m is a typical distance for an operator and 7 m for a passer by. Using this code, machines can be rated as to the effectiveness of the silencing treatment provided.

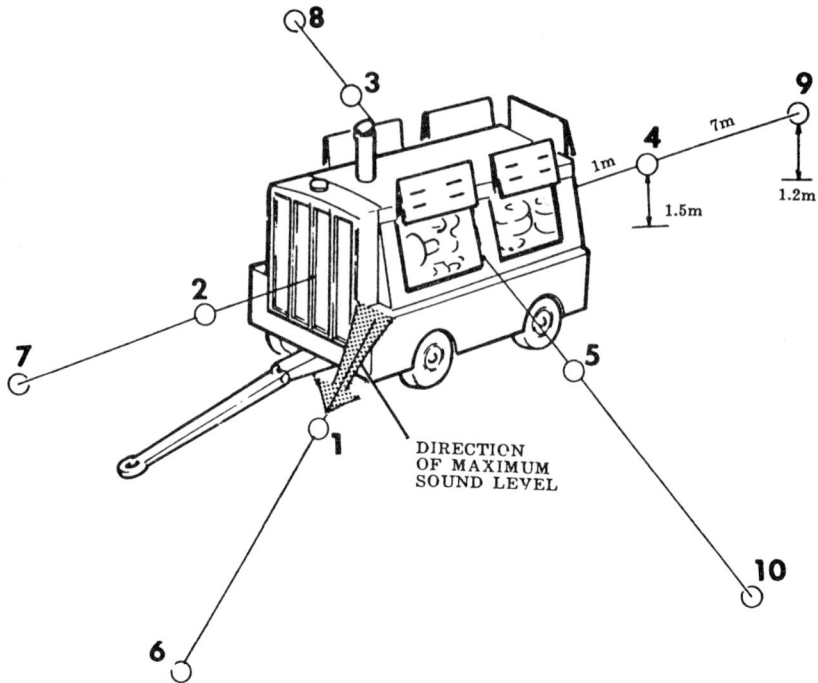

Figure 3 Sound Pressure measurement for large machines

This code, or variations of it, have been adopted by other industries because of its basic simplicity. All that is required is a suitable sound level meter and a measuring tape. A-weighted readings are taken at five or ten points and corrected for background sound; the arithmetical average of each set of five is then calculated (if the maximum variation is less than 5 dB this is acurate enough but if not, the logarithmic average can be taken instead) and that value presented as the noise rating of the machine. This gave rise to the categorisation of compressors as unsilenced (more than 75 dB(A)), silenced (less than 75 dB(A) and supersilenced (less than 70 dB(A), all measurements taken at 7 m.

The drawback of this code, or any based on sound pressure readings alone, is that while it gives some indication of noise exposure of an individual it is of little use in further calculation. Also, because of the limited number of the measurement points, some significant areas may be missed. By taking the measurement distances from the surface of the source rather than its centre, larger machines tend to have lower sound levels.

Sound Power Measurements

Most modern test codes require the determination of the sound power level of a machine, rather than its sound pressure. The distinction between the two is that sound power is a characteristic of a machine, irrespective of where the measurements are taken, whereas sound pressure is a reading taken at a specified point near the machine and varies with distance.

The easiest way to determine sound power is when the noise source is situated in a free field. The technique, in principle, consists in surrounding the source with an imaginary enclosure, taking sound pressure readings over the surface of this enclosure and then integrating over the surface area. In practice, what is usually adequate is to take sound pressure levels at a number of points approximately evenly distributed over the surface, then find their average and apply that value over the whole area, to get a final value of sound power.

The theoretical ideal of a free-field in all directions is only attainable by suspending the noise source in an anechoic chamber. For heavy equipment this is not usually practical, so the next best alternative is adopted with the equipment standing on hard ground in the way it will be used in practice. The arrangement is theoretically equivalent to two sources, the second one being the mirror image of the actual source situated under the surface. The equivalence is only exact if the surface is truly relective to sound waves, usually satisfied by smooth concrete or well-laid non-porous tarmac. If the test is done over rough ground or grass, a correction has to be made. The sound pressure level measured over a perfectly reflecting surface is 3 dB greater than for the same source suspended in a free field. The directivity index is therefore equal to 3 dB.

The imaginary enclosure is preferably a hemisphere, as shown in Figure 4. For very large sources, a rectangular parallelopiped may be used as in Figure 5. The area of either of these is readily calculated. The use of a more complex surface may lead to errors and most approved standards use one or the other. In principle large measuring distances should be chosen, which in the case of a hemisphere is achieved when the distance between the hemisphere and the outer

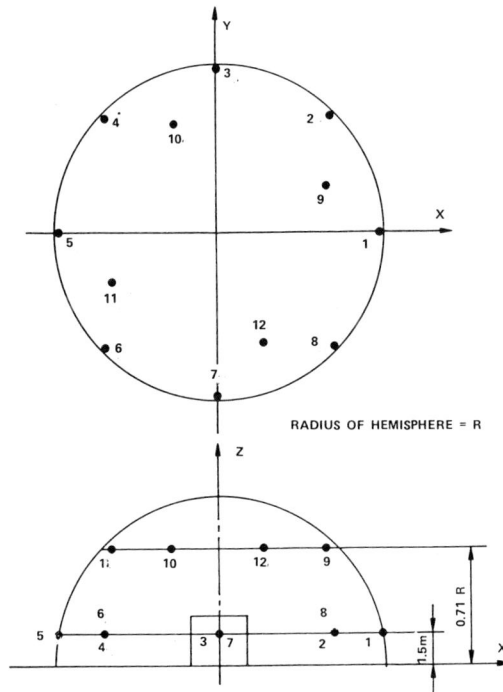

Figure 4 Hemispherical Measuring points according to EEC Directives

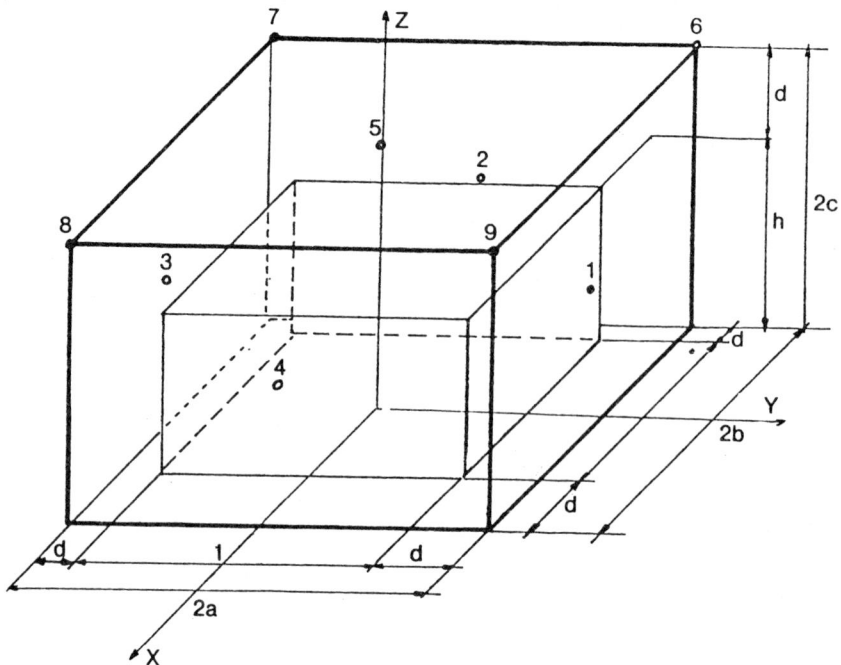

Figure 5 Parallelopiped Measuring points according to EEC Directives

surface of the source is not less than twice the largest dimension (length, width or height) of the source. For construction equipment, the recommendation given in the EEC Directive 79/113/EEC is that if no dimension exceeds 4 m, the measuring surface should be a hemisphere radius 10 m; if no dimension exceeds 1.5 m, the radius should be 4 m. The area of the measuring surface may be calculated approximately – an error of 20% produces only a 1 dB difference in the sound power.

The average sound pressure level can be found, by taking an arithmetic average if the differences are small or a logarithmic average otherwise. The sound power is found from:

$$L_{WA} = L_{pA} + 10 \log (S/S_0)$$

where S is the area of the measurement surface and S_0 is 1 m^2.

For a hemisphere with radius r = 4 m, 10 log (S/S$_0$) = 20 dB. For a hemisphere with radius r = 10 m, 10 log (S/S$_0$) = 28 dB

It will apparent that the same pressure readings taken to satisfy the requirements of a test such as the Pneurop/CAGI mentioned above can also be used to determine sound power levels. The test conditions will not be identical to an EEC test, (in particular the number of measurement stations will differ), and will not therefore meet the legislative requirements, but the calculation of sound power will be reasonable accurate. Thus for example if a noise source has a physical width of 2 m, a length of 4 m and a height of 2 m, a parallelopiped measuring surface placed at 7 m away from its sides will have an area of 2 × (7 + 2) × (14 + 4) + 2 × (14 + 2) × (7 + 2) + (14 + 2) × (14 + 4) = 900 m^2. The value of 10 log (S/S$_0$) is 29.5 dB, so this value can be added to the measured sound pressure level to provide an estimate of the equivalent sound power. This calculation has validity only if the sound pressure levels are reasonably uniform over the measuring surface. It will be apparent that the measuring points on a hemisphere can be made to correspond to those on a parallelopiped with a differing area, with a consequent doubt as to which area is to be taken. It is consideration of this kind, among others, which has led to a reconsideration of sound power determination and has led to the theoretically more satisfying method of sound intensity measurements for sound power.

Sound Intensity Measurements

There are two main purposes for which sound pressure measurements are made by a microphone: the exposure to noise of a person in the same location, and the determination of the sound power of a piece of machinery. To meet the first purpose, all that is required is a microphone with a suitable circuit and display. The second requires further calculation for which more information is required such as the properties of the environment – whether anechoic or reverberant or somewhere in between and if there are other sound sources in the vicinity which may effect the result. Until recently, the latter was the only means of determining sound power. The theoretical possibility of using sound intensity to determine sound power has always been acknowledged, but has not until recently been capable of practical realisation.

Sound intensity is the rate of energy flow across a surface and has the properties of a vector, i.e. it possesses magnitude and direction, so that if it were possible to measure the flow perpendicular to a measurement surface enclosing a noise source, the total power could be assessed. Any suitable surface can be chosen but usually a simple one is best – it can be a hemisphere centred on the ground surface, a sphere, a rectangular parallelopiped or a conformal surface. The device used to measure the intensity is known as an intensity probe. In use the probe is moved over the measurement surface and normal to it. An integration of the energy flow gives the total power of the source.

The great advantage of using an intensity probe is it only responds to the energy flowing out of the imaginary enclosure defined by the measuring surface; it is not influenced by any inward flowing energy. It does not require the use of a chamber with any particular acoustic characteristics, and in use does not need any other noise sources to be turned off. High background noise, provided it remains constant during the test, does not affect the accuracy of the method. It is claimed that sound power can be determined to an accuracy of 1 dB from noise sources that are as much as 10 dB below the background. Because of the highly directional properties of the probe it is possible to identify those parts of a machine which are radiating the most noise, so it is particularly valuable for diagnostic purposes. One of the disadvantages of the conventional method of sound power determination is the expense and trouble of designing or measuring the acoustic properties of the test environment. With the intensity method, tests can be done in any location which obviates the expense of building anechoic or reverberant chambers.

In the chapter on *Room Acoustics*, the technique for measuring sound insulation and determining the sound reduction index of panels and walls was described. The method adopted is as described in BS 2750 and uses two reverberant rooms which are independently isolated. The reverberant properties of both rooms have to be determined; this is expensive to set up. The alternative, using sound intensity, requires only one reverberant room. Inside the room, the sound pressure is measured; outside the room, the sound intensity is measured. This technique has the further advantage in that the properties of different parts of the insulating partition can be assessed, e.g. windows and doors.

Recalling the definition of the intensity as the rate of energy flow per unit area, then if the source has a power of W watts, the intensity has units of watts/m^2.

In the simplest case of a point source surrounded by a sphere, radius r, the intensity is given by

$$I = \frac{W}{4\pi r^2}$$

Intensity levels are also measured in decibels, with a reference level of 1 pW/m^2.

Sound intensity is the product of particle velocity and pressure. It obtains its vector quality from the particle velocity which is itself a vector.

Intensity = pressure × velocity

$$= \frac{force}{area} \times \frac{distance}{time} = \frac{energy}{area \times time} = power$$

The pressure is readily measured in the usual way by a single microphone. The particle velocity is the variable which has so far proved difficult to measure accurately. It can however be determined by relating it to the pressure gradient, i.e. the rate at which the instantaneous pressure changes with distance. The pressure gradient therefore requires the measurement of two accurately determined pressures separated by a known distance between them. This in its simplest form is done by two accurately calibrated microphones close together. It can be shown that the particle velocity is the time integral of the pressure gradient:

$$u = \frac{1}{\rho} \int \frac{dp}{dr} \, dt$$

where u is the particle velocity, ρ is the density and dp/dt is the pressure gradient.

An intensity probe consists of two back to back microphones separated by a short spacer. The average pressure measured by the two microphones multiplied by the integration of the pressure difference gives the intensity. The accuracy of the system depends upon the accurate calibration of the two microphones and of the signal analyser, which performs the integration. The density also neds to be measured, determined from the local barometric pressure.

Figure 6 shows a modern intensity probe complete with its analyser. In a probe design, there are strict requirements to be met: the two microphones must have identical phase response and have a flat amplitude response curve. Care must be taken in choosing the correct probe for the intended purpose, particularly the frequency range of interest. The length of spacer between the two microphones governs the frequency range. Typically, the spacing can be 6 mm, 12 mm and 50 mm. The smaller the spacer, the higher the practical frequency range. The Bruel and Kjaer recommendations for their own equipment are:

Microphone spacing	Frequency Range
6 mm	450 Hz – 10 kHz
12 mm	250 Hz – 5 kHz
50 mm	63 Hz – 1.25 kHz

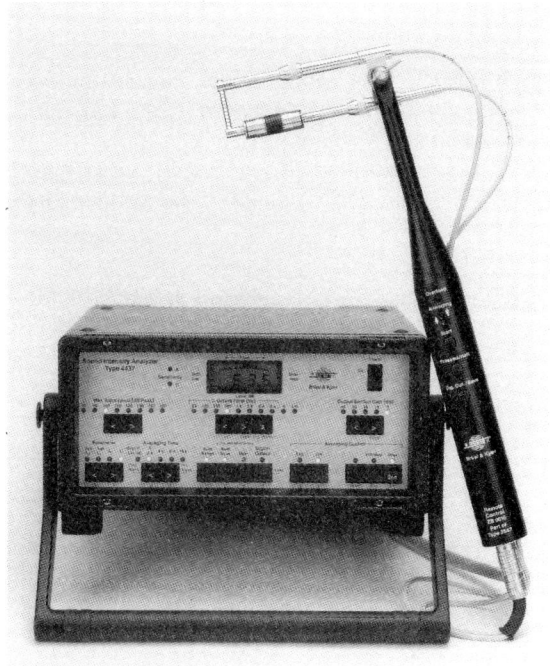

Figure 6 Intensity probe and Analyser

These values are illustrative only. The actual frequency range will depend on a number of factors such as the accuracy of the instrumentation and its calibration. The low frequency limit is also influenced by the *pressure intensity index*, which is defined as the pressure level minus the intensity level

Choice of Analysis Equipment

Different analysers are available for different purposes. For simple sound power determination, one which incorporates digital filters for A-weighting and octave octave bands (and possibly 1/3 and 1/12 octave bands) would be adequate. For laboratory use and for experiments involving narrow band determination an analyser based on FFT is available.

An essential piece of equipment when making intensity measurements is the calibrator. It will be apparent that accuracy in the determination of intensity depends upon having two microphones accurately balanced throughout the frequency range of interest. This can be assured by calibrating either each microphone in turn or ideally the two microphones simultaneously by means of a coupler. Figure 7 shows the principle of an acoustic calibrator. It consists of a sound source and special coupler with two cavities. The cavities have ports for the insertion of microphones. One of the cavities is connected to a sound source, e.g. a pistonphone, while the other is connected to it by a coupling element. The signals in the two cavities have a phase difference which is proportional to frequency. Although any well defined sound source can be used for calibration, it is desirable to acquire a complete system from the supplier of the intensity probe.

Techniques of Measurement

When using intensity measurements it is necessary to make sure that the probe is accurately tracked over the measuring surface, in position and orientation. This may require the surface to be defined by a a network of string and wire or by the use of a fixed frame. One of the standard surfaces such as a hemisphere or a parallelopiped should be tried first with the same number of measurement points as called for by the Standard. If there is much variation from point to point, the number can be increased. It is not necessary for the readings to be taken in the far-field. The measurement surface should not include any other sources of sound. One precaution to be observed is that if the measurement surface encloses absorbent material and the background noise is high, the sound power may be incorrectly determined, so absorbent material should be shielded.

At the present stage of development, there are no international standards which can be used to define a method, although ISO are preparing draft guidelines. The user may have to devise his own. It is likely to become much more accepted as a technique of the future.

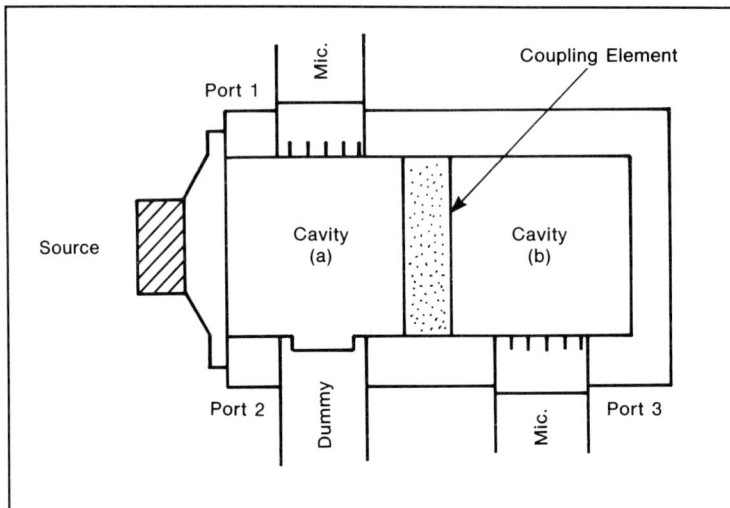

Figure 7 Principle of the intensity calibrator. The microphones are placed in ports 1 and 3 for calibration of intensity or velocity sensitivity.

Chapter 2
Sound Level Meters and Microphones

Choice of Sound Level Meters

A simple meter, which gives only a linear output of the measured sound, is of little value as a general purpose tool. Most noise test codes are satisfied by a meter giving an A-weighted reading. If one is concerned only with human exposure or with machinery characterised by a spectrum having nothing unusual in the way of frequency or impulsive noise, a meter with this limitation may be perfectly adequate. Increasingly other weightings are being used, particularly C-weighting, so one might wish to plan for future developments. In making a choice to meet specific requirements one can consider some of the features that can be incorporated in a sound level meter and decide which are the important ones. Of the possibilities, the following may serve as a check list:

Built-in or external microphone facility

Flat frequency (no weighting)

A– and C– weightings, linear (sometimes B– and D– weighting also for special purposes)

Octave and one-third octave band display obtainable by built in filters or with a separate filter set

Built-in calibrator or external calibration – acoustic rather than electronic calibration is preferred

Leq (integrating facility) for a range of integrating times from 1/8 second to several days

Sound exposure level (SEL)

'Fast' and 'Slow' response

Peak hold facility

Impulse noise facility for measuring short peaks

Digital or analogue display

Accuracy to meet specified standards

Interfacing facility for use with external recorders or processors; DC and AC output for recorders; RS232 interface for computer processing

Dynamic range required – usually from 25 to 140 dB

Battery capacity to satisfy periods of continuous use

Ability to withstand environmental conditions

Portability or wearability if used as a personal dose meter

Many sound meters can be used for the measurement of vibration as well as sound, and in a small laboratory with limited equipment it may be that a multi-purpose meter can serve in a dual capacity. In such an instrument the microphone input can be replaced by the accelerometer lead via a suitable amplifier. However since vibration measurement is becoming more and more specialised, for many applications a separate instrument may be the best answer. In deciding on the important features for a vibration meter, the following should be considered, which are in addition to the relevant ones from the list above:

Measurement ranges in dB, acceleration, velocity and displacement (e.g. 0.01 – 1000 m/s^2 etc.). In some meters, Imperial units may be available.

Human vibration filter set in accordance with ISO 2631 and ISO 5349

Triaxial input for seat acceleration measurements

Categories of Sound Meters

A portable instrument will necessarily have only a limited facility for analysis and storage and will probably not incorporate an internal calibration check. A personal dose meter is wearable rather than portable in that it can be carried by the worker in a shirt pocket or clipped to his belt throughout the whole of a working shift.

Measurement Circuits

Whatever the kind of meter that is chosen, the complete circuit of a sound measurement system will consist, in general, of that shown in Figure 1. The elements of the circuit can be separate, but are often combined in the one meter. In particular the filter networks are often incorporated in a separate module which can be attached to the main instrument; this allows for the filter networks to be chosen for the application.

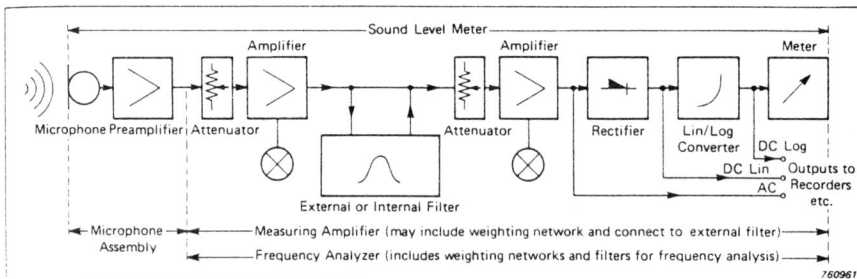

Figure 1 Block diagram of a noise measuring system

Standards for sound meters

BS 5969 (equivalent to IEC 651), following the recommendations of CENELEC HD425, designates four accuracy grades for sound level meters – Table 1. The requirements of a meter measuring peak levels is also specified.

The standard includes A–, B–, C–, D–weightings and Lin. The latter is a flat frequency response (no weighting). Not all meters are required to incorporate all weighting networks. Time weightings are also included, in particular S (slow), F (fast), I and Peak. F has an rms exponential time constant of 125 ms, S an exponential time constant of 1 s. The time constant for I is 35 ms and for Peak, less than 100 microseconds.

TABLE 1 – GRADES FOR SOUND LEVEL METERS

Grade	Type	Accuracy dB
0	Laboratory reference	± 0.4
1	For laboratory use and for field use where the acoustical environment can be controlled	± 0.7
2	General field application	± 1.0
3	Survey applications	± 1.5

The directional characteristics to be satisfied by the microphone are also specified in BS 5969. The higher the frequency, the more the measured sound pressure is likely to be affected by the orientation of the microphone. If A– weighting is the primary concern of the investigation, this frequency effect may not be important, but to avoid difficulties, the manufacturers' recommendations must be followed carefully for repeatable results. The type of microphone and the method of calibration must be chosen to match the characteristics of the sound field.

Figure 2 shows a simple Type 3 Sound Level Meter suitable for general purpose field work; it has an analogue output, linear and A– weighting filters, and a range from 40 dB to 110 dB. Figure 3 shows a rather more sophisticated meter with a wider range from 25 dB to 130 dB and a Type 2 accuracy suitable for industrial and environmental use. This instrument combines in one unit a conventional sound level meter, an integrating meter (L_{eq} meter) and a peak sound level meter.

An example of a top of the range meter appears in Figure 4. This can be used as a Type 0 meter with selected microphones, otherwise as a Type 1 meter. It incorporates A–, C– and Lin weightings and can be used for octave and 1/3 octave analysis using plug-in modules. It has a digital output for connection of a printer or computer for further processing. The integrating module shown in the figure allows a full range of integrating modes.

BS 3539 applies to meters intended for use when measuring sound from motor vehicles. The only frequency weighting required is A–, and the time weighting is as for BS 5969.

BS 6698 applies to integrating sound level meters, also known as integrating-averaging sound level meters. The grades are identical to those in BS 5969, but in addition give an output of the Equivalent Continuous A– weighted Sound Pressure Level. They may also measure Sound Exposure Level (SEL).

BS 5330 describes a test method for estimating hearing handicap and contains a brief specification of a meter able to determine the equivalent continuous sound level. It applies whether the sound is steady, intermittent, fluctuating or includes impulsive components. This standard is similar but not identical to ISO 1999. These meters can be set with a threshold level (usually 85 dB or 90 dB according to the exposure criterion they are intended to measure).

Figure 5 shows a typical dosemeter. A device of this kind can store up to 8 hours worth of 1 second readings, giving an exact profile of noise exposure during the day. A dosemeter (or dosimeter) is a special kind of integrating sound level meter, and so some manufacturers are able to combine the two functions in one instrument. A dosemeter has a clip-on microphone to allow the sound to be picked up near the wearer's ear and should be small enough for easy carrying in a shirt pocket. An integrating meter, like most sound level meters, has a microphone attached to the meter case, either rigidly or by a gooseneck.

Figure 2 Sound Level Meter
This pocket-sized Sound Level Meter is designed
for general purpose fieldwork

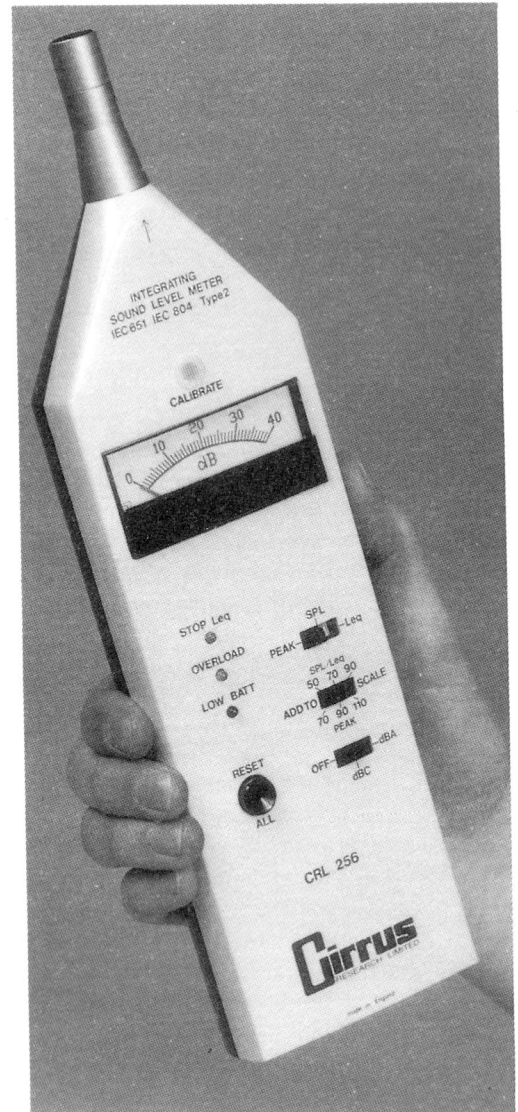

Figure 3 Combined instrument,
Type 2 accuracy

BS 6402 applies to Sound Exposure Meters, which may be worn by persons in a factory or other locations. The accuracy required is for Type 2, with A– weighting. Due to the varying directions from which sound comes and the shielding effect of the body, the readings taken from a meter used in this way may be up 2 dB different from those measured in the absence of a person. The output of a sound exposure meter to this standard is a function of the A– weighted sound pressure known as the Sound Exposure, defined as:

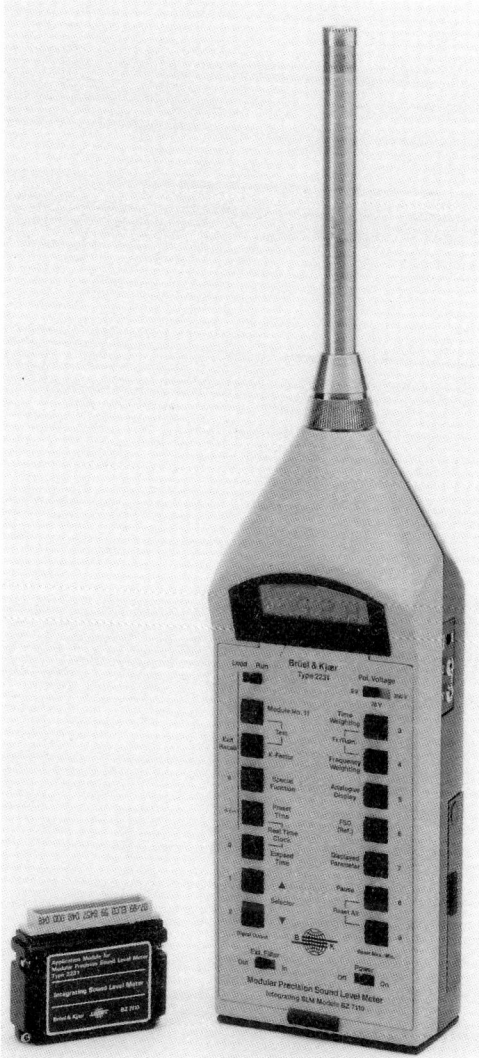

Figure 4 Sound Level Meter with Integrating
Module

$$SE = \int_0^T p_A^2 (t)\, dt$$

where p_A is the A– weighted sound pressure and T is the time interval. The output is given in Pa^2 h.

The sound exposure, when obtained, has to be converted into the Equivalent Sound Pressure Level over the working day of 8 hours; some meters incorporate the conversion and may also give a reading of the accumulated dose as a percentage of the maximum daily allowed exposure.

The relationship between the Equivalent Continuous Sound pressure Level (L_{AeqT}) and Sound Exposure (E) in Pa^2 h is:

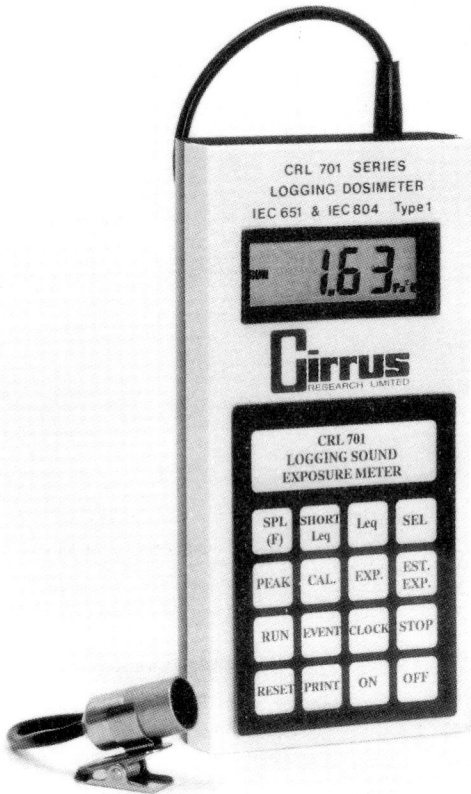

Figure 5 A dosemeter can be worn or carried in the pocket, with a clip-on microphone. It measures peak, L_{eq}, SEL, dose time and sound level. It weighs 350gm

$L_{Aeq, T} = 10 \log \{E/(Tp_0^2)\}$, where p_0 is the reference pressure 20 Pa.

For a period of 8 hours, the value of L_{Aeq} can be found from Table 2, which is a calculation of the equation above.

In addition to the international and British standards indicated, other national standards may apply. In USA, ANSI S1.4 is broadly in line with BS 5969 (IEC 651). German standard DIN 45655 also applies.

TABLE 2

E	$L_{Aeq,8}$
0.1	75
0.2	78
0.5	82
1.0	85
2.0	88
5.0	92
10.0	95

It will be apparent that, with such a large number of standards relating to sound level meters, care has to be taken to determine the most suitable for one's purpose. Clearly if it is required to perform tests to a specified test code, that code will itself specify a minimum grade of meter. Otherwise, the prospective purchaser should buy the best he can afford.

Calibration

Microphones are individually calibrated at the factory and a calibration chart is usually provided. Sound levels meters, comprising the complete circuit from microphone to display, can be calibrated by means of a sound calibrator. This is a device which generates a sinusoidal sound pressure of a specified level and frequency when coupled to a specified model of microphone. Thus when obtaining a sound level meter, the appropriate calibrator must be obtained to match the chosen microphone. Corresponding to the grades of meter (0, 1 and 2) as defined in BS 5969, there are the same grades of calibrator (0, 1 and 2), whose characteristics are specified in BS 7189 (IEC 942).

This standard applies only to acoustic calibrators of which there two common kinds. One is a pistonphone, which is an accurate reliable and simple device, employing a pair of independent pistons driven by an electric motor. A distortion-free sound pressure signal is created inside the coupler cavity of the microphone. Another form of calibrator works by vibrating a metallic diaphragm which, in a similar way, produces a pure sinusoidal signal in the coupler cavity. A common reference frequency is 1 kHz with a signal magnitude of 94 dB. Because the magnitude depends on the volume of the coupler cavity, the calibrator must be suitable for the microphone being used. There may have to be separate matched couplers for each microphone.

Electrical calibrators are also available for field calibration, but wherever possible an acoustic type is preferred for accuracy, since it checks the complete circuit from microphone to display. One type of sound meter incorporates an acoustic calibrator in the same casing as the meter (Figure 6). The ready availability of calibrators, either internal or separate, means that very frequent calibration is possible and indeed recommended; some test codes require it to be done at the beginning and end of each set of readings. It is also recommended that the sound meter is returned to the manufacturers for an overall calibration check. This should be done at least once a year, or in accordance with the recommendations of the supplier.

Microphone Types

A microphone used in sound meters is influenced by its orientation to the source of the sound. Care should be taken to ensure that the characteristics and orientation of the microphone chosen are suitable for the sound field which it is attempting to measure. Response characteristics are expressed as free-field, pressure or random incidence. A free-field microphone is one which is intended for pointing at the source of sound and is designed to compensate for the disturbance caused by its presence. A pressure microphone should be held at grazing incidence to the sound source. A random incidence microphone responds uniformly to sound from whatever direction and is used in diffuse fields.

Microphones have different types of construction according to the intended use – in sound meters, telephones, music and voice recording and so on. Three main types are used for measuring sound levels: piezoelectric, electret and condenser. Piezoresistive microphones can also be used for special purposes.

Piezoelectric microphones utilise the effect of a charge which is produced when a piezoelectric material (a ceramic or crystal diaphragm) is stressed by the fluctuating sound pressure. The

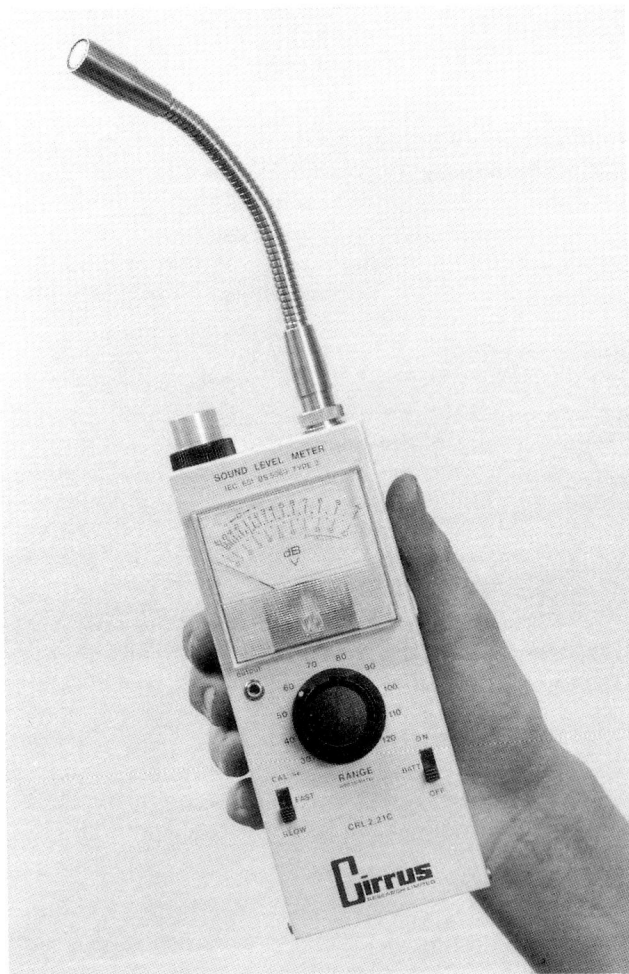

Figure 6 Sound Level Meter incorporating an acoustic calibrator

charge across the crystal is usually fed to a high impedance charge amplifier. These tend to be used in general purpose instruments of a robust and economical design.

Condenser microphones consist of a thin diaphragm close to a rigid back plate. A polarising voltage (typically 200 V) has to be applied across the plates. As the gap between the two plates change with pressure, the capacitance changes and generates an alternating voltage.

Electret are similar to condenser microphones except that, instead of an air gap, a dielectric foil is stretched over the back plate; the foil is pre-polarised so there is no need to apply the polarising voltage.

A piezoresistive microphone incorporates a diaphragm and a strain gauge bridge; it allows the measurement of static pressure fluctuations.

Each type of microphone has its own particular characteristics.

The piezoelectric type is robust and generally used in portable meters for field use; it is cheaper but tends to have a lower sensitivity than electret or condenser types. Condenser microphones have good long-term stability and have stable temperature characteristics; they have been traditionally used for precision instrumentation, and are still preferred for many applications. Construction is shown in Figure 7.

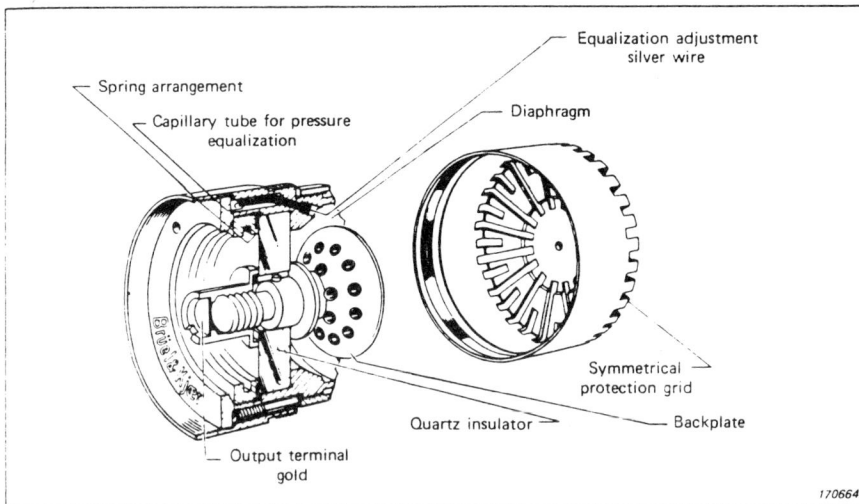

Figure 7

Electret microphones, soon after their introduction, had a reputation for not being as stable as the traditional condenser types, but today they are likely to prove as satisfactory, if not superior in this respect. In conditions of high humidity they are to be preferred. Although electret microphones are more expensive, they do not require a polarising voltage so the circuitry is cheaper.

Many of the microphones available on the market are interchangeable from company to company, so it should be possible to tailor a system to one's requirements. Microphones are graded for accuracy by the same system as for meters. Accuracy grades can be mixed.

The signal from a microphone is so small that it is of no practical use without amplification. An amplifier is needed not only to produce a usable signal but also to match the high impedance of the microphone to the low impedance of the analysis circuit. The two main types are voltage amplifiers and charge amplifiers. A microphone will possess a certain sensitivity but if a long cable is connected between microphone and amplifier, the voltage drops and the overall sensitivity changes; so a voltage amplifier needs to be close to the microphone to counter the cable effect. The charge at the microphone is unaffected by the length of the input cable, so a charge amplifier can be situated some distance from the microphone. Charge amplifiers are to be preferred on this account, although they tend to be more expensive.

The problem of cable length is not important if the microphone is permanently connected to the sound level meter but, in those applications where the signal has to be output to an octave analyser or computer situated some distance away, it can be crucial.

Chapter 3
Environmental Noise Monitoring

Environmental Noise Generally

Much of the pressure towards the reduction of environmental noise over the last twenty to thirty years has come from the public through the activities of noise abatement societies. Lawyers and engineers have each played a part in contributing to an overall reduction in noise. The various sources of environmental noise are:

> Aircraft
>
> Construction sites
>
> Mining and quarrying
>
> Activities of statutory utilities: water, gas, electricity etc.
>
> Factories
>
> Road vehicles
>
> Domestic
>
> Places of entertainment

The legal restrictions on sound levels and methods of reduction in all these areas are dealt with in other chapters, but we are concerned here with the equipment needed for measurement. Refer to the chapter on *community standards*.

A number of noise indices and procedures for analysis have been proposed over the years and instrument systems have been developed to process the sound pressure data into the favoured index. Some of the indices have fallen into disuse while others have survived. L_{eq} and its variants has survived as the primary index and so most instruments for use in the environment are capable of giving that value at least. In addition, some form of statistical analysis is often required, so L_n centile levels from L_0 to L_{100} may be required.

Aircraft Noise

In assessing the noise of aircraft flyovers there is a bewildering variety of noise measures, a few of which are dB(A), L_{eq}, NNI, L_{EPN} (EPNL), L_E (SEL), L_{NP}, N_{EF}, L_{AX} and L_{TPN}; these are defined in the chapter on *Noise Scales and Ratings*. There are others in less common use, such as NBF (Noise Burden Factor), CNR (Community Noise Rating), SENEL (Sound exposure Level for Single Events) and CNEL (Community Noise Exposure Level). The acoustics engineer may be faced with having to measure one or more of them, irrespective of the merit he may place on them, so his choice of measuring instrument will be determined by that requirement. The noise 'footprint' for a flyover, Figure 1, with a prescribed operational cycle, requires readings to be taken over a contour around the runway.

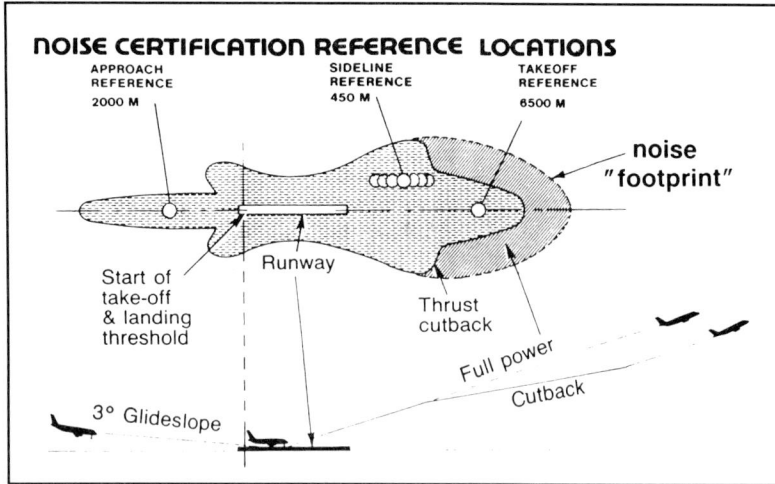

Figure 1 Noise Footprint

For specialised applications such as measuring sonic booms characterised by very short duration (300 ms) and a high sound pressure level (130 dB), a single event meter with a large dynamic
range is required (at least 45 dB). The characteristic shape of the N-curve is shown in Figure 2. See
BS 5331 (ISO 2249) for details. To record the full frequency content of this curve, a sound level
meter which starts recording at a preset threshold, or a single event recorder with the same
facility, is required. Another approach is the fitting of a single event recording module to a meter
that can accept it.

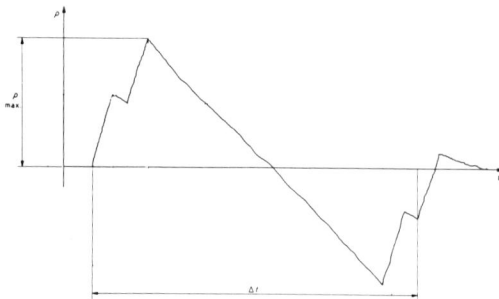

*Figure 2 Shock Wave produced by an aircraft
travelling at supersonic speed. P_{max} is the
maximum overpressure. Δt is the time interval
between the first shock and the last shock in the
signature.*

For non-aircraft surveys (most of the other noise sources mentioned above), a simpler
measure such as L_{eq} (or L_{Aeq} as it is sometimes denoted, emphasising the A-weighting) is all
that is needed. If this is combined with a facility for statistical analysis (L_n) and for recording the
time history during the day or during the event, most situations can be adequately covered. For
traffic, a Traffic Noise Index (TNI) may also be needed.

If peak values are likely to be significant, where impact, gun shot or explosions are involved,
then this facility should be available. A dynamic range of 100 dB or more is essential.

Dedicated Environmental Noise Meters

It is now possible to obtain integrating sound level meters which combine, in one hand-held instrument, the characteristics of a precision meter and a statistical data logger. An internal battery is provided, which will give a total of about 24 hours of readings. For the environmental noise engineer this may prove to be the most suitable equipment to obtain. Such a piece of equipment is capable of recording during a period of 24 hours up to 100 000 samples. The data in the samples can then be downloaded to a desktop computer for subsequent analysis, directly to a printer or via a modem. An RS-232 interface should be provided. Two-way communication via the RS-232 is also possible to control the data gathering via a suitable computer. A number of meters can be connected together via a daisy-chain cable to allow for automatic data recording and analysis. One point to bear in mind is that transmission through a serial port such as RS-232 is relatively slow; so if communication is to be established with a large number of stations, an ASCII mode of transmission may be inadequate to cope with the large amount of data. Transmission via a binary mode is to be preferred in this case. Some meters allow for both modes.

Automatic stop–start operation may also be incorporated, allowing for sophisticated test programmes. Some possible applications are:

1. A continuous noise survey spanning 7 days from midnight Sunday to midnight the next Sunday. An external power source may be required.

2. An airport wishes to count aircraft departures and measure the noise during the day but only for aircraft emitting noise above a certain noise level. Measurements to be taken during selected hours of the day over a period of several days, for example between midday and 2 pm Monday to Friday, for aircraft exceeding 65 dB(A). Total test time is 10 hours, so internal battery is adequate.

3. A transport survey of rush hour traffic is required, morning and evening during an entire month, from 7 am to 9 am and 4 pm to 6 pm. An external power source is required.

Each of these examples allow for unattended recording over the total period. Periodic downloading for analysis may also be done. Some instruments permit downloading, recording and analysis to be performed simultaneously.

If the instrument is to be used outside in all weathers, the microphone must be protected. An electret type is to be preferred for its tolerance to high humidity. A windscreen and rain cover suited to the microphone in use can be fitted. Some manufacturers offer temperature and humidity stability by offering a microphone equipped with a heater and a drying agent.

One manufacturer, Cirrus, offers a waterproof case as an extra facility. The instrument is fitted in the case and connected to extra batteries to allow for longer operation. The case can be closed and locked and an external waterproof microphone connected. This seems to be an excellent arrangement which, if not available for one's favoured meter, can be readily made up specially.

Environmental Noise Measurement with Standard Equipment

An alternative to the special purpose meter described above is a user-assembled system. A stand-alone microphone and amplifier or a simple sound meter with an interface for connection to a noise level analyser can be put together from standard items, not necessarily from one supplier. This approach will probably be the cheapest when a number of readings have to be taken at the same time when preparing a noise map around a factory or airport. Sequential sampling of an array of microphones is unlikely to be satisfactory unless the sound levels are constant or changing slowly; for aircraft flyovers, simultaneous recording is essential. A microphone and noise monitoring meter will be required at each terminal, or alternatively a multichannel tape or other recorder can be used to capture the signal for subsequent analysis.

Each system contains a microphone, monitoring terminal, data transmission link, control and computing system and a read-out or display. Typically there may be between one and twenty monitoring points, logging one or more of the noise indices discussed above.

Chapter 4
Analysis Equipment and Signal Processing

The availability of sound level meters that are able to perform advanced signal processing means that there is now not so much need for separate equipment to produce a directly usable output. It is often the case that a meter is required to satisfy the need of one particular user in one application, for example for weighted sound power or for one type of environmental measurement. It is likely, unless the need is very special, that such an instrument already exists. An advanced laboratory may require something more versatile.

Oscilloscopes

A sound level meter or a processing recorder can only give a limited amount of information (however useful that may be). For information about the wave form of the signal in the time domain, an oscilloscope is still a valuable tool. A modern storage oscilloscope can record and replay a portion of the signal for analysis or inspection. Alternatively photographs of the trace can be taken for a permanent record.

This technique is not so widely used today because of the time taken for photo-processing. Digital storage techniques allow for a permanent record to be obtained on a chart recorder or X-Y plotter.

Recorders

There will be instances in noise processing where it is not convenient to perform the analysis on site. A recorder is needed so that the signals can be processed at the laboratory. This is particularly valuable where a signal needs to be processed in several different ways. A single event such as a gun-shot, an explosion or an aircraft movement may occur relatively infrequently, and for convenience a facility for repeated processing is valuable. If an analysis needs to be done over several octave or fractional octave bands for comparison purposes, a recording is essential.

Tape Recorders

A tape recorder supplied for normal hi-fi audio recording should be suitable for analogue sound recording, if the only interest is in the audio range (say from 20 Hz to 20 kHz). Such machines do not record reliably at low frequencies, although it may be possible to extend the lower limit by special techniques. For low frequency work an FM recorder is to be preferred. Figure 1.

If one recorder is to be used both for noise and for vibration, it is worthwhile obtaining a special instrumentation recorder. The following features are available from which a choice can be made.

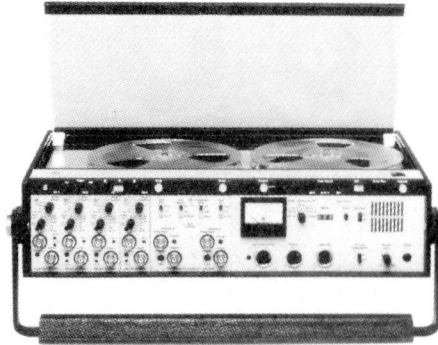

Figure 1 Tape Recorder Type 7005/06.

For some applications, the user may wish to record the vibration signals entering the three channels of the human-vibration unit. This is particularly relevant where future analysis of field data is required using laboratory equipment (e.g. 1/3-octave frequency analysis using Real-time Frequency Analyzer Type 2123). Tape Recorder Type 7005/06 is ideally suited to the task due to its ability to record all three channels simultaneously.

Multi-channel recording with a mixture of direct and FM channels; instrumentation channels with built-in charge amplifier also available. Number depends on tape width – typically 4 channels on $^1/_4$ inch tape; 14 channels on $^1/_2$ inch tape; 28 channels on 1 inch tape.

Recording level and overload indicator.

Separate input for voice microphone.

Tape speeds in the range $^{15}/_{32}$ in/s to 120 in/s, doubling speed at each step.

For field work the recorder should be robust, in a weatherproof case and unaffected by external vibration.

Operation from battery or mains.

Dynamic range to suit the application up to 90 dB. Range attenuation available.

It is possible to record at one tape speed and play back at another. Used in this way, the frequency range of the recording can be shifted into the audio range, which can be a valuable technique for short duration signals containing high frequency components. The response curve has to be flat for this method to work. Useful qualitative information can be gained by audible playback of a signal at a speed lower than that at which it is recorded – a simple procedure, but sometimes useful for diagnosis.

Calibration of the record and playback is important. This can be done by use of a sound calibrator (a pistonphone, for example) or by relying on internal calibration.

One can expand the effective dynamic range of a multi-channel recorder by using several channels simultaneously but with different gain settings, providing that the common input stage is not overloaded.

An instrumentation cassette recorder using standard cassettes can also be used. The main application when used for noise is for digital signals, taking the output from a suitable sound level meter with a digital output. Such a recorder is not suitable for sound recording direct, although for slowly varying analogue rms signals it could be satisfactory. Up to 8 channels can be recorded

Figure 2 8 Channel Analogue Recorder, using standard casettes

on standard 0.15 in tape. The recording tape speed is constant. Within these limitations, it can be an economical way of recording multiple output from sound level meters. Figure 2.

Cassette recorders using modern digital recording techniques are also available. With a 2 hour play time and possible 8 channel recording, a very large amount of high frequency data can be captured on a standard cassette. These instruments employ sophisticated analogue/digital and digital/analogue converters. The incoming signal is encoded into a digital form for storage on tape and either decoded for later use as an analogue signal or supplied to a computer for further digital processing. In this equipment, there is a trade-off between frequency of recording and number of channels, see for example Table 1.

TABLE 1 – SAMPLING FREQUENCY FOR DataRec DIGITAL RECORDER

No of channels	1	2	3	4
Sampling frequency (kHz)	9.6	4.8	2.4	1.2
Signal bandwith (kHz)	40	20	10	5

Another type of recording system is also available, based on computer technology using floppy discs and RAM. These too are only suitable for digital recording.

Digital Event Recorders

This is an instrument which records continuously in a recirculating memory, so that the stored information is always available up to the current time. When a triggering signal is applied, which can be external or at a preset threshold, the signal is stored for analysis. One advantage of this type of recording is that the information prior to the triggering signal is not lost. The trigger can be set at a high level with the assurance that the whole of the wave is available. It is useful for such applications as gun shots or sonic booms.

Other Recorders

Other recorders are available, but most have applications outside sound recording, other than in the presentation of the results of a signal from a sound level meter or from a signal processor. These include UV recorders, XY recorders and chart recorders, both thermal and pen types. Where a meter has an RS 232 serial port or there is an IEEE 488 parallel port from a signal processor, an alphanumeric graphics printer can be used to present the results of calculated sound levels such as L_{eq}.

Signal Processing Equipment

As sound level meters have become more powerful in their processing capability, the need for a separate processor has to some extent been reduced. On the other hand, there has arisen an increased demand for more complex analysis which is not easily done in a hand held instrument. Most single station noise readings and their interpretation in terms of overall sound levels and integrated values can be adequately covered in a modern meter. For complex multi-octave analysis and signal correlation a separate processor is still needed.

Noise Level Analyser

A modern noise level analyser will take the output of a meter or recorder and provide sound pressure levels, Leq and percentile levels, and many of the noise rating functions discussed elsewhere in the text; it may also incorporate octave and 1/3 octave filter sets. The output will be as a liquid crystal display, a serial or parallel interface for connection to a computer and an alphanumeric/graphic printer; AC and DC output for further analysis and recording is also available. When supplied in a weather protected enclosure, it is usually known as an environmental noise analyser.

Real-time Analysers (RTAs)

These are used for simultaneous frequency analysis of a noise signal. One often needs to obtain detailed information about the frequency content of a signal in octave and 1/3 octave bands (sometimes finer, down to 1/24 octave bands). If this is done sequentially through a bank of filters in a conventional sound level meter, the frequency range has to be set manually for each octave band (usually in no finer steps than octaves) by a knob on the body of the meter. The time taken to do this can be very long apart from the likely change in the sound level over the period. A real-time analyser, on the other hand, processes the information in parallel through a set of digital filters so that the results are immediately available. RTAs are available which are no larger than hand held sound level meters (see for example Figure 3) and the distinction between the two can become blurred with increasing sophistication of the latter. These advanced instruments may prove rather too expensive for field work.

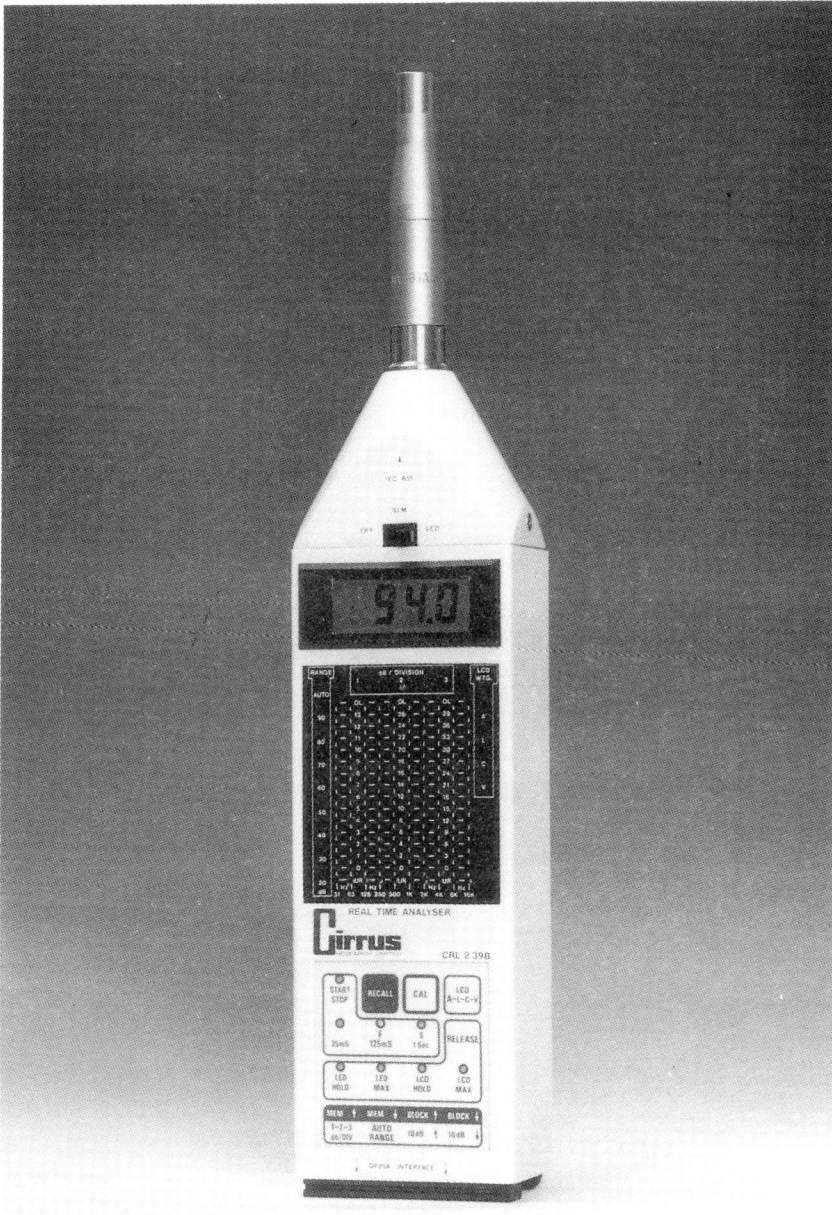

Figure 3

RTAs can be connected to a desktop computer for further processing. Usually a suite of suitable software programs are supplied with the instrument. Some RTAs have their own built-in display screen and are able to dispense with the computer. They are, in effect, a combination of a RTA with a dedicated computer.

RTAs are capable of doing much more for the acoustic engineer than just perform frequency analysis. Some of the available features (not necessarily obtainable in the same instrument) are:

Multi-channel configuration (up to eight) allowing simultaneous recording and memory and cross spectrum analysis.

Auto and cross correlation.

Spectrum weighting A–, B–, C–, D– and user defined.

A synchronised white and pink noise generator for reverberation measurements..

Separate acoustic signals using coherence power functions.

Variety of screen displays in colour and monochrome – histograms, 3D cascaded display, superimposed spectra for multi-channels.

Large internal memory for multi-spectrum storage. Additional storage on floppy disc.

Can also act as a sound intensity analyser and provide intensity and pressure spectra simultaneously.

Removes measured background noise from acoustic measurement.

RTAs are also able to handle vibration input from a linear accelerometer or from a rotational transducer. Correlation between noise and vibration signals of multi-channels can also be handled.

Hard copy from an integral alphanumeric printer or output to a separate printer or plotter

Noise and vibration analysis is a rapidly developing field. It is now possible in a portable instrument to perform many complex calculations at a modest cost. Some manufacturers offer a modular construction whereby it is possible to acquire exactly the features one requires.

Chapter 5
Audiometry Equipment

As discussed in the chapter on *Hearing Conservation in Industry*, the careful employer will consider the advisability of making audiometric tests available to those of his employees exposed to noise. There are five different types of audiometer classified in BS 5966 (IEC 645). Types 1 to 3 are intended for medical and clinical investigations. Types 4 and 5 are the ones likely to be used in industry. An audiometer made to this Standard will have the type number marked on it.

Diagnostic or Clinical Audiometers

These are used by audiologists and medical specialists for detailed analysis of hearing loss and to assist in the type of remedial or corrective action required, such as the provision of a hearing aid. These instruments will include provision for air and bone conduction, perform a variety of test procedures, such as those involving single and coincident tones randomly presented in terms of frequency and intensity above and below the threshold, and assess speech intelligibility. Bone conduction, in which sound is transmitted to the inner ear by vibration of the cranial bones and soft tissue, is not usually investigated in industrial surveys.

Screening or Monitoring Audiometers

These provide a simple indication of threshold shift and identify those subjects suffering from a hearing loss. They may be automatic, when a audiogram is produced as a permanent record on a chart, or manual when the readings require subsequent plotting by hand. The levels in an automatic recording audiometer are under the control of the subject being tested; only the recording is automatic.

Type 5 is a basic manual survey instrument with no specified frequency range. Type 4 is the one recommended in BS 6655 and is likely to be the one most suitable for routine use. It specifies a range of pure tones at 500 Hz, 1, 2, 3, 4 and 6 kHz, with a further recommendation for 8 kHz. For audiometry where the threshold is a nominal 0 dB, the levels are from −10 dB to +70 dB. A pulsed tone facility is incorporated. The levels are normally applied in 5 dB steps.

A basic audiometer incorporates an oscillator as a signal source, an amplifier and attenuator to vary the level, and a pulse-shaping network to modify the rise and fall times to remove transients. A self recording audiometer may be preferred to reduce operator errors to a minimum and for rapid screening.

Audiograms

A audiogram should be taken at regular intervals to assess the effects of noise exposure on the employee's hearing threshold. The test consists of applying a pure tone sound level at set levels and frequencies which are heard through headphones. When the subject considers that the threshold level has been reached, he presses a button and the processing module records that level. A reading is taken for each ear separately. There is a difference between ascending and descending thresholds, the former being obtained as the test signal is increased in level from below audibility, while the latter is obtained by reducing the signal from a clearly audible level. The hearing threshold is taken as the average of the two. BS 6655 recommends two methods: the ascending method and the bracketing method. The former merely records the ascending threshold, the latter determines the ascending threshold followed by the descending threshold. Whichever is chosen must be indicated on the result sheets.

A plot of a manual audiogram for a person suffering from mild hearing loss is shown in Figure 1, while Figure 2 shows a similar plot from a self-recording audiometer. Note the difference between rising and falling threshold levels on the self recording instrument.

It is convenient, when processing a large number of subjects, for a hard copy of the audiogram to be filed with each employee's records. An instrument which produces such copies is known as a recording audiometer. Any subject found to be suffering from a hearing loss, i.e. falling within one of the categories proposed by the Health and Safety Executive as calling for action (see the chapter on *Hearing Conservation*), can be referred for further investigation requiring the use of a diagnostic audiometer. Some audiometers store the results from previous tests on the subjects in the population and 'categorise' the results. Alternatively, an audiometer with a communications port (probably RS 232) can be connected to a desktop computer for further analysis and storage. It may be helpful to the employer to have available a statistical analysis of the various populations in differing risk groups so that protective measures can be taken, and a computer of this kind is ideal. Some suppliers of audiometers make available software packages adapted to this purpose. Figure 3 shows a typical modern audiometer to BS 5966, Type 4. The complete system requires a printer and a portable computer for control and recording. It allows manual or automatic testing, recall of stored results, and categorisation.

Audiograms are preferably taken in a proper booth (see the chapter on *Hearing Conservation*), but failing that, precautions have to be taken that the ambient noise level is not high. The recommendations in BS 6655 specifying the maximum permissible ambient sound level, I_{max}, taking into account the average attenuation of an audiometric earphone, are given in Table 1.

TABLE 1 – VALUES OF L_{max} WHEN THE HEARING THRESHOLD TO BE MEASURED IS 0dB

Frequency (Hz)	31.5	63	125	250	500	1k	2k	4k	8k
L_{max} (dB)	73	59	47	33	18	20	27	38	36

Audiometer Standards

The requirements to be met by an audiometer designed for determining threshold levels are contained in BS 5966 (IEC 645). Refer also to ANSI S3.6 – 1969 and ANSI S3.21 – 1978. It must be capable of producing pure tones to the absolute threshold as defined in BS 2497 (ISO 389). A limited range of fixed frequencies are provided, usually 125, 500, 750 Hz, 1, 1.5, 2, 3, 4, 6, 8, and 12 kHz, or a subset of these. For monitoring audiometers the seven main frequencies – 500 Hz, 1, 2, 3, 4, 6 and 8 Hz are considered to be sufficient. ANSI S 3.26-1969 is the US Standard.

An internal calibration of frequency and level is essential. External calibration can be carried out with a sound level meter coupled to the earphone via an artificial ear. The sound level meter

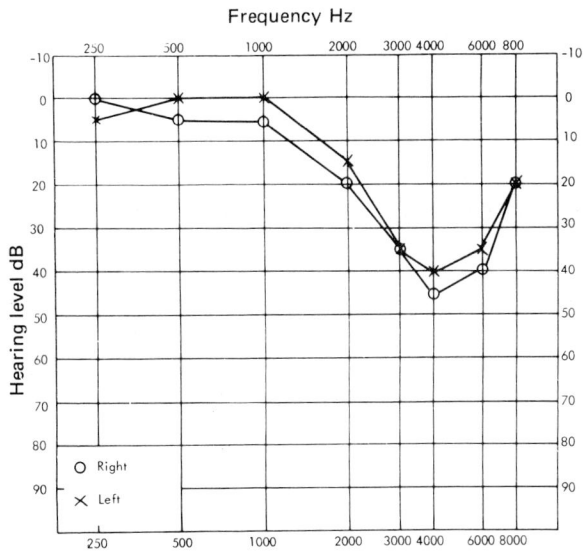

Figure 1 An audiogram chart illustrating the threshold of hearing of a person with a mild noise-induced hearing loss, measured with a manual audiometer

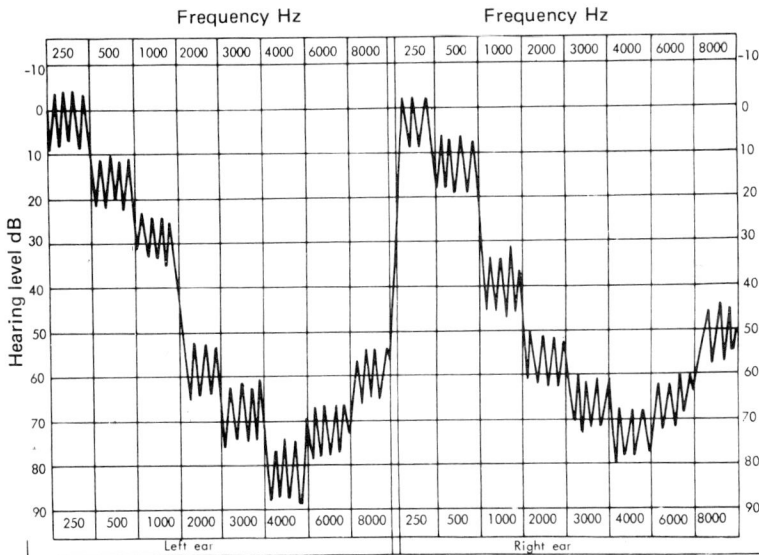

Figure 2 An audiogram chart illustrating the threshold of hearing of a person with a severe noise-induced hearing loss, measured with a self-recording audiometer

Figure 3 The ASRA Audiometer Type 4 instrument

itself can be calibrated in the usual way with a pistonphone. Calibration of the earphones is covered by BS 4668 (IEC 303) and the requirements of the artificial ear by BS 4669 (IEC 318).

The tone can be presented at a continuous level or pulsed, the latter being the appropriate mode when the subject suffers from tinnitus (persistent ringing noises in the ear). It is claimed that pulsed tones can be used to detect cheating. A continuous tone threshold which appears to be better than a pulsed tone threshold can indicate cheating. Pulse tone audiometers are better when used in a noisy room, although it is always preferable to perform the tests in a proper acoustic booth.

Presbyacusis, or the natural loss of hearing with age, starts as young as 25 years. Some audiometers are capable of making allowance for this by subtracting the natural loss (represented by standardised curves of threshold change with age and frequency) from the measured loss, so as to distinguish the change caused by industrial exposure.

Other Audiometry Instruments

Audiometry equipment has to be calibrated in the correct manner. An artificial ear can be used; this is a device intended for the calibration of earphones used in audiometry. It presents to the earphone an acoustic impedance equivalent to that of a human ear within the frequency range adopted for audiometry (80 Hz to 10 kHz); refer to BS 4669 (IEC 318). An acoustic coupler on the other hand loads an earphone with a reference impedance, which is not that of a human ear; it has a cavity of a prescribed shape and volume. BS 4668 (IEC 303) defines a particular form that can be used within the frequency range 125 Hz to 8 kHz. An artificial mastoid simulates the *mechanical* impedance of the human mastoid and has its application in the calibration of bone conduction audiometry; see BS 4009 (IEC 373).

SECTION 4

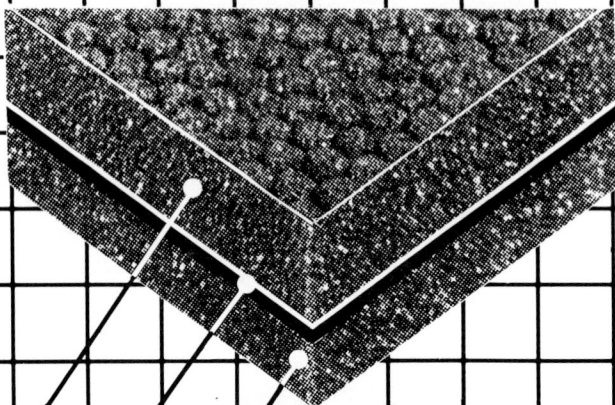

Chapter 1
Vibration Measurement and Analysis
Transducers and Pick Up Devices

Types of Pick up Device

A variety of devices can be used to measure vibration. There are three basic quantities which are relevant to the study of vibration: displacement, velocity and acceleration. It is, in principle, possible to obtain the other two from a knowledge of one of them and the frequency. They are interrelated by integration or differentiation, a process which can be performed by calculation or, if the signal is present as an electronic signal, in the appropriate circuit. The following are some of the devices that can be used to measure vibration:

> Visual displacement indicators.
>
> Stroboscopes.
>
> Non-contact proximity probes, working on the eddy current principle.
>
> Displacement transducers in a variety of types: strain gauge, potentiometer, photo-electric.
>
> Velocity sensors – seismic devices using the motion of a mass on a resilient element to produce an electrical output proportional to velocity. Moving coil and moving magnet types are available. Non-seismic pick-ups are also used where it is possible to make measurements relative to a fixed or moving point.
>
> Non-contact velocity measurement using laser Doppler technology.
>
> Accelerometers – the most commonly used vibration measuring device. They can be classified as piezoelectric, piezoresistive, strain gauge and capacitance.

Visual Displacement Indicators

These devices are seldom used today, mainly because of the ready availability of modern electronic measuring methods; however there may be occasions when they can be used with advantage. The crudest way of estimating vibration amplitude is by means of a pointer attached to the moving part; a fixed scale held against it gives an idea of the amplitude. A rather more advanced form is the vibrograph, much used for vibration studies in the past. The main body is hand-held; it acts as a seismic mass and is connected by a spring to a probe held against the moving surface. The body has a scale, against which the amplitude can be assessed, or, in the more advanced instruments, a pointer makes a mark on a moving roll of paper. In the latter form an estimate of the frequency is also possible. Although they can be useful for making a crude assessment of amplitude, they do have disadvantages: if the frequency or acceleration is too high, the probe tends to chatter rather than remain in contact with the surface; and the force applied through the probe can affect the vibration being measured.

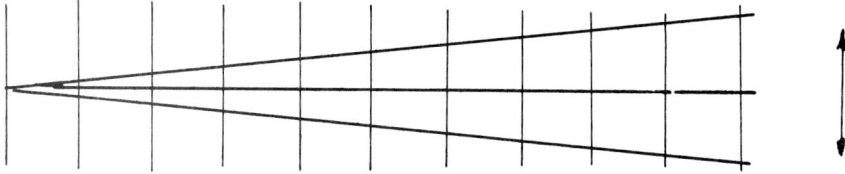

Figure 1 Optical amplitude indicator

Another visual displacement device is shown in Figure 1, a simple way of making a rough assessment of vibration amplitude, but effective only when the amplitude is constant. The figure is attached to a surface vibrating in the vertical direction. Through persistence of vision the sloping lines cause a shadow which intersects the horizontal axis. The amplitude can be assessed by noting the point where the shadow ends. With the proportions as drawn it effectively gives a 10:1 magnification of amplitude.

Stroboscopes

These can also be considered as optical devices. A stroboscope is a flashing light source which illuminates a moving surface at a set rate. If the rate is equal to that of the vibration frequency or some submultiple of it, the vibrating object appears to be stationary. A small deviation from that frequency causes the object to appear to be slowly vibrating, with an amplitude which can be readily assessed. Both amplitude and frequency can be determined but care is needed to ensure that the correct value of the frequency and not its multiple is recorded. As with other optical methods, it can only be relied upon when the vibration and frequency are uniform. Its frequency range is limited by the maximum speed of the flash, in practice about 500 Hz, although most stroboscopes have a maximum frequency of about 200 Hz. It is possible, by using sub-harmonic techniques, to accommodate higher frequencies. Figure 2 shows a modern stroboscope which can give up to 12 000 flashes per minute; it can give a digital or analogue display and can be mains or battery operated.

Non-contact Proximity Probes

A proximity probe comprises a probe, a driver and a probe-to-driver connecting cable, Figure 3. The driver produces a high frequency signal which is fed into a coil in the probe which generates a magnetic field around the probe tip. When a conductive material (ferrous or non-ferrous)

Figure 2 Hand held Stroboscope

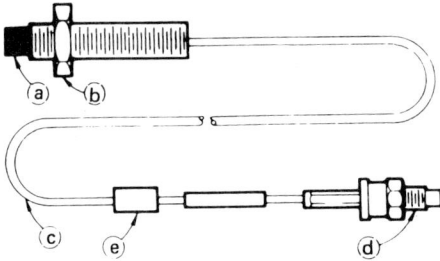

Figure 3 Eddy probe – functional diagram.
a) probe tip
b) lock nut
c) coaxial cable (flexible armour
   optional)
d) coaxial connector
e) floating sleeve

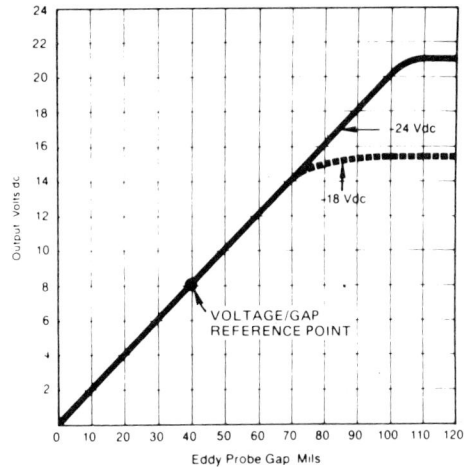

Typical response

approaches the probe tip, the field strength is affected and a voltage output can be used to assess the vibration. Proximity probes are used mainly for measuring shaft vibration and for torsional vibration. They are also frequently used for proximity sensors in manufacturing. When used for vibration measurement, they are limited to a maximum of about 2 mm with a sensitivity of 10 mV/micron. They are capable of measuring at frequencies ranging from very low ('static') up to 10 kHz. The calibration changes as the material changes, so a calibrator is essential.

Displacement Transducers

Linear potentiometers are capable of measuring very large amplitudes up to 1 metre, but have a poor resolution at small displacements. They can be wirewound or made of conductive plastic. Wirewound types are the most economical but have the poorest resolution. They give an electrical signal directly proportional to displacement, without the need for an amplifier. Potentiometers rely on contact of a moving part with the resistance element and so are prone to wear.

An alternative is the linear variable differential transformer (LVDT), and its rotary equivalent (RVDT), which have no contacts so are not as prone to wear as the potentiometer. The construction resembles a transformer consisting of a primary and two secondary coils, wound on a former. A magnetic core (the vibrating element) moves freely inside the former. The primary coil is energised by an A.C. signal. A voltage proportional to displacement is produced across the secondary which, when demodulated to remove the carrier signal, gives a D.C. output. They are linear over a limited range.

Strain Gauges

Strain gauges, bonded to a deformable structure such as a cantilever beam, can be used to measure displacement. When placed in the arm of a bridge circuit, an unbalanced D.C. voltage is produced which is proportional to displacement. They can be used to measure displacement up to 100 mm. The frequency range is limited by the natural frequency of the beam. A strain gauge displacement transducer of this form can use higher voltages (around 5 V bridge excitation) than that required for the more usual piezoelectric instruments, which simplifies the amplifier requirement.

Photoelectric Devices

Photoelectric devices are often used as proximity detectors but there are types available which can be used for measurement of displacement. Laser vibrometers are discussed below, but simpler optical techniques are also available. Optical tachometers give a readout of rotational frequency, without physical contact. For the measurement of torsional vibration, a transducer based on infrared transmission through a slotted disc is sometimes used.

Velocity Sensors

Velocity sensors are used for low to medium frequency measurements (up to 1000 Hz). Because they tend to filter out high frequency signals, they are less susceptible to amplifier overloads which can compromise the fidelity of low amplitude and low frequency signals. They are useful for monitoring and balancing operations on rotating machinery. Traditional velocity sensors employed an electromagnetic sensor to pick up the velocity signal. Piezoelectric velocity sensors are now available. Both seismic and non-seismic sensors are available, but seismic transducers are more common.

In some vibration regimes, the velocity signals of interest tend to be of equal magnitude throughout the frequency range. When measuring human exposure to vibration, it is assumed that the human body responds primarily to the energy input (which is velocity dependent). In such cases, a velocity sensor can be used; it has the advantage that the signal is less likely to overload the recording chain.

Laser Doppler Sensors

These represent a modern development which can be of use in situations where an electro-mechanical sensor has drawbacks. The principle employed is that of the Doppler effect using the coherent light from a laser, usually a helium-neon (HeNe) source. The latest systems use a low powered laser, less than 1 mW (Class 2). When light is reflected back from a vibrating target, it will undergo a frequency shift proportional to the velocity. As the target moves towards the source, the light increases in frequency; as it moves away, it decreases in frequency. The reflected beam is frequency modulated at the Doppler frequency which is proportional to velocity.

The advantages of this type of sensor are:

Non-contacting measurement, so no problems of transducer attachments, or added mass. For some systems it is not necessary even to attach reflective tape, except for large stand-off distances.

Can be used for living biological applications, e.g. ear drums.

Measurement at adjustable stand-off distances, up to 200 metres.

Measurement through a transparent surface.

Measurement of a rotating element, e.g. a shaft or computer disc. Imperfections in the shaft geometry do not influence the measurement.

Not prone to physical damage induced by high shocks or high temperatures.

Sensitive to a wide range of magnitudes of vibration, typically 1 μm/s to 1 m/s.

Frequency range from 0 to 10 MHz.

Low noise floor.

Does not suffer from resonance frequency limits or non-linearities as in a mechanical transducer.

Wide dynamic range (80 dB)

The disadvantages to be borne in mind are:

High cost compared with conventional sensors, although cheaper industrial systems are now becoming available.

Safety radiation aspects, although with low-powered lasers, the only danger is when staring directly into the beam.

Employs a fixed laser so not suitable for seismic measurements.

One system is illustrated in Figure 4. The HeNe laser is split into two parts by using a beam splitter. One of the beams is directed at the test object. The other beam (the reference beam) is directed at a rotating disc via one fixed and one moveable prism. Back scattered light from the rotating disc and the test object return on-axis with the incident beam and are mixed in the beam splitter. This light is then directed towards a photo-detector. The frequency-tracking electronics first measure the Doppler frequency shift caused by the vibrating surface and then produce a voltage proportional to the surface velocity in the direction of the beam. Because during vibration there are moments of zero velocity and therefore no output, there is ambiguity over the direction of the velocity. The problem is overcome by frequency shifting the reference beam. In this design, the prisms and the rotating disc add a constant Doppler shift to the reference beam. The rotation frequency can be changed to optimise the dynamic range of the unit. Figure 5 shows the laser mounted on a tripod.

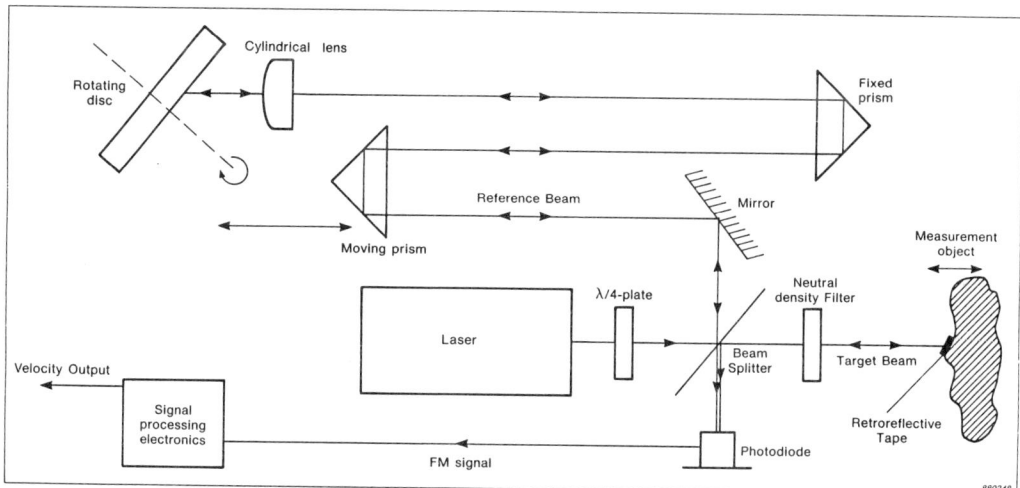

Figure 4 Laser Doppler Velocity sensor using a rotating disc

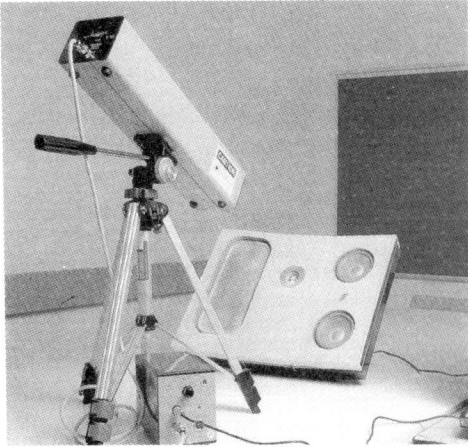

Figure 5 Laser Doppler Velocity transducer, measuring vibration of a speaker diaphragm

An alternative method of introducing a frequency shift is shown in Figure 6. This behaves in much the same way as in Figure 4, except that the rotating disc is replaced by a Bragg Cell (an acousto-optic modulator), with a 40 MHz frequency. BS1, BS2 and BS3 are beam splitters; D1 and D2 are opto-electronic converters.

A rotating diffraction grating can also be used for frequency shifting. It is claimed that the rotating disc method is more robust for industrial use, probably at the expense of increased bulk. In practice it is not necessary to understand how the equipment works. The output is a velocity signal which can be processed by a spectrum analyser or a FFT.

By combining fibre-optics with laser technology it is possible to reduce appreciably the size of the optical heads, making the system more versatile. Figure 7 illustrates an extension of Figure 6 where two optical fibres have been added. In this alternative technique the reference beam is not internal but external. It exits the system and can be used to determine the relative displacement of two points. The use of fibre optics makes it possible to site the sensing head in confined places which the laser itself would not be able to reach.

The laser Doppler principle can also be used for non-contact measurements of angular velocity and torsional vibrations of rotating shafts.

Accelerometers

The seismic accelerometer is the workhorse of vibration measurement. A wide range is available to meet most industrial applications.

The most common type in use today is piezoelectric, whose construction is shown diagrammatically in Figure 8. In its basic form (configuration a), the transducer consists of two piezoelectric discs on which is resting a heavy mass. The mass is loaded by a spring and the whole is mounted in a metal housing. When the accelerometer is subject to vibration, the mass exerts a force on the piezoelectric material which is proportional to the acceleration and a voltage is generated across the discs. The combination of mass and spring is a resonant system. At frequencies well below resonance the motion of the mass is virtually identical to that of the housing and the

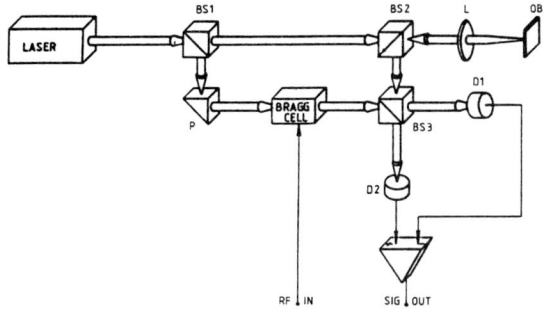

Figure 6 Optical layout of a single-point vibrometer

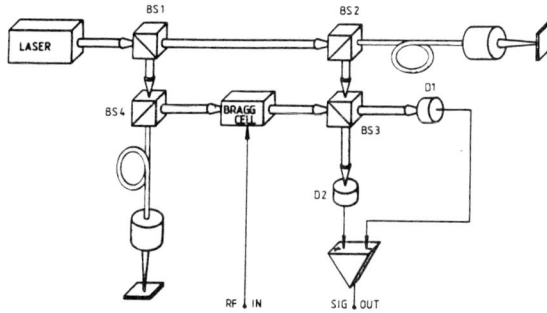

Figure 7 Optical layout of a fibre-optic vibrometer

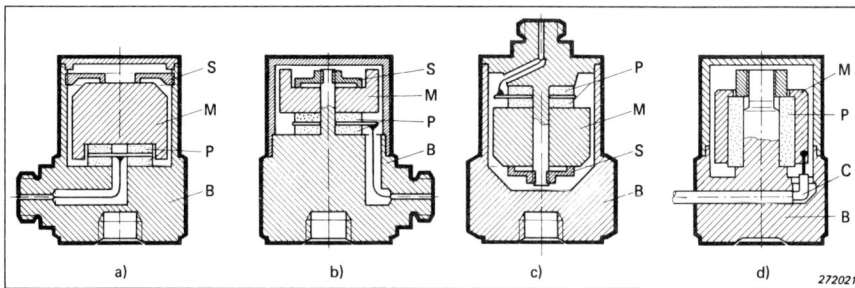

Figure 8 Schematic drawing of four accelerometer configurations
a) Peripheral mounted compression design   c) Inverted centre mounted compression design
b) Centre mounted compression design   d) Shear design
S = spring. M = mass. P = piezoelectric element. B = base. C = cable

sensitivity is constant; at frequencies approaching resonance, the sensitivity varies and measurement is not possible. This represents an upper limit to the use of this kind of accelerometer. The sensitivity of the accelerometer, i.e. the voltage generated across the output terminals when subject to unit acceleration, depends on the properties of the material and is proportional to the mass. The smaller the accelerometer, the lower the sensitivity but the wider the useful range of frequencies that can be measured. The frequency response and the sensitivity are the two important factors to be considered when choosing an accelerometer. Another important factor is the cross sensitivity, or the response of the instrument to accelerations normal to the measurement direction. Typically the response curve of a piezoelectric accelerometer will be as in Figure 9. Some suppliers, when presenting a calibration curve, ignore the region below the useful range; it should be understood that with a generating accelerometer there is a lower and an upper frequency outside of which the device should not be used.

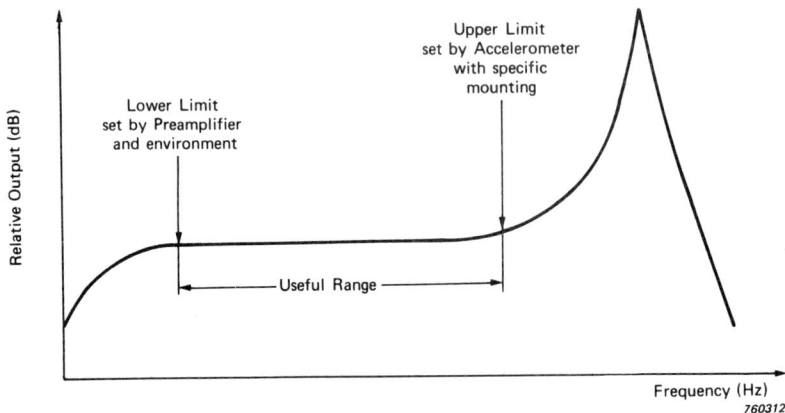

Figure 9 Frequency response of an accelerometer when subjected to constant acceleration

Quartz crystals are naturally piezoelectric when properly cut, but have very low sensitivity, so man-made ferro-electric ceramics or lead zirconate titanite are more commonly used. Figure 8 shows several basic constructions for descriptive purposes but other configurations are possible, with the crystal placed in shear or as part of a cantilever beam. Different piezoelectric formulations are available for different applications, for example in extremes of temperature.

When choosing a piezoelectric accelerometer, the following factors should be borne in mind:

Sensitivity – in the range 2 to 100 mV/g for voltage sensitivity or 2 to 100 pC/g for charge sensitivity.

The desired usable frequency range which can vary from a minimum of about 1 Hz to a maximum of about to 25 kHz.

Maximum permissible vibration level up to 10 000 g, or higher (200 000 g) for special shock accelerometers

Environmental conditions – high or low temperature, or exposure to water or oil.

Mass of accelerometer may affect the dynamics of measured system. It can vary from 1 to 100 gm.

Attachment method – usually through a mounting stud where that is permissible, otherwise by an adhesive. A hand-held probe may also be used.

Need to measure motion in several axial directions simultaneously – special triaxial accelerometer mounts are available.

Preamplifiers

Piezoelectric accelerometers are characterised by a high output impedance and a weak signal, so a preamplifier is required to transform the impedance to a lower value and to amplify the signal. Two types are available – voltage and charge amplifiers. A voltage amplifier is based on the accelerometer as a voltage source; it has a high impedance and a non-capacitive input. The capacitance effect of the connecting cable cannot be ignored, so unless the correct cable is used, which normally means the one supplied with the accelerometer, the calibration can alter. Voltage preamplifiers are only recommended for use with short cables.

A charge amplifier is based on the accelerometer as a charge source and produces a voltage output proportional to the change in input charge. A high capacitive feedback is used to obtain a high input capacitance so that long connection cables can be used without affecting the sensitivity. For this reason charge amplifiers are to be preferred although they tend to be more expensive. Some suppliers now offer accelerometers with an amplifier or impedance converter incorporated in the main body, which is an ideal solution for some applications, however the mass is greater and they are more prone to shock and high temperature damage. An extension to this idea is the incorporation of a conditioning circuit in the amplifier which can tailor the frequency response for specific applications or can suppress the accelerometer's resonant frequency. If a high impedance sensor is used, the connecting cable has to have high insulating resistance and low noise; with a low impedance sensor, a standard connecting cable is adequate. Accelerometers are also available with an integrating circuit incorporated into the electronics, where the output is proportional to the velocity. A further extension is a transducer which has a dual output of either acceleration and velocity or velocity and displacement.

Piezoresistive Accelerometers

Piezoresistive strain gauge elements are solid state silicon resistors which change their resistance in proportion to the applied stress. When used as an element in an accelerometer, they are connected in pairs to a cantilever beam and form part of a bridge circuit (Figure 10). They are particularly suitable for measurement at low frequencies down to D.C. but are normally limited to a maximum frequency of 1 kHz. Because they are energised by an external electrical source, they have an inherently low output impedance, so a preamplifier is usually unnecessary. Silicon resistors are superior to metallic gauges in that they are virtually free from hysteresis and have a higher sensitivity. They are however prone to temperature changes, unless a compensating circuit is incorporated. The bridge can be excited by a constant voltage or constant current. The latter has a degree of self-compensation with temperature.

Variable Capacitance Accelerometers

These also have the ability of measuring from D.C. to about 1 kHz. The sensor element is manufactured from a single silicon crystal, which is electrostatically bonded to form a parallel plate capacitance device. The result is a sensor which has stable damping characteristics and a ruggedness to withstand high shock loading, which gives these sensors the ability to measure low accelerations in a high shock environment. As with the piezoresistive type, they have to be externally energised.

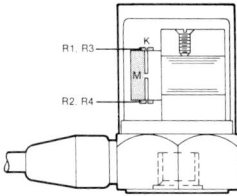

Piezoresistive accelerometers use solid state gauges R located on a cantilever beam.

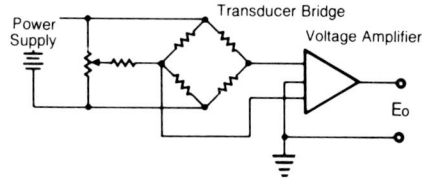

Equivalent circuit for a PR sensor, power supply and amplifier.

Figure 10

Accelerometer Comparison

Table 1 summarises the suitability for the various accelerometer types of one manufacturer, but it can be assumed to apply generally.

Mounting of Accelerometers

To obtain reliable results from acceleration measurements it is imperative to mount the sensor properly. Insufficient attention to this may lead to gross errors in readings. Refer to BS 7129 for detailed recommendations on mounting.

TABLE 1 – COMPARISON OF ACCELEROMETER CHARACTERISTICS

Characteristic	Piezo-electric	ISOTRON™	Piezo-resistive	Capacitance
Self-generating	yes	no	no	no
D.C. response	no	yes	no	yes
Sensitivity to to non-vibration environments	low/very low	very low	low	very low
Low impedance output	no	yes	yes	yes
Cryogenic temperature	yes	no	no	no
Turn-over or shunt calibration	no	no	yes	yes(1)
Subminiature designs	yes	yes	no	no
Zero shift at high shock	(2)	(2)	no	no
Rugged, high sensitivity	yes	yes	no	yes
Availability of damped resonance designs	no	(3)	yes	yes

Notes: ISOTRON™ is a piezoelectric accelerometer with a built-in preamplifier
(1) Variable capacitance offers capability for active self-test
(2) Depends on design
(3) Electronic filters can effectively change frequency response range

Some of the possible causes of error are:

1. Signal distortion. There are two main forms, both of which can be avoided by proper selection of the transducer. The first is harmonic distortion which occurs when a large amplitude signal approaches the clipping amplitude; the distortion appears at integer multiples of the clipping frequency. This can be cured by making sure that the measuring range of the accelerometer is much wider than the likely signal and rejecting any results which are close to the limit. The second form is sometimes called washover distortion; it occurs when low frequency components are to be measured in the presence of high frequency noise. This is difficult problem, which may require the use of a shock accelerometer or a velocity transducer. It can also result in D.C. shift (see below).

2. Electro-magnetic interference. This is an unwanted signal from another electrical source, which can be caused by transients from ignition systems, mains hum used by poor grounding, RF pickup from radio transmissions. It can usually be solved by the use of screened cables.

3. Electrostatic discharge caused by a charged body coming into contact with the sensor or its cables. It may be in the range of kilovolts and can cause irreversible damage to the sensor or amplifier. The solution is to take precautions that any apparatus close to the accelerometer or its cables is earthed to the same potential. Special shielded accelerometers are also available.

4. High acoustic fields may show as a mechanical vibration.

5. Operation outside the approved temperature range. All accelerometers should be supplied with a temperature response curve in addition to the frequency response curve.

6. Some instruments are particularly sensitive to rotational vibration or have a large cross sensitivity, rendering them unreliable when the vibration is other than along the measurement axis.

7. Incorrect mechanical attachment. Unless the accelerometer is firmly and uniformly attached to the vibrating surface, its base may bend, putting a stress on the piezo crystal, which may generate a false reading. Some of the acceptable methods of attachment are by: a screwed stud, adhesives, beeswax or thin pressure sensitive tape. When using an adhesive, a hard setting or a thermosetting cement is preferred to one relying on solvent evaporation, which remains soft internally. Cyanoacrylate adhesives are recommended where subsequent removal is desired.

When electrical isolation of the accelerometer from the vibrating surface is necessary, an isolating stud and a mica washer may be used. All methods may modify the frequency response of the unattached sensor (usually the resonant frequency is lowered); so, if there is any doubt, a calibration check should be carried out.

Figure 11 shows a recommended way of attachment using a mounting stud. A thin layer of silicon grease between sensor and surface secures a good contact. Cable attachment is also important as shown in Figure 12. It is always good practice to clamp the cable to the vibrating surface or use tape or adhesive to eliminate noise through mechanical motion of the cable.

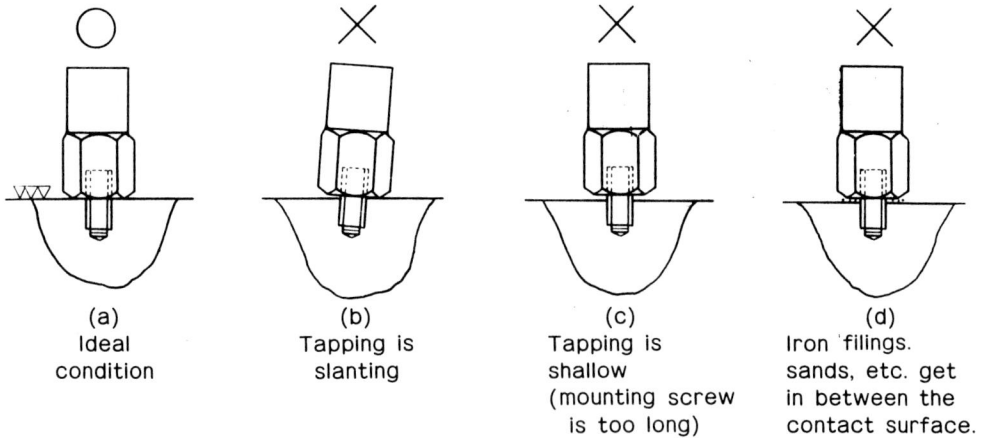

(a)
Ideal
condition

(b)
Tapping is
slanting

(c)
Tapping is
shallow
(mounting screw
is too long)

(d)
Iron filings.
sands, etc. get
in between the
contact surface.

Figure 11 Right and wrong ways of attaching an accelerometer

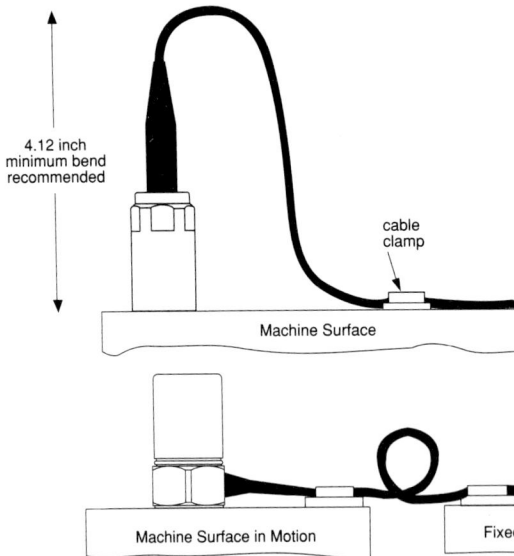

4.12 inch
minimum bend
recommended

cable
clamp

Machine Surface

Figure 12 Cable attachment of accelerometer

Machine Surface in Motion   Fixed

High Shock Environments

It is easy to damage an accelerometer; merely by dropping it on a bench can expose it to 1000 g or more. When measuring the vibration from a high shock-producing machine, such as a pneumatic concrete breaker, a shock of the order of 100 000 g or more may be generated. Unless one knows the magnitude of the highest occurring shock, it may not be possible to decide on the specification of the accelerometer. A 'shock' accelerometer may be better than an industrial one, yet may still not be adequate. A common result of shock damage to the accelerometer is 'D.C. shift'. This phenomenon, a sudden change in the D.C. signal level, can be caused by a number of mechanisms, some of which can be external to the accelerometer such as amplifier overload and tribo-electric effects in the cable. Internally, a severe stress of the piezoelectric material can produce D.C. shift. The phenomenon can be permanent or temporary; in either case, wild results may be produced. The problem is particularly difficult when the accelerometer has to measure low vibration levels at low frequencies which calls for an instrument having a high sensitivity, yet one which does not suffer damage from shocks.

One method of avoiding exposure to high shocks is by the use of a mechanical filter, which effectively cushions the accelerometer from high level shocks. It consists, in its basic design of an elastomeric material sandwiched between two washers and inserted between the vibrating surface and the body of the accelerometer, Figure 13. The combination becomes a welldamped springmass isolator, with a resonant frequency lower than that of the accelerometer alone. Provided that the sensitivity remains linear within the frequency range of interest, reliable readings can be taken. In many cases this may prove to be a better solution than the use of a special shock accelerometer which has an inherently low sensitivity.

Another method of avoiding D. C. shift is by the use of a laser sensor as described above. If damage caused by high shocks is suspected, frequent calibration or measurement by two accelerometers placed back to back may be called for.

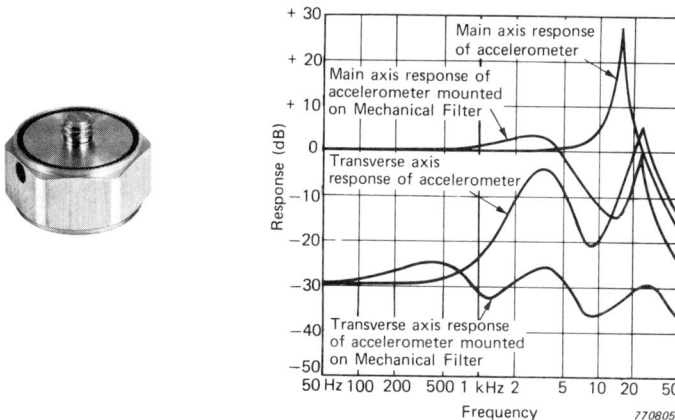

Figure 13 Mechanical filter and curves showing the modified response of an accelerometer fitted with the device

Calibration

All accelerometers are calibrated at the factory and are sent out with a calibration chart, which gives the sensitivity as pC/g or mV/g, as well as the frequency response . It should also give the temperature sensitivity. This calibration chart can be used for first use, but it needs to be confirmed regularly either by returning the unit to a laboratory for recalibration or by use of an in-house calibrator. For reliable results, the whole measurement chain consisting of the accelerometer amplifier readout should be calibrated.

A portable reference source, generating a standard rms level at a fixed frequency, can be useful for field checking. For more accurate checking a calibration exciter, capable of producing a known force level through a range of frequencies, combined with a reference accelerometer and amplifier, can be used.

For a piezoresistive or a capacitance accelerometer (i.e. one which is suitable for D C applications), inverting it in a gravity field will give a 1 g signal, and is therefore useful as a check. This method is useful when investigating the temperature characteristics of a sensor: it is an easy test to perform in a hot or cold chamber.

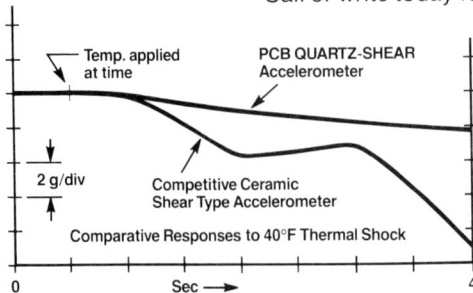

Chapter 2
Machinery Health Monitoring

Predictive Maintenance

A study by the UK Department of Trade and Industry of 2000 companies established that they could save £150 million annually by applying a comprehensive predictive maintenance programme. There are three categories of maintenance – run to breakdown; preventive maintenance which seeks to avoid breakdowns by replacing machines or components at regular intervals whether or not they require it; and predictive maintenance which seeks by condition monitoring to detect potential failures, so that only those items that require it are replaced and then only at a time convenient for the running of the factory. Vibration monitoring is a key element in predictive maintenance, although not the only one.

Machinery Health Monitoring

An important application of vibration measurement is in predicting the failure of machines. If one can assess the likely time of failure well before it occurs, not only is it possible to avoid catastrophic breakdown of the machine, but also maintenance can be properly planned, so that the failing component can be replaced at the next most convenient scheduled period. Bearings, gearboxes and shafts are the components most likely to require monitoring. Significant monitoring parameters relative to all types of machines are summarised in Table 1. Vibration level is normally the most significant quantity to be measured but it needs to be related to the other parameters in the table. In particular, acoustic signature analysis can be equally valuable, because of the interrelation of noise and vibration. Temperature of bearings can give a warning of danger before a significant change in vibration level is measured.

Condition monitoring systems are either intermittent (periodic) or permanent. Intermittent systems are those in which instrumentation is applied to selected points on the machine being studied, either at predetermined intervals or when a potential failure is suspected. Analysis of the measurements usually requires the employment of skilled personnel. A permanent monitoring system is one in which the measurement transducers are left in place and the magnitude of the measurements are constantly compared. Machinery diagnostic techniques have developed around the need to keep all critical components under permanent review. A number of companies offer complete monitoring systems which are capable of being used by unskilled personnel, but they tend to be expensive (although perhaps when considered against the cost of a complete breakdown, they may well be looked on as a worthwhile investment). There is little in the way of standardisation yet in this area; but reference might be made to German Standard VDI 2056 on Acceptability Criteria and USA National Standard ANSI/SAE AIR 1839A on Aircraft Turbine Engine Vibration Monitoring Systems.

TABLE 1 – MONITORING PARAMETERS

Parameter	Measuring Instruments/transducers	Remarks
Temperature	Thermocouples, resistance type temperature detectors (RTDs)	Monitoring points should include: (i) bearings (as distinct from lube oil temperatures). (ii) lubricant. (iii) stator and rotor windings (in electrical machines). (iv) machine casing. (v) any other significant parts.
Vibration (rotor)	(i) proximity probes. (ii) velocity pick-ups (iii) accelerometers	Measurement should cover all three phases of: (i) shaft motion relative to bearings. (ii) shaft motion relative to free space.
Vibration (non-rotating parts)	Seismic transducers	Monitoring points should include: (i) bearings. (ii) bearing housings. (iii) casings. (iv) foundations or mount. (v) connected ancillaries.
Rotational speed		Shaft acceleration should be considered as an auxiliary measurement for torsional vibrations
Shaft phase angle	Keyphasor probe	Can be used directly for balancing
Position	Various sensors	Monitoring points should include: (i) shaft axial position. (ii) shaft radial position (eccentricity ratio). (iii) casing expansion. (iv) shaft and casing alignment.
Process variables	As appropriate	e.g. temperature, pressures and flows of fluids handled by a machine.
Sound levels	microphones	Often used in conjunction with vibration readings
Electrical power consumption	voltmeters, ammeters.	To give warning of excess power consumption
Engine parameters	Oil pressure gauges thermometers, torque meters	As above
Particle effects	filter examination, particle monitors	metal particles indicate bearing breakdown

Monitoring Techniques

Most machinery failures give a warning in advance of their occurring. This warning is known as a potential failure, defined as a physical condition which indicates that a functional failure is about to occur or is in the process of occurring. A functional failure is defined as an inability of an item to meet a specified performance. The time between the two is known as the P-F interval; clearly the frequency of inspection or the monitoring interval depends on the P-F interval. Figure 1 shows a typical P-F curve, where 'Condition' represents any measurable physical characteristic, which for the purpose of this chapter may be the magnitude of vibration (displacement, velocity or acceleration). The ability accurately to determine a potential failure is crucial to the task of diagnostic monitoring.

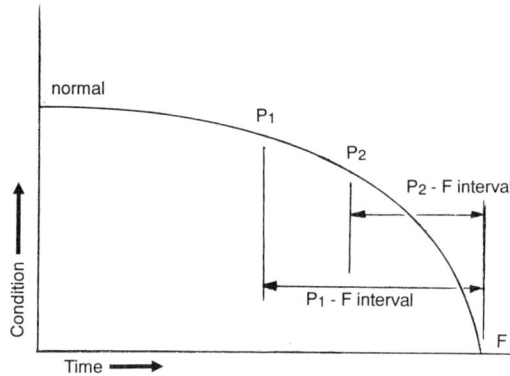

Figure 1 P-F Intervals

It can be seen from Figure 1 that, to secure a long P-F interval, potential failure must be deter-mined at a point higher up the P-F curve. But the higher up the curve, the smaller is the deviation from the normal, the more difficult is the task of identifying it and the more sensitive the equip-ment has to be. There is a partial trade-off between measurement accuracy and monitoring inter-val, although it is clearly of benefit to maximise the P-F interval, so that maintenance can be prop-erly scheduled. Implicit in this requirement is the need for as accurate a determination as the equipment allows. By the use of computer analysis on stored data, it is now possible to examine trends in the values of the parameters of interest, i.e. to predict the shape of the P-F curve. Such an approach is clearly an improvement on a system which merely gives a warning when the chosen parameter exceeds a preset trigger value. Software which is able to do this is generally available in association with comprehensive diagnostic systems available from a number of suppliers.

A typical diagnostic system will consist of a selection of the transducers listed in Table 2, the output of which are fed into a data acquisition and analysis system and a host computer with a communication link, so that trends and warnings can be conveyed to the plant managers.

Such a system is not limited to detecting deterioration in service: a new or rebuilt machine may need to be checked to ensure that no potential fault has been incorporated in the build.

The extent of the monitoring, and whether it be continuous or applied only intermittently, depends on the significance of a particular machine. In general, machines can be sub-divided into three groups.

(i) *Critical machines* (e.g. compressors, turbines, motors or generators) that cause a plant to shut down in case of failure.

Here permanent, continuous monitoring systems are required for full benefit, monitoring all the significant parameters. The cost of a protection system is more than paid for if the system pro-tects the machine against major damage only once during its lifetime. In this critical category it is therefore vital that the optimum choice of transducer/monitor for a specific machine has been made, (see Figure 2 for typical system).

(ii) *Semi-critical machines* which, if they fail, may cause part of a plant to be shut down but still allow production to proceed (if only on a reduced scale).

Comprehensive permanent vibration monitoring systems are not essential for machines in this group. The best approach for such machines is often to install a comprehensive set of trans-ducers but use a reduced number of permanent monitors to cut the overall system cost. (See Figure 3 for typical system).

A very good approach is to use intermittent preventive monitoring, as a complementary

TABLE 2 – RECOMMENDED MONITORING FOR MACHINES

Machine	Monitors	Parameters monitored
Electric motors	X-Y proximity probes Keyphasor probe Temperature indicators	(i) axial vibration. (ii) position measurements (periodically). (iii) casing vibration. (iv) speed, phase angle and timing. (v) bearing and oil temperatures. (vi) rotor and stator winding temperatures.
Pumps	X-Y proximity probes Keyphasor probe	(i) axial vibration. (ii) shaft motion relative to bearings. (iii) shaft phase angle (unless directly coupled). (iv) bearing and oil temperatures. (v) casing vibration. (vi) casing temperature.
Fans	X-Y proximity probes	(i) shaft vibration. (ii) bearing housing vibration. (iii) casing vibration.
Gears	X-Y proximity probe at each bearing	(i) axial vibration. (ii) input shaft. (iii) output shaft. (iv) thrust loads (axial probes). (v) gear teeth interaction. (vi) casing vibration. (vii) bearing and oil temperatures.

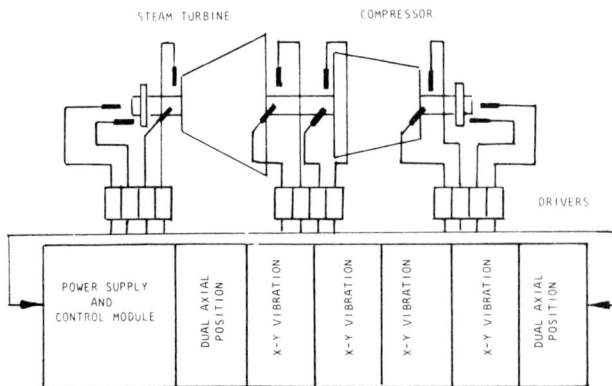

Figure 2 Typical system for a Group 1 machine

back-up of the limited permanent monitoring system. A well organised and meticulously execut-ed preventive maintenance programme has proved to be very successful, achieving a very high degree of plant reliability.

(iii) *Non-critical* machines whose failure only represents material cost and increased work load for maintenance personnel, but does not affect production significantly.

These are usually checked for vibration on a periodic basis (or when the operator feels that a machine is vibrating more than normal). Since the machines in this category are not critical, a simple portable vibration monitor based on overall level is a good cost-effective approach.

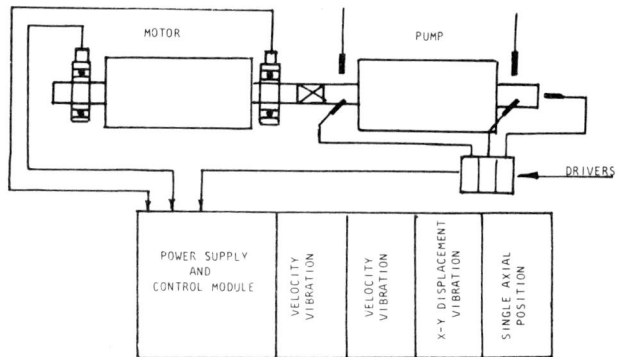

Figure 3 Typical system for a Group 2
machine

The other major consideration is scale, since there comes a point at which the burden of manual logging – no matter how simple a method is used – becomes too big to handle. At this stage the merits of a computer managed system become self-evident.

The recent advent of computerised monitoring systems provides a very economical means of continuously monitoring many non-critical machines which have previously gone unmonitored.

Basic Systems

The two basic forms of vibrations monitoring systems are:

(i) Built-in instrumentation providing a measure of vibration in terms of overall level, coupled to alarm and shutdown devices in the event of the vibration level rising to a certain level. Such a system of monitoring provides continuous protection but no specific information other than that vibration has increased, indicating wear or a potential fault.

(ii) Vibration measurement by analytical instruments to provide a vibration signature. This would normally be done periodically but, in the case of extremely critical machines, could be continuous, although continuous monitoring with built-in instrumentation plus periodic check with analysis machines would be more realistic. Signature analysis provides a much more sophisticated preventative maintenance capability since the deterioration of specific machine components can be isolated whilst the machine is running.

Dynamic Monitoring Techniques

Much attention is being given today to condition monitoring, but it should not be considered as separate from general planned maintenance procedures. Because of their complexity, there is a temptation to treat vibration measurements and analysis as a separate or stand-alone system, running in parallel with other scheduled maintenance. This temptation should be resisted or it will add to the cost. As far as is possible, all the separate systems should be integrated into one comprehensive system, even taking account of the relationships between different machines or production lines. The goal of a successful monitoring programme is to predict the time of breakdown and, unless it is simple enough and clear enough in its warnings to do that, effort spent on it will be wasted.

It is not always obvious exactly what vibration characteristics should be measured. Broadly vibration can be measured as displacement, velocity or acceleration. It will be recalled that displacement sensors are more suitable for low frequencies, velocity sensors for the middle range of frequencies and accelerometers for high frequencies, but for most purposes an accelerometer is employed. Subsequent analysis of the measurements is of crucial importance for interpreting the readings.

Broad-band analysis detects overall changes in vibration level without discriminating the frequency spectrum. It can be used to detect wear, imbalance and mechanical looseness, and is applicable for general surveys. The information provided by a vibrometer is limited in scope, and the P-F interval is uncertain. Its advantage is that it can be used by unskilled personnel without further interpretation. A vibrometer is normally supplied with hand-held probes; it is not intended for a permanent installation.

Octave band Analysis is rather better than broad-band analysis because it gives some indication of the frequency spectrum. A sound level meter, capable of accepting an accelerometer input, is a suitable instrument for analysis. The octave bands of interest can be consecutively or simultaneously recorded. One-third octave band readings may also be taken. This equipment has the advantage that it is portable and operated by the maintenance technician.

Narrow-band analysis (constant bandwidth or constant percentage bandwidth) is of advantage where the the critical vibration frequencies are known in advance or through experience. It is possible to examine just the frequencies of interest (e.g. the harmonics of the operating frequency or the gear meshing frequency), so it is valuable for rotating machines which run at a limited number of speeds. Changes to bearings, gears and rotating shafts can be easily spotted. Narrow band analysis by manual search is not really practical, so some form of automatic scanning equipment is required; it is not usually portable. Interpretation depends on a skilled operator or a computer.

Real-time analysis is a sophisticated technique, leading to early detection of a potential failure. The resolution can be as fine as experience has indicated. Results can be presented in real time or in the frequency domain. Without subsequent computer interpretation the required skill of the operator is high.

Proximity analysis employs a proximity probe which measures displacement; it is suitable for checking imbalance, misalignment and bent shafts. By the time that a imbalance is shown, the machine is near to failure, so the P-F time is short.

Shock pulse analysis is useful when the generating mechanism is in the form of a shock such as gear hammer, broken balls in a bearing or valve impact.

Acoustic analysis using a sound level meter can give an indication of mechanical failure. A crack in a structure such as a gearbox can cause a distinct change in the sound quality of the noise, but interpretation can be difficult

A combination of several different kinds of analysis techniques may prove to be the ideal solution.

Hand-held Instrumentation

Figure 4 is an example of a vibration meter intended for checking bearing quality. The signal is measured by an accelerometer probe and the output can be either audible, through earphones, or by a meter reading. In this instrument, the value of the meter reading has little absolute significance, but for comparison purposes, provided that manual records are kept, it can be used to give an early warning of failure; only trends when recorded on a particular bearing have meaning. This kind of instrument is for small factory installations, and is the simplest one that can be of practical use.

**TABLE 3 – PERMISSIBLE VIBRATION LEVELS FOR
ROLLING BEARINGS SPECIFIED IN US MILITARY
SPECIFICATION MIL–B–17931D**

Bearing bore (mm)	rms vibration limits (μm/s)		
	50–300	300–1800 Hz	1800–10000 Hz
Hz			
10 – 12	120	90	75
15	150	120	75
17 – 20	150	120	120
25 – 30	180	150	120
35 – 40	180	150	150
45	240	150	180
50 – 55	240	180	180
60 – 65	240	180	240
70	300	180	240
75 – 85	300	240	300
90 – 95	300	300	300
100	300	300	380
105 – 115	380	300	380
120	380	300	455

Note that these vibration magnitudes are readily met by new
bearings. Higher quality bearings with lower vibration limits
may be specified for precision purposes.

Hand-held instrumentation providing a much higher quality of information is also available. It
may incorporate features of storage and analysis and an interface to a printer or computer. The
use of such instruments enforces a discipline on the maintenance engineer and necessarily

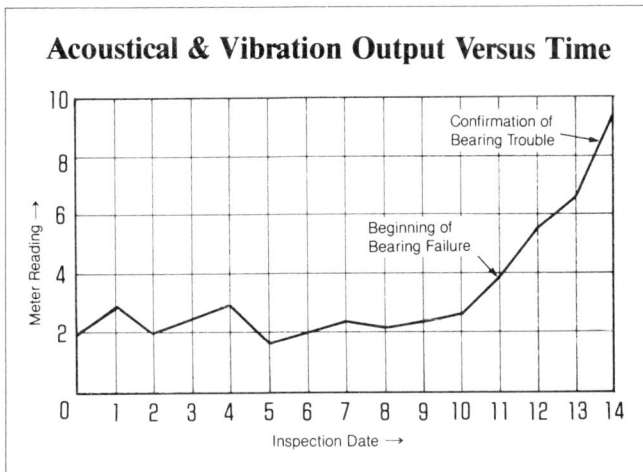

Figure 4 *Bearing checker, showing a typical record*

extends the P-F interval by the inspection interval. A stage intermediate between one based on a hand-held probe and a permanent installation has the accelerometers permanently attached to the machine which the maintenance engineer then visits on a scheduled tour with a plug-in recorder.

Shock Pulse Method (SPM) for Rolling Bearings

This is based on monitoring the mechanical impacts created in a damaged bearing. It can be distinguished from conventional acceleration measurements in that a piezo-electric accelerometer is mechanically and electrically tuned to a specific resonant frequency (usually 32 kHz); it responds to the shock pulse created by the impacts, rather than the actual vibration signal in the gearbox or housing. The compression wave sets up a damped oscillation in the transducer at its resonant frequency, which can be analysed in the associated equipment. Because the the dampened transient is well defined and of a constant decay rate, it is possible electronically to filter out all the other vibration signals.

Typically the shock pulse generated by a damaged bearing can be 1000 times that of a new bearing, so the intensity of the shock pulse is expressed as the decibel shock value (dB_{SV}). A new bearing has a small but measurable initial shock value (dB_i), which depends on the rotational speed and the bearing diameter. The increase above this initial value is the normalised shock value (dB_N).

$$dB_N = dB_{SV} - dB_i$$

The instrumentation gives dB_N as a readout and defines three zones:

$dB_N < 20$ Good operational condition

$dB_N > 20 < 25$ Caution zone

$dB_N > 25$ Bad operational condition

Refer to Figure 5, which shows a method of estimating the initial shock value and the relationship between dB_N and the percentage bearing life. If periodic measurements are taken and recorded as in this figure, an estimate of when to replace the bearing can be readily made.

Shock pulses generated in the bearing are propagated ultrasonically through the housing. When selecting a suitable test point for location of the probe, the signal path should be as short as possible, there must be only one mechanical interface (that between the bearing and the bearing housing) and the pick-up point must be as close as possible to the load zone of the bearing.

When used as a manual monitor, a portable test instrument is used in conjunction with a pair of earphones, Figure 6. The necessary measurement, analysis and readout is quite straightforward. Before a meter is used, the bearing diameter and bore are dialled in to assess the dB_i value, which is then automatically subtracted from the transducer output. The analysis performed by the instrument can be extended also to give an indication of the quality of the lubrication of the bearing as well as detecting damage. A continuous monitoring system is also available which can handle up to 50 bearings, giving a visual or audible warning when a pre-set value of dB_N has been exceeded. The SPM has been developed primarily for identifying bearing failures, but it also has applications in which structural shocks are generated such as cavitation in pumps or mechanical impact external to the bearing. Typical amplitude patterns are shown in Figure 7.

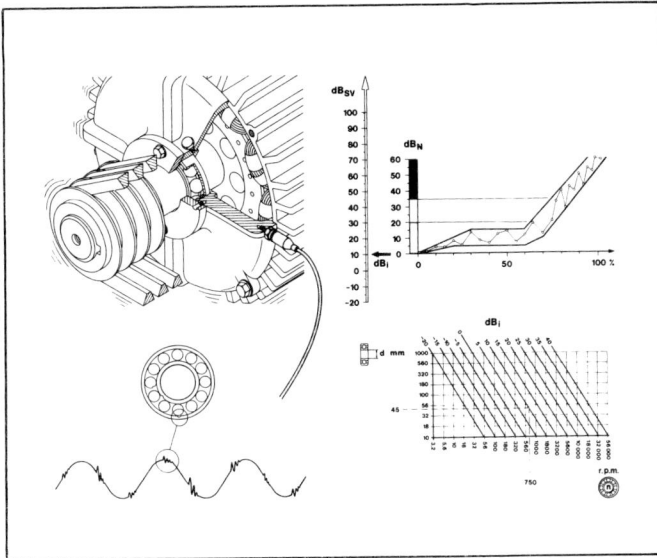

Figure 5 Illustration of the Shock Pulse Method for Producing Bearing Failure

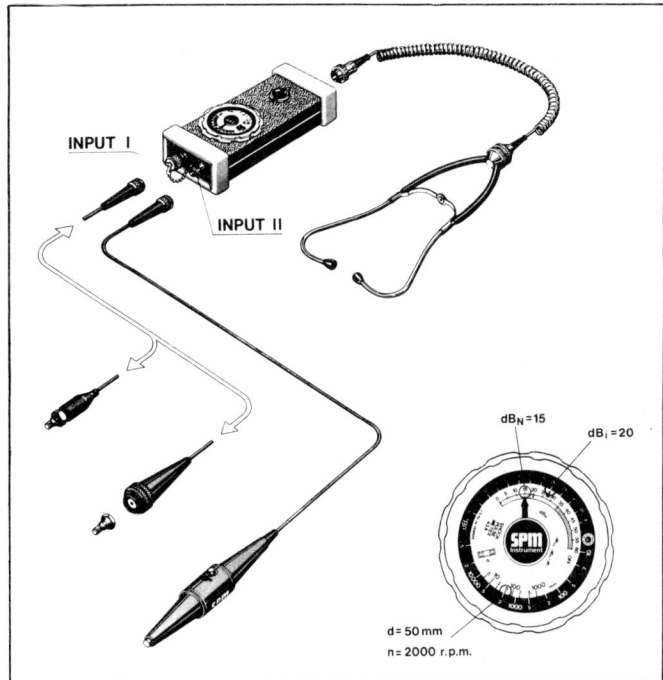

Figure 6 Monitoring shock pulses with a portable instrument

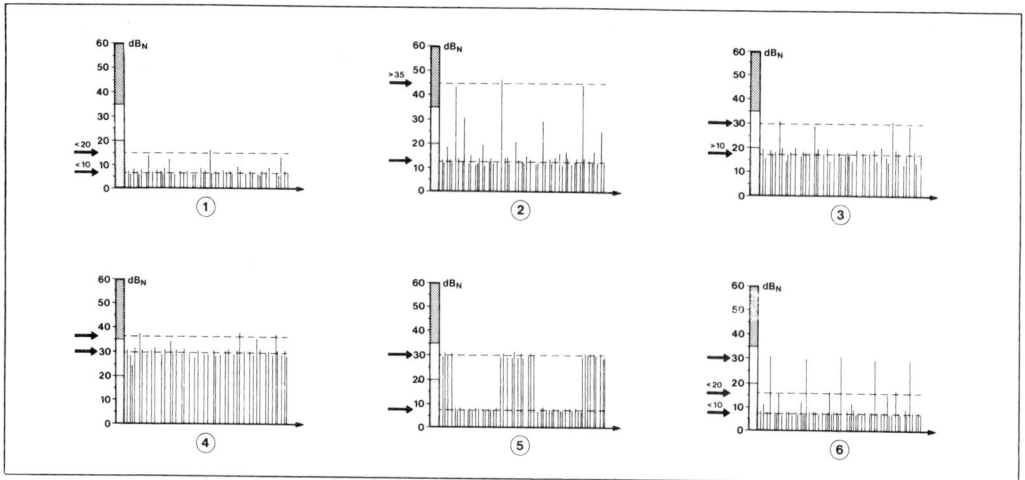

Figure 7 Typical patterns found in SPM

1 Good bearing properly lubricated – carpet value under 10dB$_N$, peak value under 20 dB$_N$

2 Damaged bearing – raised carpet value caused by an increase in overall surface roughness with high random shocks

3 Improper installation or lack of lubricant exhibited by high carpet value

4 Pump cavitation – high carpet values with little difference between carpet and peaks

5 Mechanical rubbing causing rhythmic shock bursts

6 Machine cycle load shocks

Permanent Monitoring Systems

In a permanent installation, an accelerometer is the preferred type of transducer and it is wise to choose one which is designed for that purpose. It should be robust with a sealed construction for use where exposed to water or oil flooding. It should be capable of sustaining a high shock over-load. Internal amplification and/or filtering can be incorporated provided there is no compromise on robustness. Types for use in areas of high radiation or in explosive environments are obtain-able. The connection cable and its attachments must be chosen to suit the same conditions as the accelerometer. Cable with a stainless steel outer shield is desirable for full mechanical protection. The installation of a permanent monitoring accelerometer has to be done to a higher standard than for laboratory investigations, with particular attention to cable routing and attachment. When some intermediary equipment is incorporated in the system, e.g. preamplifiers, junction boxes or barrier boxes, the same standard of protection must be adopted as for the monitoring equipment itself.

The value of a permanent monitoring system lies in its ability to give a clear indication of potential failure, without the intervention of a skilled specialist. The analysis equipment may be of an advanced design, but unless it is capable of giving an unambiguous guide to the maintenance engineer, it will be of limited value. The output of the signal processor must be analysed in a com-puter with the appropriate software, tailored to the application. For a simple installation it is pos-sible that the installation can be done in-house; in more complex cases it may be wise to bring in a specialist company.

Another advantage of a permanent monitoring system is the possibility of transmitting the signals, after the initial processing stage, to one central processor through a digital communications network. If a high degree of skilled interpretation is required (as might be the case in the initial stages of the installation), that work can be left to one trained expert.

A complete condition monitoring system consists of:

A measuring device, which may incorporate preliminary processing (a filter set or weighting network) or recorder

An analyser with a RTA, FFT, correlation or other processing circuits

A computer with the appropriate software which manages the measurement program, compares the vibration spectra and gives a warning of the changes, stores the data and interprets trends.

An example of a system developed by Stewart Hughes consists of the machine under surveillance, the Data Acquisition and Analysis System (DAAS) and the Host Unit (a PC), Figure 8. The complete system can be used either for online monitoring or for testing of new and rebuilt machinery. This system provides for:

A visual annunciation of machine condition via a mimic display

A plain English statement describing the affected component, the fault type and the estimated fault severity. At this stage, no specialist knowledge is needed.

A range of analogue and digital outputs.

Automatic or manual setting of alarm thresholds, which can compensate for variations in machine speed and load.

Database of past history for detecting trends.

In addition to conventional detection techniques based on signal levels, pattern based methods are available for gears and gear boxes, shafts and bearings. Identifiable gear faults include gear mesh quality, alignment, damaged teeth, pitch errors, damaged shafts, casing misalignment and out of balance. Rolling contact bearing faults include Brinelled raceways and damaged elements or raceways.

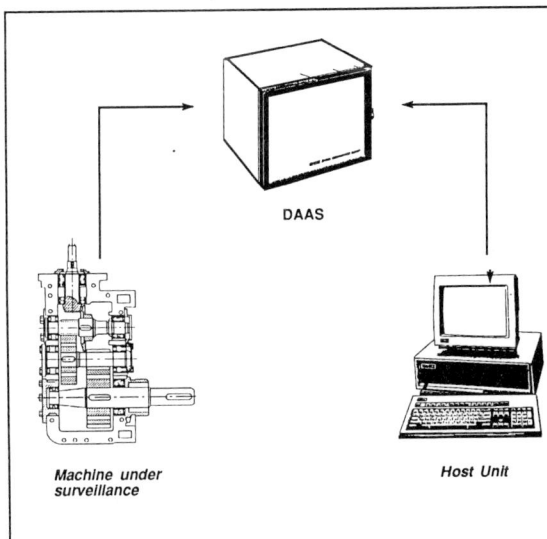

DAAS

Machine under
surveillance

Host Unit

Figure 8 Permanent Monitoring System

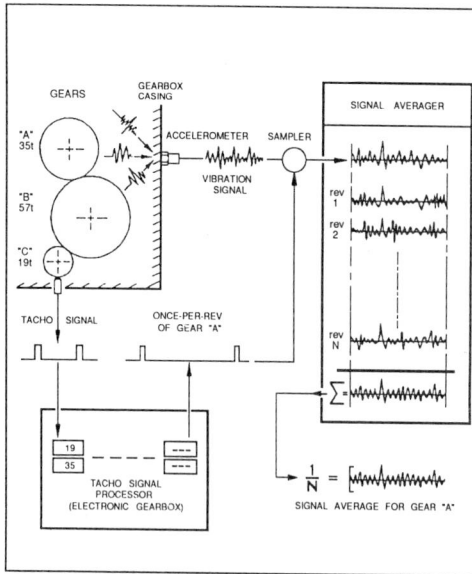

Figure 9 Analysis of a gearbox monitoring system

In order to identify the failing gear in a gearbox, information about the tooth numbers and the meshing frequency has to be provided to the programme.

Figure 9 shows a typical system taking a signal from an accelerometer mounted on the casing, and a tachometer on a suitable shaft. Three stages of analysis can be defined.

In the first stage of analysis, the signal is sampled at a frequency related to the rotational speed of the appropriate gear, enabling the relevant vibration characteristics of that gear to be obtained.

In the second stage, the primary analysis from the first stage is processed to measure energy levels for each vibration signature and to check for specific diagnostic patterns. Figures of Merit (FM numbers) are produced.

In the third stage, the results of the second stage analysis are combined with the historical data and other evidence to give the final presentation on the computer screen or printer, which can be as straightforward as a message warning of a pitted gear or a misalignment of a shaft. Although the warning is clear, the implementation by the software requires a specialist knowledge.

In a damaged gearbox, the level of the impulses relative to the gear noise is low and the changes are masked by the gear noise, so the above procedure is necessary to distinguish the signal caused by the damage.

A similar procedure can be adopted when analysing bearings. A normal bearing manifests only white noise, but a damaged bearing exhibits impulsive events related to the ball-pass frequencies, which stand out above the white noise levels. Refined techniques can be used not only to detect the faulty bearing but also to identify the specific element of the bearing (e.g. the inner or outer race or a damaged ball).

For such applications as helicopter gearboxes and aircraft turbines, this degree of sophistication may be called for, but for other less sensitive applications a simpler analysis will be adequate.

Spectrum Comparison

A useful technique, only possible when narrow band analysis by a FFT is available, is spectrum comparison. The first stage is to record a series of constant percentage bandwidth spectra when the machine is known to be in good condition. To these spectra are added a tolerance, the value of which must be determined by experience; the spectrum plus tolerance gives a reference mask which is really another spectrum defining the limits of acceptable vibration level. When the mask is penetrated by the measured spectrum, an alarm is triggered (see Figure 10). Provided that sufficient storage capacity is available to hold a sequence of spectra, trend analysis can be performed, either at the critical frequencies or over the whole spectrum. This allows a prediction of when the levels are likely to become critical. To achieve the optimum in fault prediction, continuous monitoring can be combined with periodic spectrum comparison.

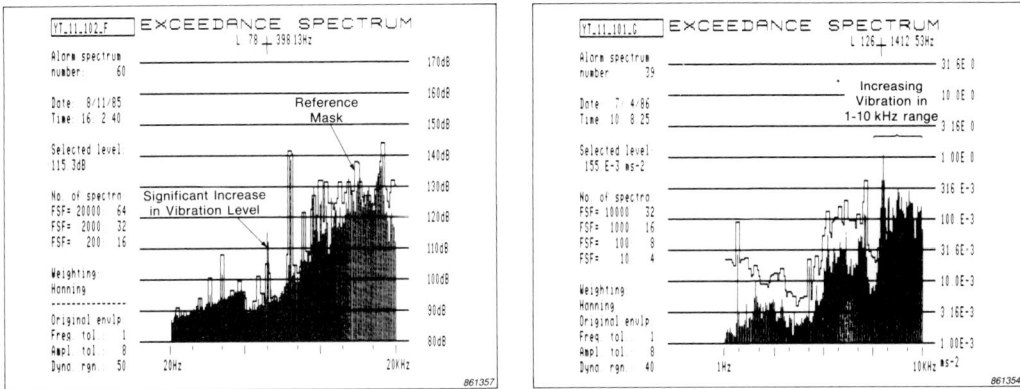

Figure 10 Principles of spectrum comparison. Any frequency components that exceed the reference mask will cause an alarm. The spectrum to the left shows increases at discrete frequencies, whilst the spectrum to the right shows increases in a broad band of frequencies

Chapter 3
Torsional Vibration Measurement

General Principles of Torsional Vibration Measurement

The theory of torsional vibrations has been dealt with in a previous chapter. The measurement of vibration in a rotational sense presents a different problem from that of measurement in a linear sense. When studying torsional vibrations, it is possible to use an accelerometer mounted on the bearing housing or, when studying unbalance or whirling, a proximity gauge mounted near to a critical shaft. But torsional vibrations may be present in a rotating system without there being any measurable vibration at the bearings or an unbalance signal from the shaft. Measurements then have to be taken directly on the rotating component itself.

There are severe difficulties in attempting to measure torsional vibrations using an accelerometer of a conventional type:

> An accelerometer mounted on a shaft rotating at a constant speed would be subject primarily to centripetal acceleration, which would tend to overload it.

> The balance of the component would be disturbed by the additional mass of the accelerometer.

> There would be problems in transmitting the signal to the analysis equipment – both slip rings and telemetry present problems.

The traditional ways of measuring torsional vibrations directly have involved the use of complex rotational accelerometers driven from a suitable shaft or off a convenient gear (a starter gear or an intermediate gear). They rely on a seismic mass driven by a system of springs from a light pulley wheel which exactly follows the motion of the wheel under investigation; a linkage between the two is transformed into a linear axial movement which can be picked up as a trace. Such a device has a short life and is incapable of phase discrimination.

Modern methods do not rely on contact between the rotating wheel and the transducer. Vibrations are sensed by a magnetic or opto-electronic pick-up or by use of a laser.

Torsional Measurements with a Magnetic or Optical Sensor

A magnetic pick up creates a magnetic field which, when interrupted by a moving ferro-magnetic object, induces a voltage pulse in a pick-off coil. Much the same kind of signal pulse can be produced by the use of an opto-electronic sensor using an infrared transmitter in conjunction with reflective tape or a slotted aluminium disc, Figure 1. If reflective tape is used, care must be taken that the spacing between the segments is uniform. A non-contacting probe has the advantage of all such probes, that they are not subject to mechanical damage if excessive amplitudes are experienced.

Mathematically, the presence of torsional vibration of a rotating shaft is analogous to a frequency modulated wave form, where the carrier frequency represents the steady speed of the shaft and the modulating frequency is the electrical equivalent of the vibration. The waveform is generated by positioning the transducer in close proximity to an existing gear wheel, if one is conveniently available, or to a special toothed disc attached to a shaft. The carrier frequency, f, is proportional to the number of teeth on the wheel, T, and the rotational speed, N. Thus

$$f = \frac{TN}{60} \; Hz$$

The input waveform is squared and the pulse train is averaged by a low-pass filter. When the shaft speed is constant the A.C. output is zero, but when a vibration is present, an A.C. signal is produced at the frequency of the vibration and proportional to the amplitude. Figure 2 shows the system.

The Econocruise type of instrument is capable of detecting torsional vibrations at frequencies up to 500 Hz; the peak amplitude before a meter reading occurs is 0.05° for the range 40 to 500 Hz, rising to 0.2° at 10 Hz. The maximum amplitude is 5° peak. The output signal depends on the number of teeth on the wheel or disc (proportional to the number of teeth). Although an existing gear wheel can be used for pulse generation, it may not be possible to find one with a sufficiently large number of teeth, so it is preferable to use a special slotted disc provided with the equipment; this, typically, will have 60 or 120 teeth. A further advantage possessed by a special toothed ring is that it provides a sharp pulse, giving optimum resolution with small vibrations. The rotational speed can be up to 20 000 rpm. A multichannel unit can be used to determine phase differences between different sections of the shafting.

Torsional vibration measurements made with this equipment can also be an integral part of a condition monitoring system (see previous chapter).

Torsional Measurements Using a Laser

A laser can be used to measure torsional vibrations in those situations where it is not possible to mount a measuring disc. It is particularly valuable when making measurements at a distance. As with all laser measurements, a coherent light beam is directed at a moving target (in this case a rotating shaft or disc), and the velocity determined by the Doppler effect. Figure 3 shows the system. All that is required in the way of attachment to the shaft is a strip of reflective tape. Figure 4 shows the complete equipment.

Laser measurement has a rather better potential for discrimination than the magnetic probe method described above. It can measure an angular velocity from 0.3 to 10 000°/s peak, equivalent to an angular displacement of 0.01 to 17° peak for a bandwidth of 1 Hz. After initial processing through the special meter, the signal can be passed through a tracking filter to give a frequency spectrum of the vibration. It can be processed by any standard analysis system, e.g. an RTA and, recorded and further processed in a PC.

The laser can be hand held and is therefore easily used without even the necessity of stopping the machine, but the equipment can be expensive, particularly if simultaneous readings at several stations are required, because each station requires its own laser.

Often torsional vibration measurements are taken in conjunction with a modal analysis, using the Holzer technique, so that a measurement made at one station enables the complete mode to be studied, leading to the estimation of displacements and stresses.

It may sometimes be required to estimate the critical frequencies when running up or down through the speed range. A laser meter is probably best able to do this.

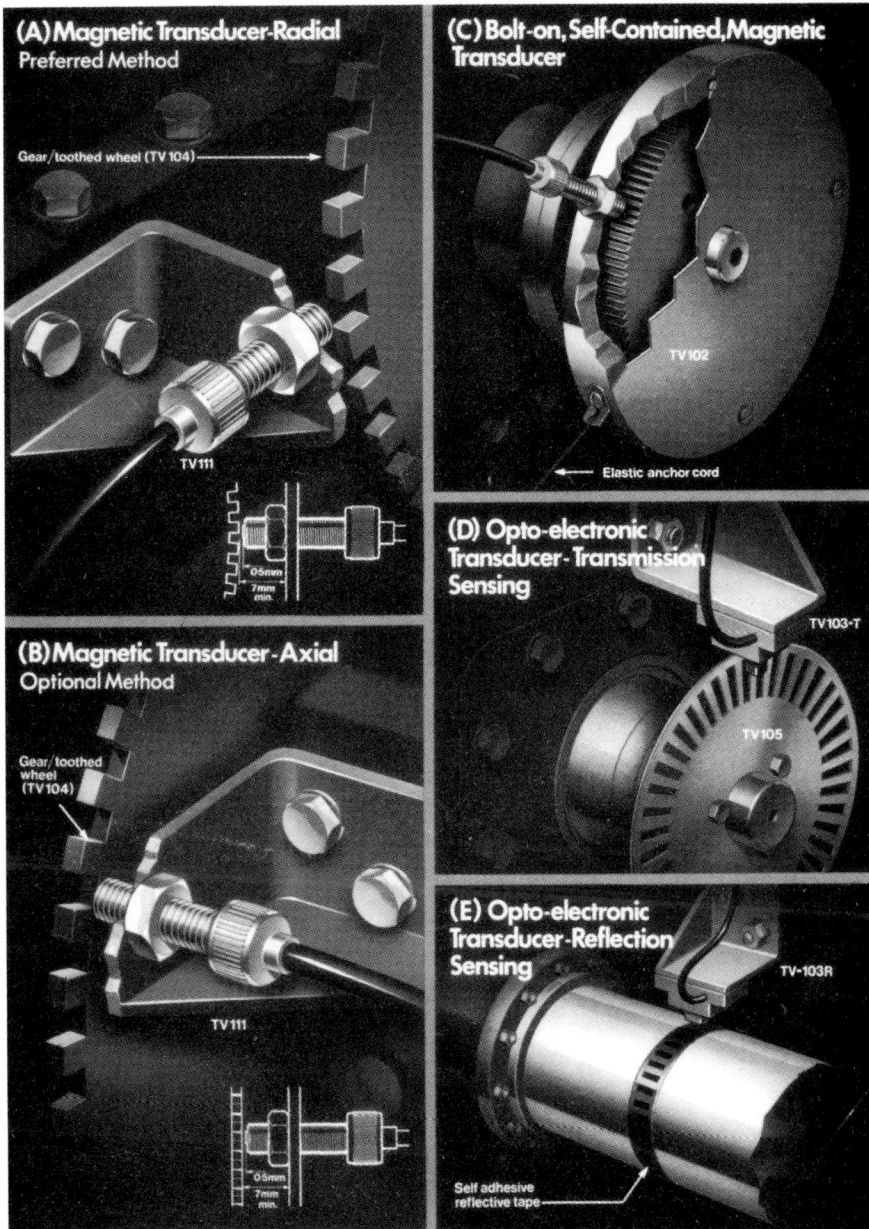

(A) Magnetic Transducer-Radial
Preferred Method

Gear/toothed wheel (TV 104)

TV 111

Ø5mm
7mm
min.

(B) Magnetic Transducer-Axial
Optional Method

Gear/toothed
wheel
(TV 104)

TV 111

Ø5mm
7mm
min.

(C) Bolt-on, Self-Contained, Magnetic Transducer

TV 102

Elastic anchor cord

(D) Opto-electronic Transducer-Transmission Sensing

TV 103-T

TV 105

(E) Opto-electronic Transducer-Reflection Sensing

TV-103R

Self adhesive
reflective tape

Figure 1

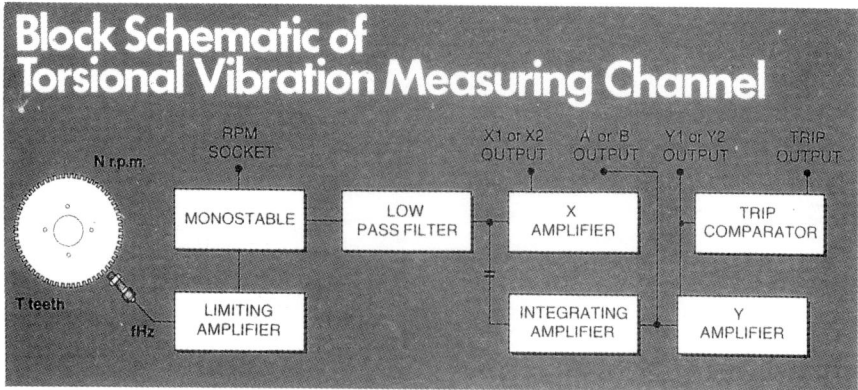

Figure 2

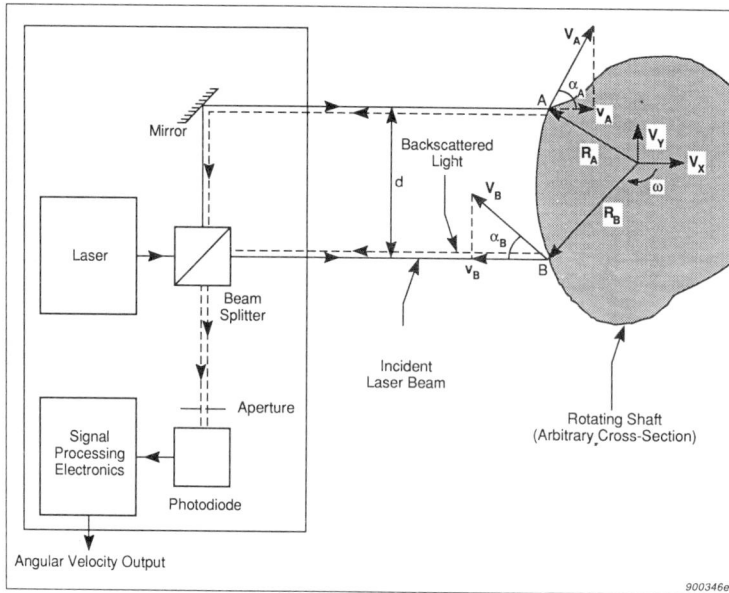

Figure 3 Measurement of torsional vibrations with a laser

Figure 4 Laser used for torsional vibration measurements, complete with meter

Chapter 4
Modal Analysis

Analysis of Structural Vibrations

Any real waveform can be represented by the sum of much simpler waveforms. In a similar manner, any real vibration in a structure or a body can be represented by the sum of much simpler vibration modes. Modal analysis is the tool for determining the shape and magnitude of the structural deformation in each mode. From this it is usually apparent how the overall vibration can be changed. When studying a system idealised by a finite number of lumped parameters, there are a finite number of modes. In any real structure, there are an infinite number, and part of the skill of the analyst is in assessing which of them are significant; these are usually the lowest frequency ones, but attention must be given to the excitation that will exist in practice. Modal testing in its modern form originated with the need to test aircraft structures. It proved to be such a powerful tool that it is widely used now for such diverse structures as turbines and bridges.

There are two basic experimental methods of determining the modes of vibration in complex structures.

Exciting only one mode at a time by sweeping through the frequency range of interest.

Exciting several modes simultaneously by the application of a random excitation followed by a computation of the separate modes from the total vibration.

There is a third 'hybrid' method

A sweep through the frequency range, with a random excitation.

Take a tuning fork as a simple example. To excite the first mode, a shaker or vibrator needs to be placed at the ends of the tines as in Figure 1. It is a common experience that in a structure such as this, the mode can effectively be excited by a single shaker, although with heavy structural

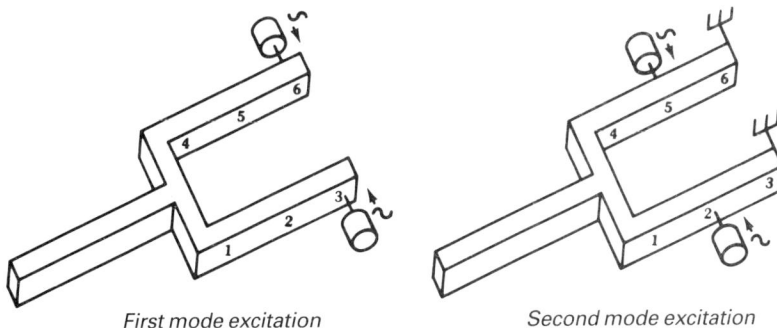

Figure 1

First mode excitation   *Second mode excitation*

damping, multi-point excitation is necessary. By varying the excitation frequency around the point of resonance, it is possible to determine the damping in that mode. In the second mode, the ends of the tines are nodes and the exciter has to be placed at or near to a node of the vibration mode. The deflection has to be measured at several points on the structure to fully define the mode.

In complex three dimensional problems, it is necessary to add more exciters in phase to ensure that only the mode of interest is excited. This already indicates one of the difficulties of performing this kind of testing.

Types of Exciters

There are several kinds of exciter available. The types that are used mainly today are electro-magnetic, electro-hydraulic, piezo-electric and impact hammers. Contra-rotating eccentric masses have been used historically, and for specific purposes gun recoil devices and step relaxation methods can be used.

Typical electro-dynamic exciters are shown diagrammatically in Figure 2. Permanent magnet types are available for small excitation forces and electro-magnetic types for larger forces. Electro-dynamic exciters are effective from 10 Hz to 1 kHz and, in the more powerful models, can generate a maximum force of about 20 kN with 25 mm displacement. Below 10 Hz, an electro-hydraulic type is necessary, which theoretically works down to DC, and can generate higher forces and amplitudes. Above 1 kHz, a piezo-electric type is necessary. A dual exciter, combining an electro-dynamic and a piezo-electric type, is also available; this can be chosen to cover a frequency range wide enough for most purposes. Exciters can be grounded, i.e. the case is effectively attached to earth, or seismic in which the reaction is provided by the mass of the casing.

When acquiring an exciter, it is worth first referring to BS 6140 for electro-dynamic exciters and to BS 7285 for hydraulic exciters. The primary criterion for selection is the force rating, which determines the amplitude at which a given structure can be excited. A test object can be mounted on a table, which itself may have a substantial mass which has to be taken into account in assessing the effectiveness of the applied force. Thus through the equation

$$F = ma$$

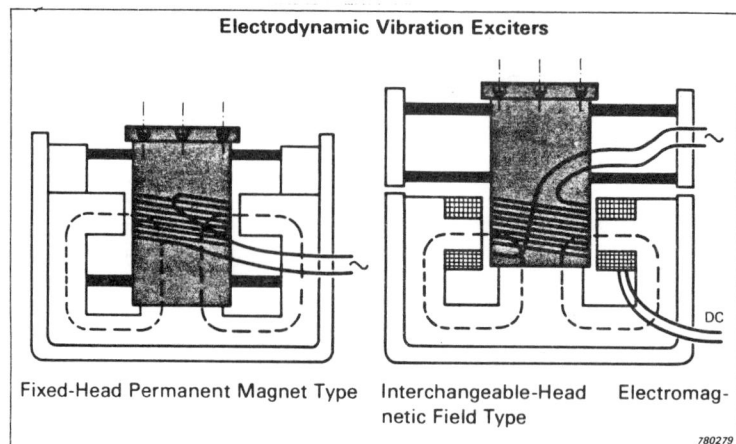

Figure 2 Electrodynamic Vibration Exciters

where F is the force, m is the mass of the exciter table plus the fixture plus the test object, and a is the acceleration, it is possible to estimate the vibration amplitude. The force is usually given as the blocked force output, which is the force generated by the exciter against a mass of infinite mechanical impedance, Figure 3.

Electrodynamic exciters can consume a great deal of power and often require cooling; both water and air cooling are employed. The power efficiency is an important characteristic, so the complete unit, Figure 4, comprising power supply and signal control equipment has to be properly matched. An electro-dynamic exciter has a maximum cut-off frequency approaching the resonance of the moving element and a minimum cut-off frequency, where the suspension resonance

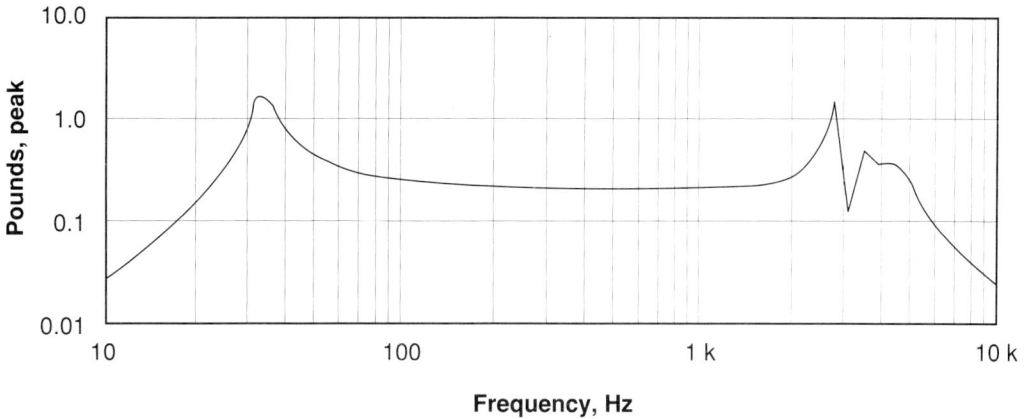

Figure 3 Typical Blocked Force Output for an Exciter

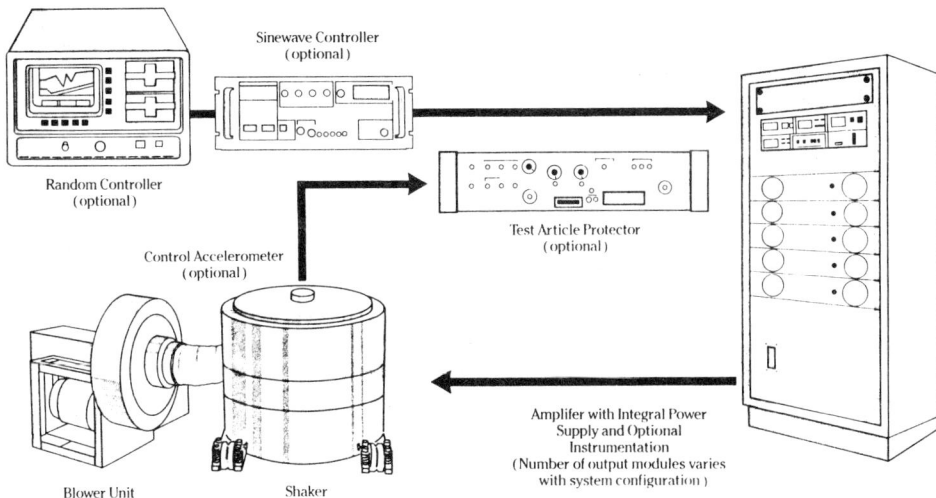

Figure 4 Typical Vibration Test System

is significant. The waveform can be distorted by non-linearities in the generator/amplifier system. Typically the working region is from 10 Hz to 1 kHz.

An electro-hydraulic exciter is one in which the vibratory movement results from the variable flow of oil, ensured by an electro-hydraulic control device, fed by an hydraulic power system, and acting on an actuator using one or more control loops. It is basically a servo-driven hydraulic ram.

A piezo-electric exciter is the converse of a piezo-electric accelerometer. A piezo-electric crystal changes its thickness in proportion to an applied voltage. It is capable of excitation up to a frequency of about 60 kHz. The displacement is very small, but by using multiple discs and high voltages, large forces can be produced at high frequencies. While an electro-magnetic exciter can be driven by an audio power amplifier, a piezo-electric one requires an impedance matching network after the power amplifier. A piezo-electric exciter presents a capacitive load to power amplifiers, therefore the electrical impedance decreases with increasing frequency. A large power amplifier is required to drive such an exciter at maximum voltage to its maximum frequency; however a smaller amplifier can be used to drive it at its maximum output at lower frequencies.

Excitation Methods

Test excitation can be a simple sine wave applied either at a number of fixed frequencies or sweeping automatically through a range of frequencies. Random vibration is the alternative choice. If mode shapes and frequency response are the main interest, sine-wave excitation is used. Random excitation is used to simulate vibrations resulting from random processes, such as wind forces, acoustic noise and road surfaces. This is a more common form of excitation of structures than pure sine waves.

The simplest kind of excitation is sine wave generation with manual control of the frequency; it is easy to identify the mode shapes by watching the output from a suitably placed accelerometer and measuring the mode shape at maximum response. Sine wave excitation with automatic sweeping through the frequency range is used for frequency response plotting. The main advantage of this form of testing is that it enables high input forces to be fed into the structure and the force level can be controlled. A swept sine test can be very time consuming, so a so-called 'chirp' signal can be used to shorten the test time. A chirp signal is a logarithmically swept sine wave that is periodic in the analyser measurement window.

Random vibration tests can be done much more quickly than a sine wave test and have the further advantage that all possible resonances can be excited simultaneously, accounting for possible interactions. The generation of random vibration requires a random noise generator combined with a spectrum shaper. When applying random vibration to a structure, the response also has a random character (modified, of course by the dynamics of the structure), so the accelerometer output has to be analysed by a suitable narrow band spectrum analyser.

Random excitation can be pure random or pseudo-random. Pure random excitation has a Gaussian distribution of the amplitude of the excitation force. The frequency spectrum is continuous and flat and has a uniform spectral density through the applicable frequency range (which may be broad band, narrow band or some form of composite banding). An applied random excitation may not result in a uniform spectrum at the exciter table due to exciter resonances, so it may be necessary to modify the applied excitation to achieve the desired spectrum through a spectrum shaper, which can be provided by the vibration exciter. Since the excitation cannot be periodic, it is necessary to use a Hanning window in the analyser to reduce 'leakage'. This however, reduces the frequency resolution of the analyser. After the time records have been transformed, the response in the frequency domain is ensemble averaged in order to reduce the non-linear effects, noise and distortion. The averaging produces a better measure of the linear least-squares response of the structure. The energy is distributed over the whole of the frequency range of interest in random excitation but not all frequencies are equally present in each record. To get a reliable result, a large number of averages need to be taken; it is suggested that more than 25 are needed.

A pseudo-random or burst random waveform is periodic with the same record length as the measurement window. This improves the resolution at the expense of converting a true random signal into one with periodicity, with a limited frequency content. There is of course no leakage, so rectangular weighting in the analyser is used. Each record length is the same, so there is no advantage in averaging since the response will be the same each time. Non-linearities, if present, will not be suppressed as they would be for truly random signals.

Shock Excitation

Shock testing is becoming common in research and production. It can be performed either with an electro-dynamic exciter modified to produce a single application of a shock pulse or by an impact hammer having a calibrated force and impulse duration. The shape of the shock pulse determines the frequency content of the applied signal. When using an exciter it can be whatever the pulse shaping circuit is capable of producing, which is usually either a single half sine wave or a single saw tooth shape. A vibration controller capable of giving the required signal must be used in conjunction with the power amplifier. A shock produced by an exciter has the advantage that the spectrum shape is well under control, but has the disadvantage that it adds an extra complication to what is basically a simple technique.

Another form of transient testing, similar to shock application, is step relaxation, where the structure is loaded to an acceptable strain level and suddenly released, triggering the measurement. This is more repeatable than impact testing and can be used on fragile specimens. When applying the shock through an impact hammer, the waveform produced is a short duration transient with maximum amplitude at low frequency and a decaying amplitude as the frequency increases, see Figures 5 and 6. There is an effective cut-off frequency, when the amplitude has dropped to about 10 dB below the maximum. The shape of the impulse depends on the construction of the hammer, in particular the material from which the impact surface is made and the

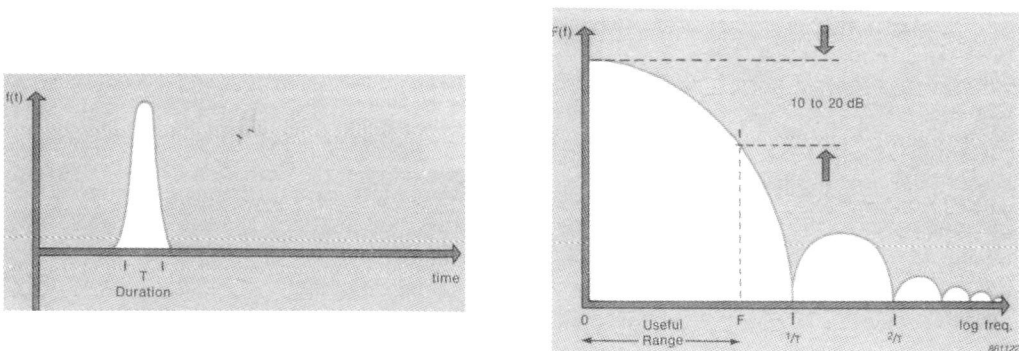

Figure 5 Left – waveform produced by an inpact device
Right – frequency spectrum of the same impact

mass of the head. Any one hammer can be fitted with a range of different heads in two ways: the effective mass can be added to and the impact tip can be made of a soft or a hard material, ranging from steel through plastic to soft rubber. These have the effect of modifying the rise time and the impulse width. A steel tip covers a wide frequency range and has a high peak force with a narrow spectrum. A soft tip has a lower frequency range and a lower peak, but a broader spectrum.

Figure 6   Impact hammer, capable of applying a force of 5000 N and a frequency range 0–6.5 kHz. The weight of the hammer head is 250g.

The force is measured by a piezo-electric disc incorporated in the body of the hammer. It is possible to construct hammers from a few grams up to several kilograms (or higher for rig mounted hammers), covering a frequency range up to 10 kHz for the smallest and up to 10 Hz for the largest; the duration of the pulse is of the order of a few milliseconds.

The advantages to be gained from performing shock testing by impact hammer are:

    inexpensive – the equipment is very simple

    very rapid testing

    no mass loading of the structure

    portable, so it allows field testing

The disadvantages are:

    limited energy input, otherwise there is a danger of local damage to the structure

    not suitable where structure behaves non linearly

    covers the whole frequency range up to the cut-off, so not suitable for zoom analysis

Certain precautions have to be taken when carrying out testing with an impact hammer. If the hammer is too heavy in relation to the response of the structure, there is a risk of double hitting where the structure rebounds and hits the hammer. This can be avoided with a skilled operator. Readings taken when a double hit is observed cannot be used. The force can be windowed by a transient window which takes the unweighted data during the period of contact and sets it to zero elsewhere.

Since the duration of the impact is usually short compared to the record length, particularly for lightly damped structures, the response also has to be windowed. A Hanning window is not suitable since that would destroy the signal at the beginning of the record, so an exponential window has to be used. This has the effect of adding an extra amount of damping to the measured signal, which has to be taken into account when determining the modal damping.

Comparison Between Types of Excitation

The following table compares the various forms of excitation:

Waveform	Analysis speed	Leakage error?	Linear approx-imation	Crest factor S/n ratio	Spectrum control	Zoom analysis?
Sine	very slow	yes	no	good	high	yes
Random	slow	yes	yes	fair	high	yes
Pseudo-random	fast	no	no	fair	high	yes
Impact	fastest	no	no	poor	limited	no
Multiple impact	slow	yes	some	fair	limited	no

Modal Analysis

It is usual in modal analysis to want to know the frequency, the response of the structure, the corresponding modal shapes and the damping. In some cases, vibration testing is done to confirm (or indeed fail to confirm) the predictions that have been made at the design stage. It has to be admitted that the prediction of vibratory behaviour of complex structures is expensive and as yet not all that successful, requiring very large computers capable of solving coupled differential equations by a fine element method. In other cases, vibration testing is carried out to make sure that there are no undesirable resonances on a structure about which no theoretical analysis has been done, and to give some assistance to the designer in taking corrective measures.

The mathematical theory underlying modal analysis is beyond the scope of this text. Fortunately it is not necessary to understand the theory in depth to be able to use the results of a vibration test. There are frequency analysers available which are capable of taking the records of accelerometer readings and processing them into a form that can be used by the practical engineer.

As has already been discussed, for linear systems the most effective way of looking at the response of a structure is in the frequency domain. The input force spectrum can be directly related to the output displacement spectrum via a function known as the Frequency Response Function (FRF), defined as:

$$H(\omega) = \frac{X(\omega)}{F(\omega)}$$

$H(\omega)$ is the frequency response function for a structure, relating the input forcing function $F(\omega)$ and the response function $X(\omega)$ at a frequency ω.

It has a complex representation, i.e. it has real and imaginary parts, corresponding to a magnitude and a phase angle, which can be interpreted in this way: at a particular frequency a sinusoidal input can be multiplied by $|H(\omega)|$, (its absolute value), to give a sinusoidal output, X, with a phase shift. For some purposes, particularly when the damping is light, it may be sufficient to deal with absolute values only and just look at the peak amplitudes, but it should not be forgotten that the functions are always complex and the analysis equipment has to handle them as such, even though in practice we may not use all the information.

The basic purpose of a signal analyser is to generate the frequency response function, given the signals χ (t) and f (t) in the time domain corresponding to X (ω) and F (ω) in the frequency domain. The presence of noise in both the input and output measurements makes the matter more complicated. Ensemble averaging has to be done to remove noise.

Vibration in the time domain can be expressed in terms of displacement, velocity or acceleration. Usually one requires to know the displacement, although the actual readings are of acceleration. The frequency response functions corresponding to displacement, velocity and acceleration are known as compliance, mobility and accelerance.

Single Sine Analysis Techniques and Tools

The Fourier Transform of a complex periodic time function, x(t), with period T, reduced to a discrete complex frequency spectrum $X(\omega_k)$.

$$X(\omega_k) = \frac{1}{T} \int_{-T/2}^{T/2} x(t) \left[\text{Cos } \omega_k t + j \text{ Sin } \omega_k t \right] dt$$

where j is the complex operator

It can be shown, by calculating the Inverse Fourier Transform, that the complex periodic function x(t) may be expressed as the General Fourier Series:

$$x(t) = \frac{b_0}{2} + \sum_{n=1}^{n=\infty} \left(b_n \text{ Cos } (nt) + a_n \text{ Sin } (nt) \right)$$

where

$$b_n = \frac{1}{T} \int_{-T/2}^{+T/2} x(t) \text{ Cos } (nt) \, dt$$

and

$$a_n = \frac{1}{T} \int_{-T/2}^{+T/2} x(t) \text{ Sin } (nt) \, dt$$

Substituting

$$X(\omega_k) = a_k + jb_k$$

where

$$a_k = \frac{1}{T} \int_{-T/2}^{+T/2} x(t) \text{ Sin } (\omega_k t) \, dt$$

and

$$X(\omega_k) = \frac{1}{T} \int_{-T/2}^{T/2} x(t) \, [\text{Cos } \omega_k t + j \text{ Sin } \omega_k t] \, dt$$

Consequently it is seen that by evaluating the integrals, the real and imaginary components respectively of the complex Fourier Series at the single frequency ω_k are determined.

Essentially all single-sine Frequency Response Analysers (FRAs) are practical implementations of the last two equations and are used to resolve the real and imaginary components of the complex Fourier series at the selected single measurement frequency ω_k. By incorporating the facility to vary the measurement frequency over a number of decade ranges, the FRA is made a useful tool for evaluating the Fourier series of a time signal x(t) over a wide frequency spectrum.

Further examination of the equations reveals some additional features of FRA operation of significant interest to the vibration engineer. Firstly, it can be seen that if the signal x(t) contains, in addition to a fundamental component, any harmonic components of the measurement frequency ω_k, the effects of these harmonic components will be averaged to zero by virtue of the computation of the integrals over a complete number of periods of the measurement frequency. Consequently, the single-sine FRA technique is seen to exhibit excellent harmonic rejection characteristics. Clearly, if the FRA is set to measure at an harmonic of the fundamental frequency, then the instrument will act as a very high Q resonant filter and reject all other harmonic components, including the fundamental.

A second interesting feature of the implementation of the equations in the single-sine FRA concerns measurement noise. Most signals derived from systems or structures contain, in addition to the desired frequency response information, some random noise component. Again, the integrating operation performed in the implementation of the equations serves to 'average out' this random noise component such that, if the integration period is chosen sufficiently long (i.e. a large multiple of integer periods of the measurement frequency ω_k), the effects of the noise on the measured component values can be reduced to an insignificant level. The diagram (Figure 7) illustrates the improvement in frequency selectivity to be gained by increasing the integer number of periods of the measurement frequency ω_k over which the integration operation is performed.

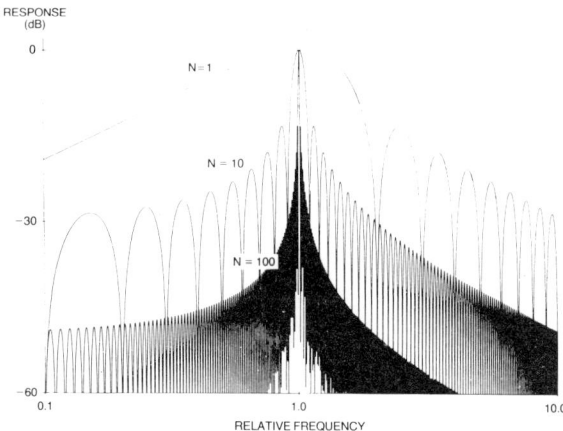

Figure 7 Harmonic and noise rejection characteristics of a single-sine frequency response analysis

To summarise, the single-sine FRA has two particular merits, good harmonic rejection and, provided a relatively long measurement period can be tolerated good noise rejection. On the minus side, the single-sine FRA only measures the components of the Fourier series at one frequency at any one time. Consequently, the evaluation of the Fourier series components over a sizeable spectrum may require many individual measurements to be taken, involving a considerable amount of time. The vibration engineer must therefore assess the trade-offs between achieving a suitably high level of measurement accuracy and the time available in which to reasonably complete the measurement programme. Such trade-offs become all the more pertinent in situations where the parameters of the system or structure under investigation are time-varying, e.g. the vibration characteristics of a turbine casing during run-up or run-down.

The twin (or multi-channel) single-sine FRA can be of value to the vibration engineer in many different ways. One very obvious application is the determination of the resonant modes of a structure using an experimental set-up such as that illustrated in Figure 8. The structure is being excited sinusoidally by the generator of the FRA and monitored at several points of interest simultaneously using the multi-channel analyser facility. Using the processing capability of an FRA, it is possible to compute the transfer characteristic between the generator input and any monitor point on the structure, or alternatively, the relative transfer characteristic between any two prescribed monitor points on the structure. Thus, by sweeping the measurement frequency through the spectrum of interest, it is possible to determine the relative frequency response characteristics of the nodes of interest, (typically as shown in Figure 9), from which the damping and stiffness of the elements coupling the node points can be determined by conventional modal analysis techniques.

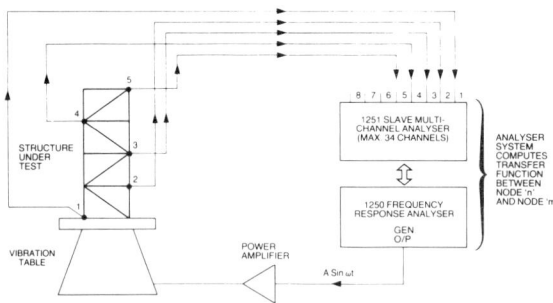

Figure 8 Modal analysis using 1250 frequency response analyser

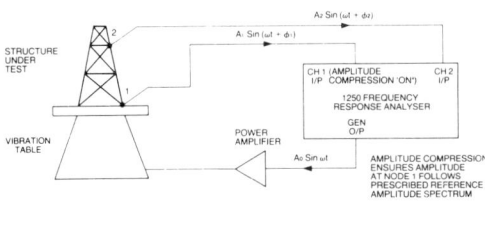

Swept sine vibration testing using a frequency response analyser

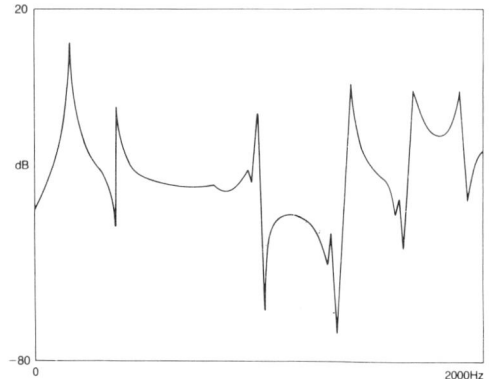

Figure 9 Typical response of lightly damped structure

Swept-sine vibration testing has proved a useful method for determining the behaviour of components of structures subjected to prescribed power spectra. Often there is a requirement with this type of testing to excite a certain point on the structure under investigation with a controlled spectrum and to monitor the response resulting from the excitation at a second point located elsewhere on the structure. A problem arises however, either because the vibrator used to excite the structure has an inherent frequency response characteristic which modifies the spectrum input at the driving point, or it is not physically convenient to excite the structure wth the prescribed spectrum at the defined driving point. Hence the spectrum input at the actual driving point is modified by the inherent transfer function between the actual driving point and the desired driving point. Either situation results in failure to achieve the required power spectrum at the desired driving point, thereby invalidating the measurement.

In applications concerning the vibration analysis of rotating machinery, there is often a requirement to analyse the vibration response of a component of the machine under investigation at a frequency synchronous with the fundamental frequency of rotation (or a prescribed harmonic thereof) of a given shaft. A typical example, the determination of the vibration characteristics of a turbine during run-up/run-down, is depicted in Figure 10

The applications cited here are but a few of the many ways in which single-sine frequency domain analysis tools can assist the vibration engineer in determing the behavioural characteristics of vibrating systems or structures.

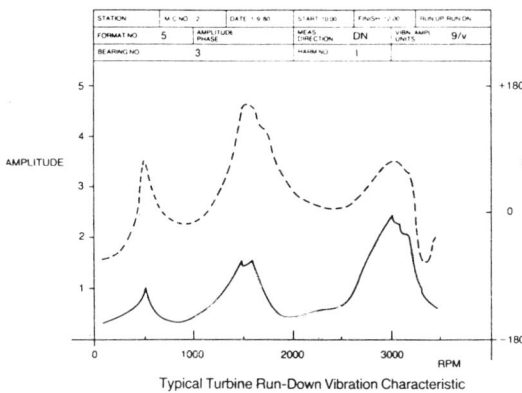

Figure 10 Typical turbine vibration characteristics

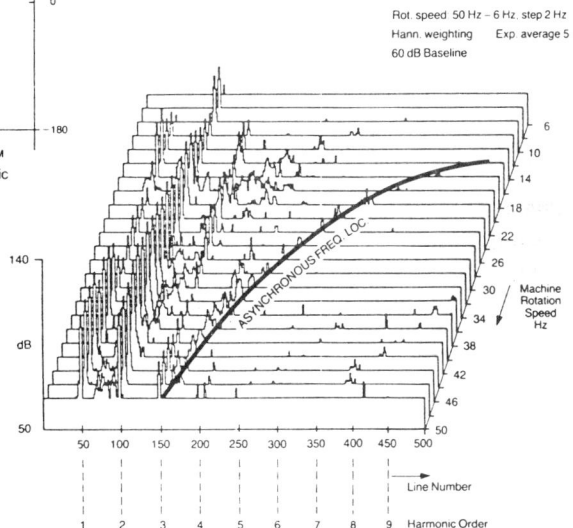

Typical Turbine Run-Down Vibration Characteristic

Typical result of tracking frequency analysis of a turbine.

FFT Analysers

A FFT analyser suitable for generating a FRF takes the input signal (the excitation) and the output signal (the response) and produces the FRF spectra. It is able to do this whatever the kind of excitation, but it must be synchronised with the excitation. The signals are first filtered, then sampled and digitised to give a series of records. The sampling rate and the record length determine the frequency range of the analysis. Each record has to be windowed as discussed above and then its Fourier transform taken. After averaging, the FRF is determined. A modern analyser does this without the user having to be involved in understanding the theory of the process. Sequential analysis is necessary to determine the modal shapes from a number of accelerometer readings. Analysers offer a variety of post processing options for the presentation of the FRF spectra.

As has been indicated, a two-channel capability is required for obtaining FRFs, but it is also possible to obtain four channel instruments which are capable of more sophisticated cross-channel processing. An analyser has to be synchronised with the excitation, particularly when random excitation is chosen. When a limited frequency range is under investigation, a 'zoom' facility is of advantage.

Some of the features which are worth considering when choosing a modern dual channel analyser are:

> Complex signal processing
>
> Wide frequency range and zoom facility
>
> Output interface for connection to a computer, printer or plotter
>
> Wide dynamic range (80 dB or more)
>
> Generation of all relevant FRFs – accelerance, compliance and mobility
>
> Range of windowing capability
>
> Waterfall displays (a form of 3-D display used to emphasise amplitude and frequency trends)
>
> Nyquist display
>
> Mode deflection pattern display
>
> Built-in storage with a floppy disc

A FFT signal analyser can be used for much more than structural response calculations as has been mentioned in previous chapters. It has a powerful ability in machinery monitoring applications and for waveform analysis in acoustics and electronics. Figure 11 shows two examples of available equipment.

Operational Deflection Measurements

It is not always necessary to determine mode shapes for a generalised excitation. If one wishes to determine the vibration characteristics of a structure in detail, one must excite all the modes so as to be able to predict its behaviour under any excitation. But for many, if not most, structures the excitation that occurs in practice is over a limited range of frequencies and amplitudes, and the only information that is available to the analyst is a measurement of the vibration of the structure when in service. Although in practice it may be possible to simulate service conditions in a laboratory, as for example when studying vibration of a motor vehicle under specified road conditions, in many instances the actual excitation is unknown. Other examples are aircraft flutter and wind forcing of tall buildings.

The concept of operational deflection shapes (ODS) has been introduced. An ODS is the shape assumed by a vibrating structure at a specific frequency under specified operating conditions. It differs from a modal shape in that it is a linear addition of several modes. ODS determination does not rely on the measurement of the vibrating force, which is usually unknown. The procedure involves the measurement of the vibrational displacement of the structure relative to a reference point. One accelerometer is chosen as the reference and the ODS found from measurements taken at the others. Analysis of the accelerometer outputs requires a two-channel FFT as for modal shapes. Although an ODS can be obtained from the modal shapes, this information is usually not available, so measurement of ODS is mainly used for trouble shooting.

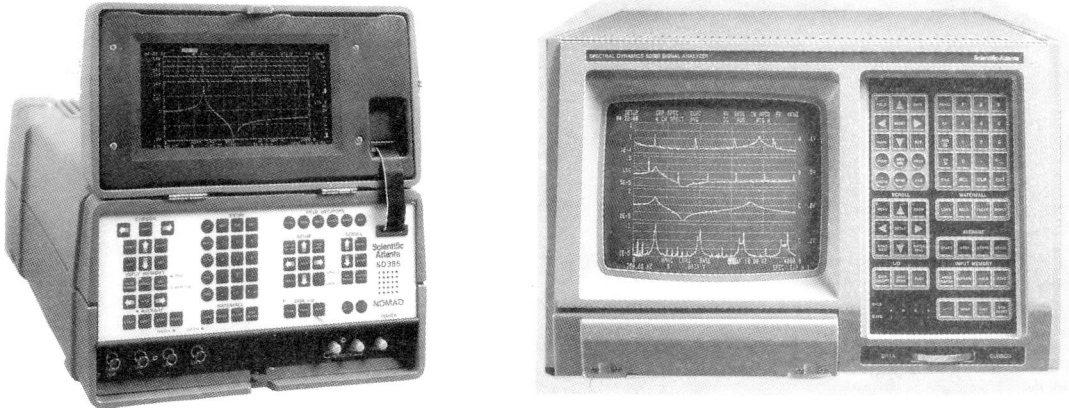

Figure 11 Two types of FFL analysers. Left is a portable instrument with ability to store field results on a floppy disc. Right is a top of the range instrument able to handle four channels with a wide range of processing options

SECTION 5

Chapter 1
Engine Noise

Legislative Background to Engine Noise Reduction

It has been recognised for some time that the most serious nuisance affecting the public is vehicular noise, see the chapter on *Community Standards*. In Europe this fact has culminated in a series of Council Directives, a current list of which can be found in the European Community Directives section of the chapter on *Standards*. The directives are concerned, in the main, with noise from the vehicle as a whole, but the main thrust of noise reduction techniques has to be directed to the principle source of noise – the internal combustion engine. The intermittent nature of the combustion gives rise to noise and vibration. The noise can be directly radiated from the body of the engine and from the inlet and exhaust systems. Vibration of the engine structure is transmitted to the main body of the vehicle. This vibration is in turn converted into acoustic noise which is very efficiently radiated by the light weight construction of the vehicle, Figures 1 and 2.

Although there are noise sources in the vehicle such as the transmission (gearbox and drive shaft), brakes, tyres, wind noise and warning noises, the main one is the engine, and it is that to which most attention has been given. Engine noise sources are

> combustion noise
>
> fuel injector noise

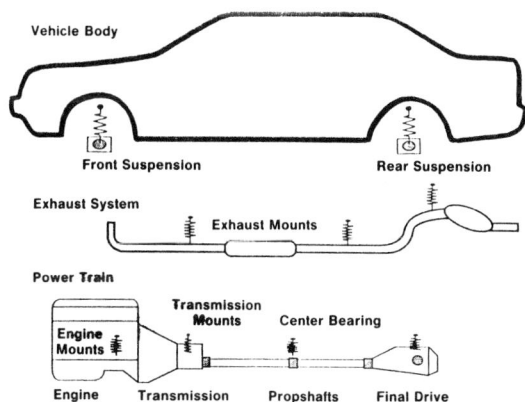

Figure 1 Vehicle noise sources and paths

| FORCE GENERATION | FORCES APPLIED TO STRUCTURE | VIBRATION TRANSMISSION | RADIATORS OF NOISE |

AIR CHARGE
Temperature
Pressure
Swirl
Turbulence
FUEL INJECTED
Spray Velocity
Distribution

FUEL AIR MIXING
IGNITION DELAY
HEAT RELEASE
CYLINDER PRESSURE

CYLINDER PRESSURE PULSES

CYLINDER HEAD → ROCKER COVER

PISTON CONNECTING ROD AND CRANKSHAFT → INTAKE MANIFOLD

COMBUSTION NOISE SOURCE

CRANKSHAFT PULLEY

GAS FORCES

INERTIA FORCES

PISTON SLAP IMPACTS

CYLINDER WALLS WATER JACKET DECKS

CRANKCASE PANELS

WATER JACKET PANELS

SIDE COVERS

SUMP

CYCLIC VARIATIONS in
Valve gear Torques
Fuel Injection Pump
Drive shaft Torque
Crankshaft Rotation

TIMING GEAR IMPACTS

FRONT OF CRANKCASE

TIMING COVER

MECHANICAL NOISE SOURCES   STRUCTURE RESPONSE

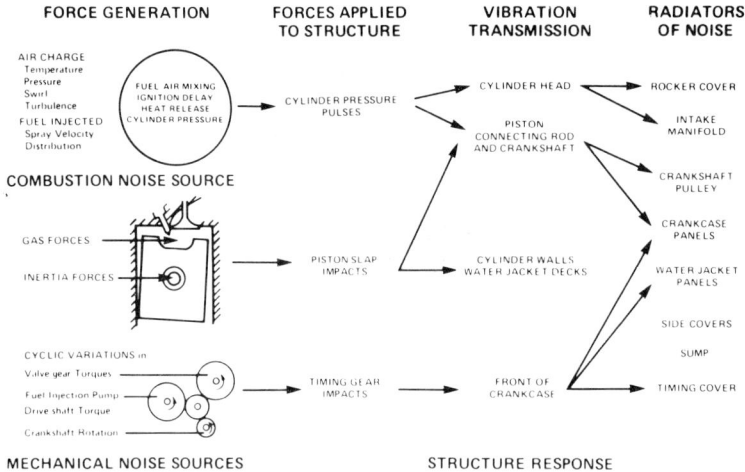

Figure 2 Noise generated in diesel engines

mechanical noise

inlet and exhaust noise

cooling fan noise

noise from the ancillaries such as the generator or compressor

When dealing with each of these sources, the operational regime of the engine also has to be taken into account (speed, load, air pressure and temperature) as well as the possibility of having to operate at a non-optimum design condition, e.g. with a badly set ignition timing. Many of the legislative codes that have to be satisfied prescribe operation in the drive-by mode, and while it is unlikely that the engine operation *per se* is likely to be affected by the driving circumstances, it is worth bearing in mind the possibility of an interaction.

Identification of Noise Sources

The frequency of combustion at a given engine speed is the forcing frequency of the noise. The design of the engine block controls how much of the combustion noise is converted into radiated noise. Mechanical noise from the piston and valve gear has its own forcing frequency. Techniques are available for improving the structure of the crankcase to make it less responsive at the important combustion frequencies. The improvements consist of both modifying the flexibility of the crankcase and increasing the natural damping.

There are two main methods of assessing which are the critical areas of the engine – measurements on the finished engine and predictive methods at the design stage. Noise measurement methods in the past have not been particularly good at identifying the critical regions, but modern sound intensity measurement techniques are much better at doing this (see the chapter on *Measurement Techniques*). They are likely to become more important in the future but work still needs to be done to develop appropriate techniques.

Finite element analysis and experimental modal analysis are increasingly used to assess the vibrational behaviour of the engine to predict the noise radiating regions. Such methods can be used to determine which parts of the structure can be stiffened to reduce the radiation.

One practical technique for identifying mechanical noise is the motored strip-down test, where the radiated noise is measured as the motored engine has parts progressively removed.

However, the removal of the combustion process has such an effect on the loading in the engine that the results of such a test should be treated with reserve.

Another traditional technique involves shielding and cladding the surfaces of the engine with a tailored lead cover lined with an absorptive mineral wool. The various surfaces are progressively treated and sound pressure measurements taken at each stage. This is a time consuming process, and has a number of technical drawbacks, such as health risks from the materials in use, the difficulties in ensuring proper sealing of the shielding components and maintaining the same degree of surface attachment for each test.

The magnitude of the exhaust noise is fairly easily determined, since it usually possible to duct away the exhaust gases so as to isolate that source from any other noise, Figure 3.

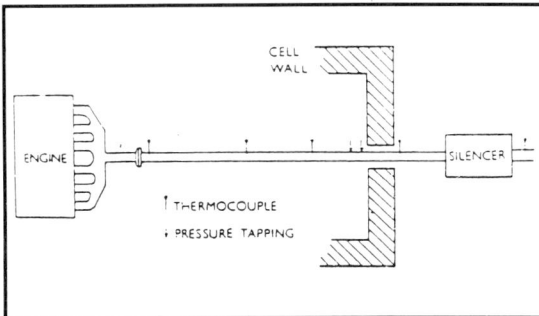

Figure 3 Experimental rig for on-engine tests

Exhaust noise is dominated by pressure pulsations reaching the atmosphere from the tailpipe, with about 15% being radiated from the body of the silencer; it is helpful to be able to distinguish between the two so that the appropriate treatment can be applied. Body vibrations can be effectively suppressed experimentally by lead cladding or enclosure by a lagged box. Research into exhaust noise can usefully employ a cold blowing facility which effectively simulates the pulsations in the manifold, but caution is advised, since higher gas temperatures result in higher noise levels.

Noise from the fuel injector mainly comes from the impact of the moving valve on its seat. Its magnitude is governed by the change in momentum of the moving parts and by the areas of the body and the inlet manifold which radiate the noise.

Variations in engine design make it difficult to give a universally applicable ranking of vibration-induced engine noise, but experience indicates that the following are the important ones, with the noisiest first:

    sump
    timing and valve covers
    induction manifold
    crankshaft pulley
    cylinder block
    exhaust manifold
    injection pump
    ancillaries

Combustion and Mechanical Noise Reduction

The cylinder and cylinder head, being designed to resist the combustion pressures without movement, are usually very stiff compared with the lower crankcase, see Figure 4. It is usually found that the areas which respond best to stiffening treatment are the lower crankcase, flat panels on the upper crankcase, and sub-structures such as oil sump (which is often made from pressed steel) and the valve cover. These last two are usually non-structural and on that account may be thought not to merit much attention, but being attached to the engine block they can be very effective noise radiators.

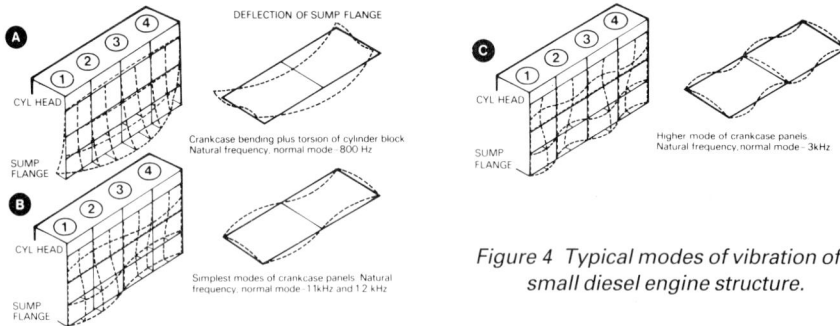

Figure 4 Typical modes of vibration of a small diesel engine structure.

Once the modes of vibration of the engine block have been determined, experimentally or theoretically, it is then a matter of putting in the appropriate extra stiffening to reduce the movement. it is often found that extra ribbing applied to the crankcase walls can make a marked contribution to the noise reduction.

It has been estimated that the total engine noise can be reduced by as much as 3 dB by effectively silencing the covers (sump, valve cover, etc.), so this is clearly an area where good results can often be achieved at a more modest cost than changes to the engine block.

There is usually a gasket between the cover and the body, and this has an effect on the total radiated noise. It has been found that, contrary to expectations, a highly damped joint between a light valve cover and a rigid cylinder head may increase the coupling between the two components and increase the energy radiating from the cover, the assumption being that high energy levels can be transmitted more easily across such a joint. When considering the joint between a rather more flexible lower crankcase and the sump, the situation is a little different in that damping and stiffness in the sump can affect the energy levels in the crankcase, and in this case increased damping is beneficial.

A change to the construction of the covers by the use of laminated steel, or where appropriate a high density plastic, can make a large change to the radiated noise.

Some further conclusions can be drawn from work at ISVR, University of Southampton:

Energy transfer is highest when modal characteristics of the cover flange are similar to those of the engine component to which it is attached.

Vibrational energy of the valve cover is lower with a central fixing.

A damped steel sump (a steel sandwich construction) reduces the energy in the sump without affecting energy transfer and reduces crankcase energy.

Isolating the cover from the crankcase can increase the radiated noise in certain frequency bands.

An example of structural modifications, carried out as an exercise by Ricardo Consulting Engineers, is shown in Figure 5.

It was found that the ideal solution was to tie the crankcase walls and the main-bearing caps together to form a ladder type structure.

A water-cooled engine is quieter than an air-cooled one of the equivalent power, because water acts as an insulator and absorbs the vibration generated in the combustion chamber. The fewer the number of cylinders on a crankshaft, the greater the difficulty in balancing and so the higher tends to be the vibration level. Air cooling is most often found in single and twin cylinder engines, so this contributes further to the reputation that air cooled engines have for being noisy.

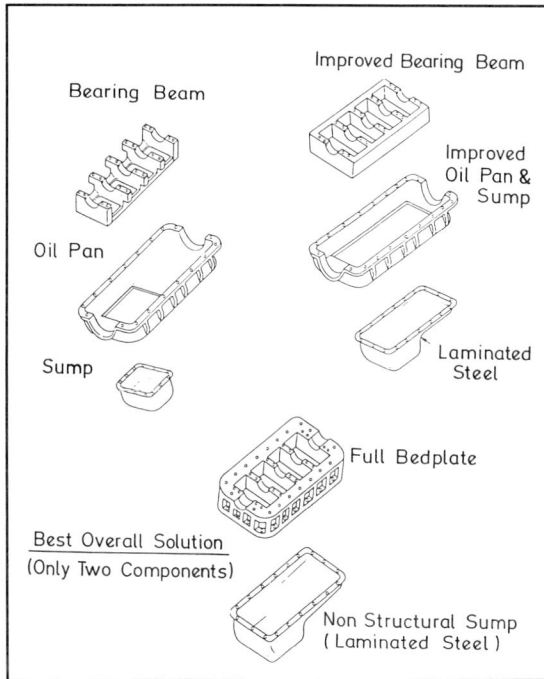

Figure 5 Example of Engine block modification

Intake and Exhaust Systems

The exhaust is the main sources of noise in internal combustion engines; the intake system is less significant. The noise is produced by the periodic expulsion of gases through the exhaust manifold. The gases are ducted away in such a way that they seem to be designed more to protect the passengers than the nearby residents.

The lowest frequency of the noise spectrum is given by the number of exhaust charges per cylinder per second, corresponding to the firing frequency. Combining the exhausts of several cylinders into one exit does provide some cancellation which increases with the number of cylinders. The spectrum might be expected to show a series of harmonics peaking at integral multiples of the firing frequency, which is indeed the case. However, other intermediate peaks may be present which modify the quality of the sound. The evaluation of exhaust noise, as with other kinds of noise, is probably most often done on the basis of the dB(A) value. There has been some attempt recently to treat exhaust noise differently; this approach has some validity in that exhaust noise is dominated by resonant peaks, which are not easily handled by dB(A) as a measure. So far the identification of a quality index, specific to exhaust noise, has proved elusive, but it is worth bearing in mind that a subjective assessment (by a jury panel, for example) of the noise quality of a finished design should always be carried out.

The only practical method of controlling the exhaust noise is by fitting a silencer. A reactive type is usually chosen but this may be combined with dissipative elements as well as resonator (Helmholtz) chambers. A silencer can serve two purposes: reduce the overall sound pressure, and by selective filtering improve the quality of the noise. Size is no criterion of efficiency. Automobile engine silencer volumes may range from less than engine displacement to five times displacement (or twice this value for twin exhausts). The quality of the silencing will vary with gas temperature and engine speed, which means that a silencer will only give its best performance at one operating condition. The spectrum, being dependent on the firing frequency, will shift with speed, so that any attempt to suppress a particularly undesirable peak in the spectrum is doomed to failure in an engine which works through a range of speeds; a resonator for example is effective over a very narrow band of frequencies. All this points to silencer design being a compromise and explains how so much of the design has in the past been largely empirical. A warning must be given about over-design: it is in theory possible to reduce the noise level to any required value at the expense of engine efficiency.

The acoustic performance of a silencer is best characterised by the *insertion loss*, which is the difference between the sound pressure levels at some defined measuring point with and without the silencer in place. As a measure it is superior to either *transmission loss* which is the ratio of the incident acoustic power to the transmitted acoustic power, or the *noise reduction* which is the difference between the sound pressure upstream and that downstream. Insertion loss is easy to measure and takes account of any possible modification of engine performance caused by fitting the silencer, but it is less easy to predict because it has to include the effect of both pressure pulses and radiation. The trial and error design methods of the past are now being superseded by the advent of modern analytical tools. The acoustic properties of the various elements in a silencer system, see Figure 6, can be simulated in a computer model. It is possible, by making a sequence of small changes to the computer model, to optimise the final design.

The best method of analysis is that based on plane wave linear theory, which seems to be satisfactory insofar as it successfully matches theory and practice, with not too heavy a burden on computer facilities.

It has to be admitted that there is still a degree of empiricism involved in exhaust design; what the simulation does is to speed up the theoretical evaluation of the changes suggested by the designer. It follows also that a model of the proposed design ought to be made to validate the theoretical predictions. One convenient model simulates the exhaust pressure pulses in a cold air rig, using the known exhaust pulsation spectrum. The advantage of using cold air rather than the hot gases is that the dummy system can be made of any convenient material, for example wood or plastic.

The exhaust system cannot be treated in isolation. A particular design may be effective as a silencer but may degrade the performance of the engine to which it is attached, by introducing a back pressure in the manifold. It is well known that a 'tuned' exhaust may enhance the performance of the engine and at the same time give an acceptable degree of silencing. This leads to the

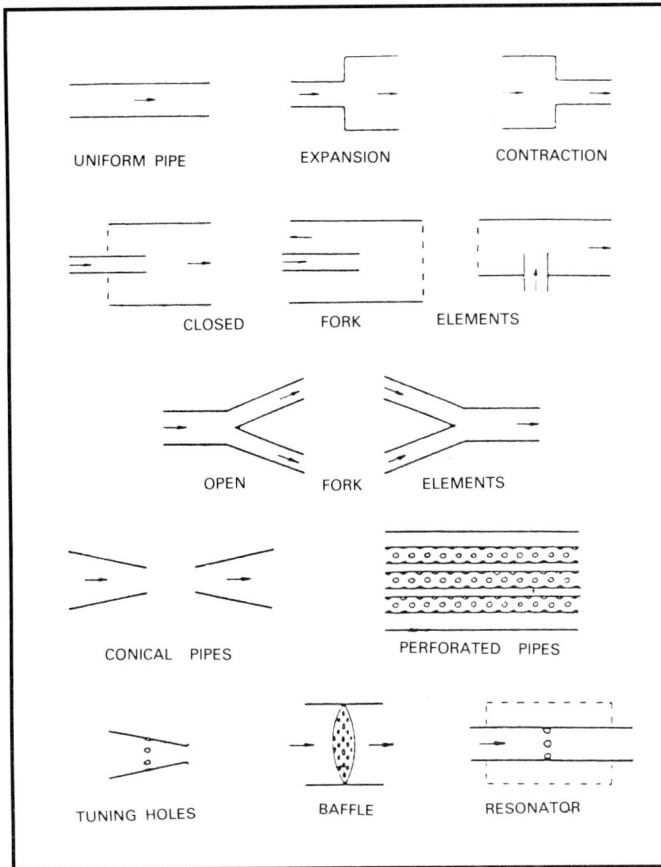

Figure 6 Acoustical elements available in the computer program

further development of simulation programmes wherein the engine, its inlet and exhaust systems are treated together, and all interactions studied. Examples of the sort of improvements that are possible can be seen in Figures 7 and 8.

It sometimes happens that the silencer provided with the engine is not sufficient to reduce the exhaust noise to a satisfactory level. It is common practice to use a secondary silencer of the absorptive type fitted after the primary silencer. This is most effective at medium to high frequencies (250 Hz upwards), but has little effect at lower frequencies, see Figure 9. It also helps to suppress tail pipe resonances. A Burgess silencer is a straight through type with a perforated inner tube; the space between the inner tube and the outer casing is filled with a heat resisting fibrous material, Figure 10.

There is little directly applicable to exhaust noise in the available standards, but reference might be made to BS AU 193A which is specific to motorcycle and moped exhausts, specifying limits and test methods, and to ANSI/SAE J1207 for measurement procedures.

Figure 7 Prototype exhaust system
designed by Ricardo

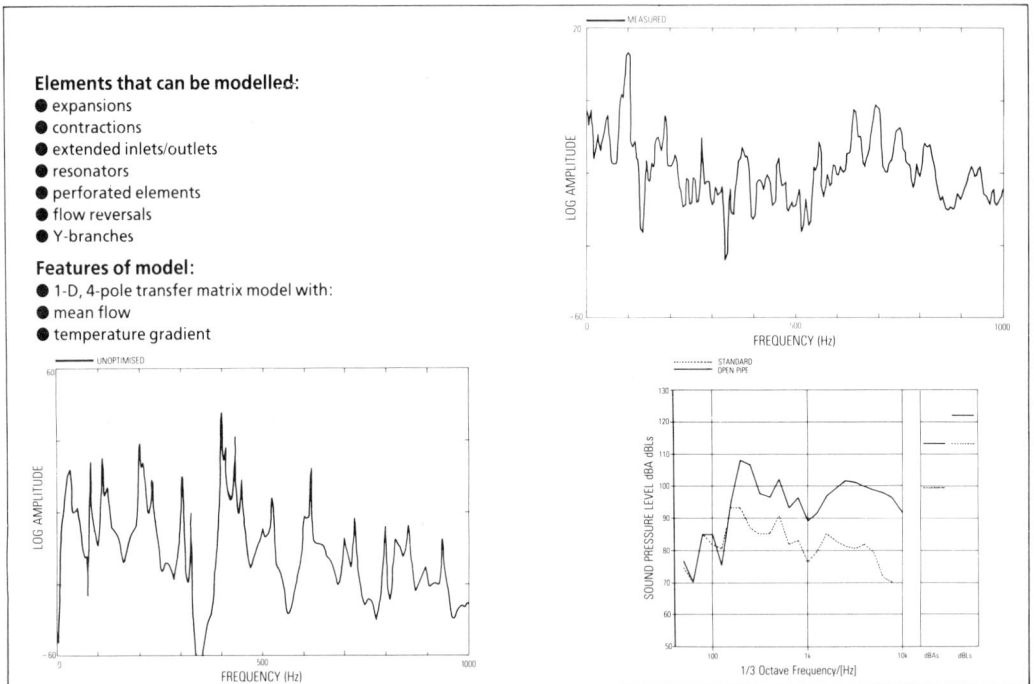

Elements that can be modelled:
● expansions
● contractions
● extended inlets/outlets
● resonators
● perforated elements
● flow reversals
● Y-branches

Features of model:
● 1-D, 4-pole transfer matrix model with:
● mean flow
● temperature gradient

Figure 8 TOP RIGHT: Acoustic modelling. Measured and predicted insertion gain of exhaust system with two mufflers LEFT: Automatic muffler design optimisation. Predicted acoustic performance showing comparison between designs RIGHT : Complete intake system designed by Ricardo for single cylinder DI industrial engine

Mechanical Noise Reduction

Piston slap and gear noise are two sources of mechanical noise which merit attention. It is rarely possible to reduce the clearance between piston and bore under full-load conditions, and since the piston and cylinder are made from different materials, the clearance increases at lower loads. Devices which distort the piston skirt as it cools offer a means of preventing the increase of piston slap under cold start conditions. Other devices which have been tried include articulated pistons, offset gudgeon pins (Figure 11) and pressure pads in the skirt.

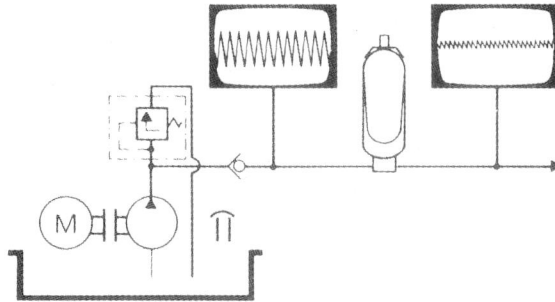

Figure 9 Typical insertion loss of an absorption type silencer

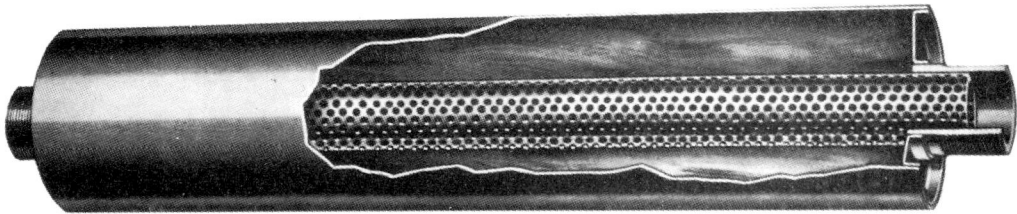

Figure 10 Burgess absorption silencer

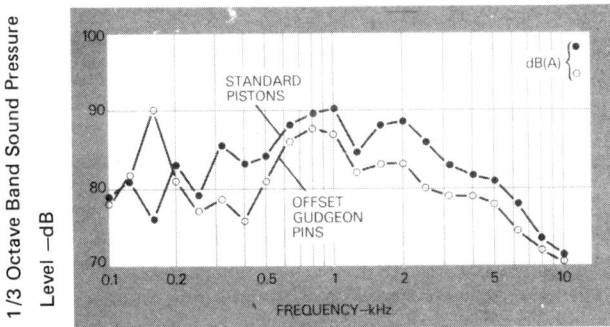

Figure 11 Reduction of external noise by fitting offset gudgeon pins to diminish piston slap. Engine running at 1500 rev/min on light load

Gear noise is covered in another chapter, but a few specific points can be made here. If backlash can be eliminated by making sure that the constant driving load is always greater than torque variations from the fuel pump, valve gear, etc., noise will be reduced. For smaller engines, a toothed belt with proper wrap-around generates lower noise.

Engine Mountings

As illustrated in Figure 1, vibration can be transmitted from the engine to the body of the vehicle, which is then changed into acoustic noise, affecting not only the occupants but also the outside observer. Transmission from the exhaust suspension and the drive train attachments is also possible. This topic is dealt with in the chapter on *Vibration Isolation*.

Other Means of Noise Reduction

The solutions offered so far have dealt with the reduction of engine noise alone. This can very often prove an expensive and time consuming operation, even with the assistance of the newer and more powerful analysis techniques becoming available. Before embarking on work to reduce engine noise at source, one should investigate the machine that is to be driven by the engine, and assess the contribution that it makes to the overall noise. It makes little sense to take extraordinary steps to quieten an engine only to find that the noise level from another part of the drive train is so great that any improvement is hardly noticeable. The obvious rule, which is so often overlooked, is to tackle the noisiest component first; this applies whether that component is a part of an engine or a complete generator or compressor.

Even if the engine is the main contributor to the overall noise level, it may prove much more economical to put the whole drive unit into an acoustic enclosure. The design of these is dealt with in the chapter on *Acoustic Enclosures*. It should be mentioned in passing that very large noise reductions are possible at modest expense by a properly designed enclosure. 20 dB is easily achieved, where such a reduction at source would be practically impossible.

Chapter 2
Gear Noise

Causes of Noise

Conjugate gears, by definition, transmit constant angular velocities, so if each pair has perfectly rigid, equally spaced and accurate teeth, with good lubrication, the angular velocities will be constant and the noise minimal. Deviations from the ideal cause vibration and noise. The noise is likely to be obtrusive since tooth meshing frequency in most gear drives lies in the audio range. The frequency for simple countershaft gearing is:

$$f = \frac{N\,n}{60} \qquad \text{Hz}$$

where n is the rotational speed in rev/min and N is the number of teeth on the gear. Gearboxes that contain more than one pair will exhibit frequencies corresponding to each gear pair. For planetary gears:

$$f = \frac{N_r\,(n_r + n_c)}{60} \qquad \text{Hz}$$

where N_r is the number of teeth in a reference gear, n_r is the rotational speed of that gear and n_c is the rotational speed of the cage. The sign in the bracket is positive for opposite direction of motion and negative for same direction of motion.

Table 1 lists some of the possible sources of noise. In the majority of cases, the dominant source of noise is vibration caused by transmission errors (geometric inaccuracies) introduced during manufacture. The most frequent manufacturing fault is caused by periodic error in tooth spacing, which will be aggravated if two gears, each with a similar flaw, mate together. Errors in the gear-hobbing machine, which has its own gear train, may be reflected in tooth spacing errors in the manufactured gears; this is particularly difficult to diagnose, since in many cases the characteristics of the machine tool are unknown. It is claimed that it is possible to identify the specific machine tool used to manufacture the gear from an examination of the noise spectrum. It is not possible to eliminate manufacturing errors entirely, so noise will always be present from this source. Indeed, each gear transmission system possesses its own particular vibration signature, which has proved to be be of military significance in identifying enemy submarines, for example. Ground gears are normally quieter, more on account of the accuracy of the tooth form than of the better surface finish. Gearboxes, gear teeth and shafts distort under load and the transmission error may change.

TABLE 1 – SOURCES OF GEAR NOISE

Cause	Form of Excitation	Remarks
Geometry – (i) Inaccuracies	Impact	Due to surface irregularities Due to unbalanced radial forces generated
(ii) eccentricity	Impact	Unbalanced radial forces
(iii) elasticity	Deformation	Elastic deformation of teeth under cyclic stressing
(iv) friction	Tooth contact	Independent of accuracy
Stress loading	Compressive stress waves	Magnitude dependent on gear geometry and tooth load
Oil pocketing	Hydrodynamic shock	Excess of oil trapped between pressure surfaces
Oil splash	Turbulence	More noticeable at higher speeds with splash lubrication
Air pocketing	Compression /expansion of trapped air	Only likely to be experienced with high speed gearing
Bearing noise	Resonance	Amplification of basic gear-generated noise
Casing	Resonance	Amplification of basic gear-generated noise
Rattle	Impact	Caused by combination of backlash and off-load condition

Gear Rattle

Gear pairs can, under some driving conditions, run unloaded, resulting in rattle between the teeth, shown in Figure 1. This is a problem which has become of greater importance recently in automobile transmissions, where idle or light load conditions are increasingly common, brought about by the higher torque and lighter weight of modern gearboxes. In the past a cure has been achieved by tuning the torsional stiffness of the clutch, but this has involved a trial-and-error approach, which can be time consuming. A computer simulation model of the complete transmission is one approach that has been tried. The analysis is complicated by the non-linearity of the equations. Gear rattle is observed as a vibration of the gearbox, which is then transmitted throughout the drive train and then into the body of the vehicle. The same effect can be produced when torsional vibration is present; this has the effect of off-loading the drive at the critical frequency of the system. In this work as in other gear investigations it is helpful to make measurements related to engine cycle, rather than to time. This makes the identifying of excitation modes easier to visualise.

Experimental Investigations

Figure 2 shows the kind of spectrum that might be expected, that is, a clustering around the integer multiples of the meshing frequency. It is not always possible to identify the source of gear noise solely by examining the frequency. Noise peaks occur elsewhere in the spectrum, as 'ghost' components. Harmonics of the meshing frequency are common as are side bands resulting from gear eccentricities and geometric errors introduced during the manufacturing process. There may also be resonances in the gearbox itself, excited by broad band vibrations. The noise spec-

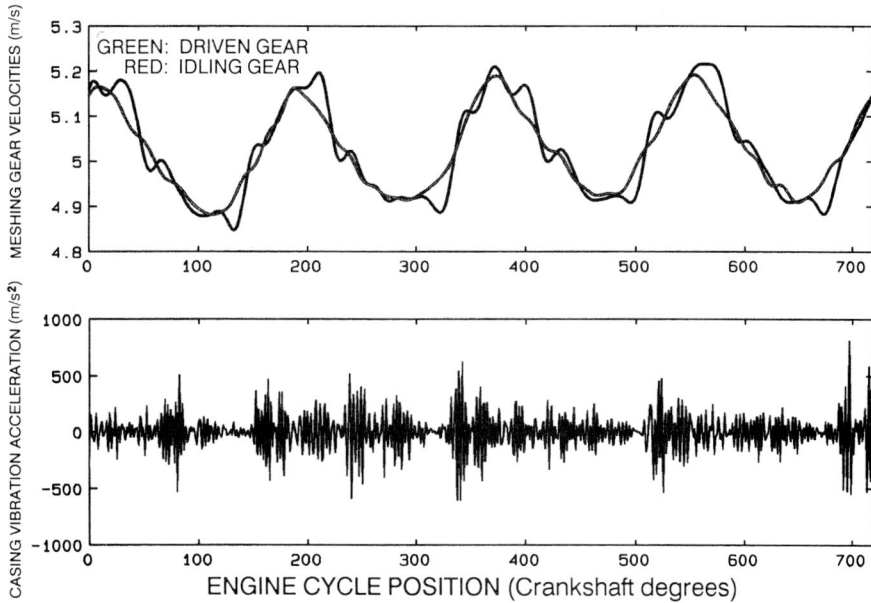

Figure 1 Impacting gear pair under rattling conditions

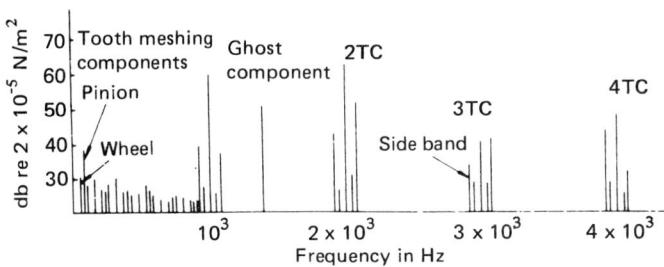

Figure 2 Schematic diagram of gear noise analysis

trum is shifted as the speed of rotation and the drive load are changed. The spatially averaged noise output is a contour plot, with noise, speed and load coordinates, Figure 3. There is likely to be considerable variation between different gearboxes off the same production line, so a number of them should be checked.

The vibration signal from a gearbox, whether it be acoustic noise or vibration, needs to be processed in order to extract the information that can be related to specific gear wheels. The signal when examined may appear to be random, and indeed there will be a degree of randomness, but it is essential to relate the measured level to the various teeth meshing frequencies. One technique that can be used for this is autocorrelation. The autocorrelation function, $\psi(\tau)$, is defined as

$$\psi(\tau) = \text{Lim } T \to \infty \frac{1}{T} \int_0^T f(t)\, f(t + \tau)\, dt$$

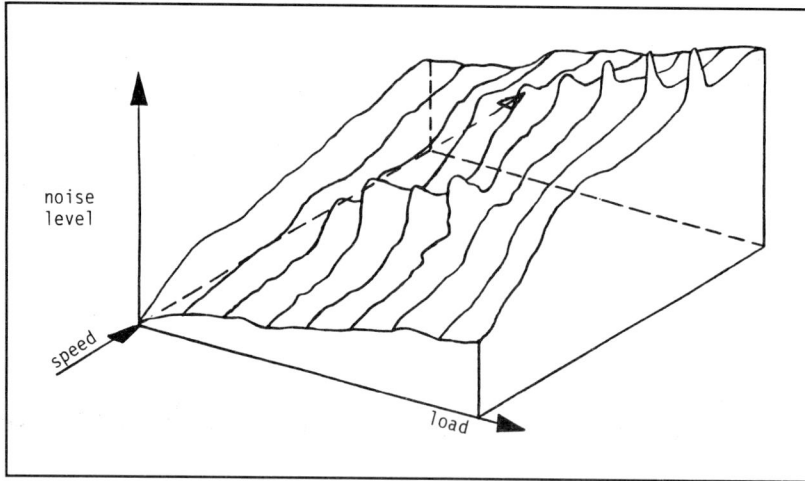

Figure 3 Contour map of noise against speed against load

where f(t) is the magnitude of the signal at a time t and f (t + τ) is the magnitude of the signal τ later. Using this function, it is possible to distinguish between random signals and a regular waveform. Autocorrelation suppresses the random signal and leaves a clean sine wave. If there are several sine waves present they will appear in the final spectrum. Autocorrelation is found from the FFT.

One of the first steps to take when examining gear noise is a geometric measurement of the gears, i.e. profile, helix and pitch, for which the standard measurement techniques are suitable. Dimensional errors, detected in this way, can sometimes point to vibration later, although the work involved in relating vibration to geometrical errors is often very involved and the results may be unreliable. Transmission error (TE) gives a more direct connection between geometry and vibration. Transmission error is defined as the difference between the position of the output shaft if the gearbox were perfect and its actual position. It is helpful to check the TE firstly in the static, no-load condition, which disregards any effect caused by vibration or tooth deformation, and secondly in the loaded but low speed condition, which is able to bring out errors caused by poor tooth loading.

Single Flank Testing of Gears

There are two methods of determining gear quality, double flank and single flank testing, Figure 4. In double flank testing, the gears are placed in a fixture with both flanks in contact so there is no backlash, the error being determined by measuring the change in centre distance. This has the drawback that data from the two flanks are averaged, so there is no way that the error can be attributed to a single flank. Single flank testing is a superior technique, in which a pair of meshing gears with backlash (a master and a work gear) are run together at the correct centre distance. The angular motion of the pair is measured by use of optical gratings. Photodiodes on the input and output shafts, with suitable circuitry, give a direct reading of transmission errors, Figure 5. Single flank testing can be performed on spur, bevel, helical and worm gears. It is not suitable for lead and spiral angle errors, but these are unlikely to contribute to the vibration.

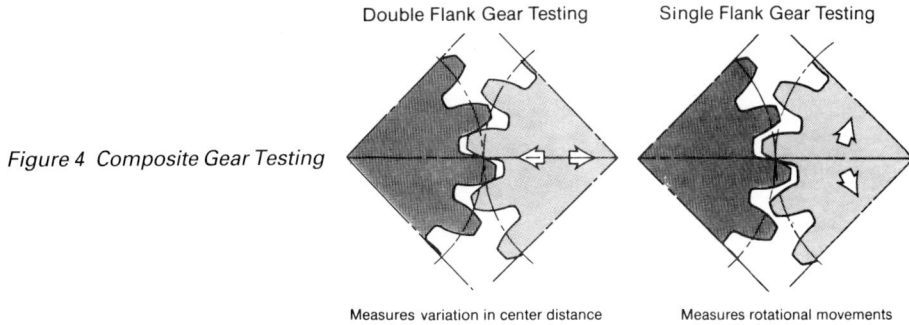

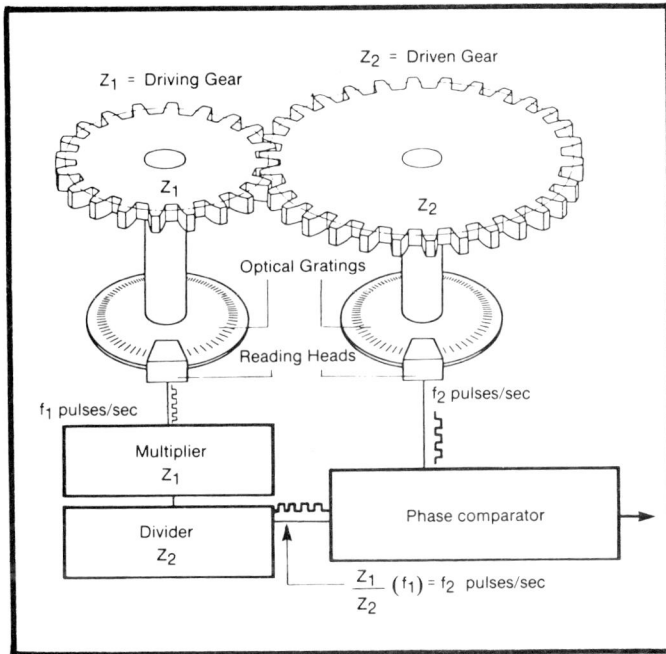

Double Flank Gear Testing   Single Flank Gear Testing

Figure 4 Composite Gear Testing

Measures variation in center distance   Measures rotational movements

Z_2 = Driven Gear

Z_1 = Driving Gear

Z_1

Z_2

Optical Gratings

Reading Heads

f_1 pulses/sec

f_2 pulses/sec

Multiplier
Z_1

Phase comparator

Divider
Z_2

$$\frac{Z_1}{Z_2}\,(f_1) = f_2 \text{ pulses/sec}$$

Figure 5

Using this method, each flank can be measured individually. The input shaft is driven at a constant speed and minimum loading, so as to avoid the risk of dynamic effects distorting the results. The Gleason equipment as shown in Figure 6 can run at less than 10 rev/min up to 300 rev/min and is accurate to 1 second of arc.

The simplest analogue output is the strip chart shown in Figure 7, indicating some of the possible errors. The TE is characteristically a once-per-tooth error superimposed on a once-per-revolution error. Vibration is usually directly related to the former pattern. Analogue data can be further analysed by the use of plug-in, high-pass and low-pass filters, leaving the once-per tooth error signal. As stressed elsewhere in the text, modern digital analysis techniques using FFT on

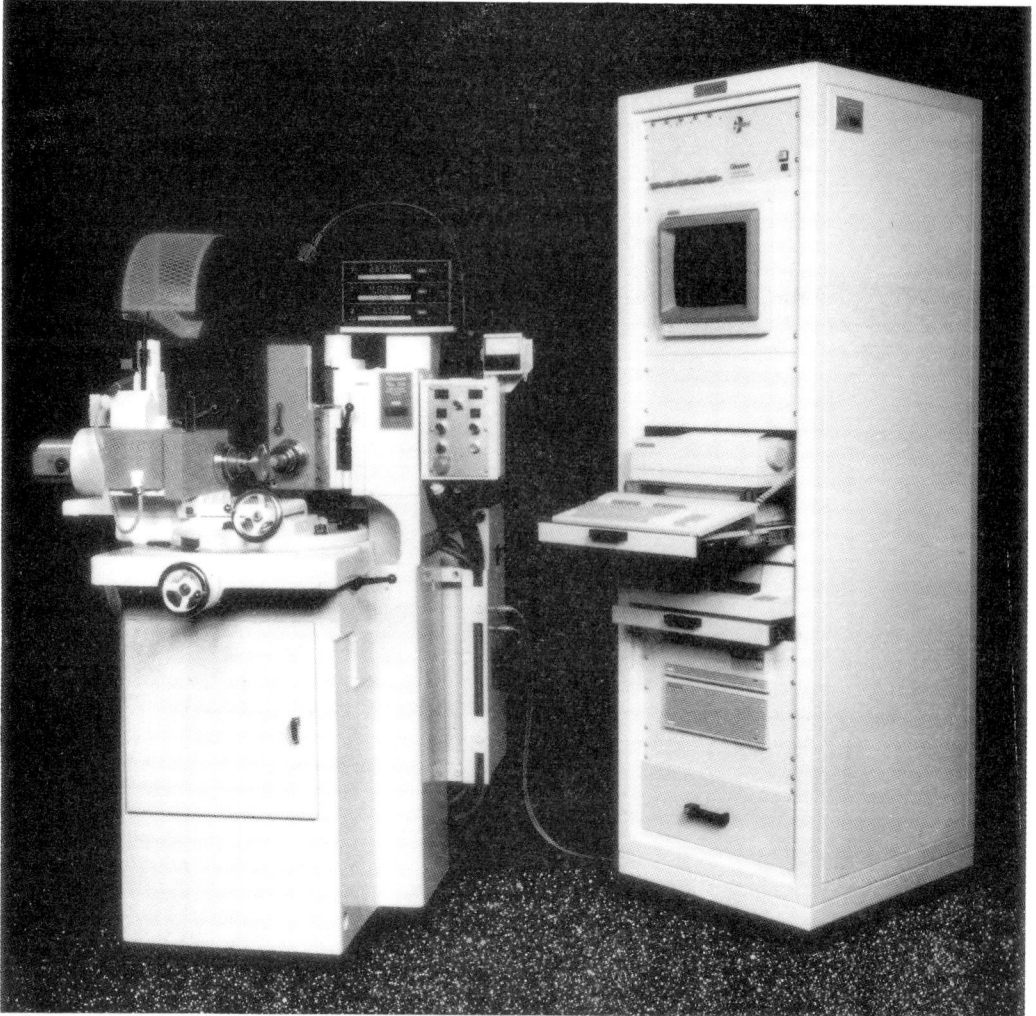

Figure 6 Single Flank Angular Gear Tester with Automatic Data Analysis System

the analogue signal will be more satisfactory. The spectral analysis, as in Figure 8, can be related to the gear geometry and the measured noise. The spectrum of Figure 9, when plotted against mesh harmonics, makes it easier to identify the tooth errors, particularly when the spectrum contains ghost and side-band frequencies. Ghost harmonics are generally caused by flats or facets on the normal tooth form and by such things as cutter run-out or non-uniform motion of an element in the gear train of the machine that generated the tooth form.

Equipment for the Measurement of Gear Noise

A twin channel FFT can be used for analysis of the vibration signal from the complete gearbox, but its ability in this area is limited in that the analysis is performed in real time. Real time analysis is valuable for structural work, but for rotating machinery it turns out not to be ideal. A better time

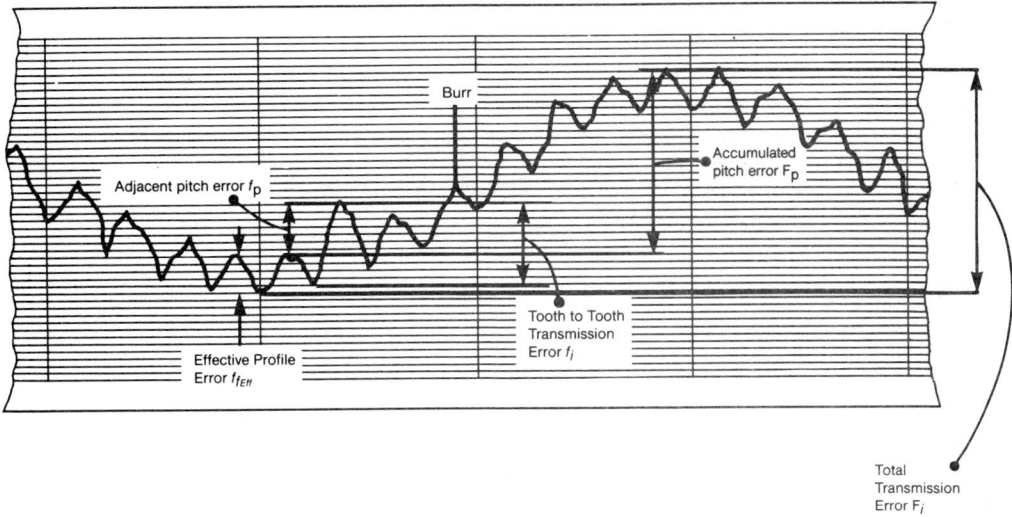

Figure 7 Individual errors revealed by single flank testing

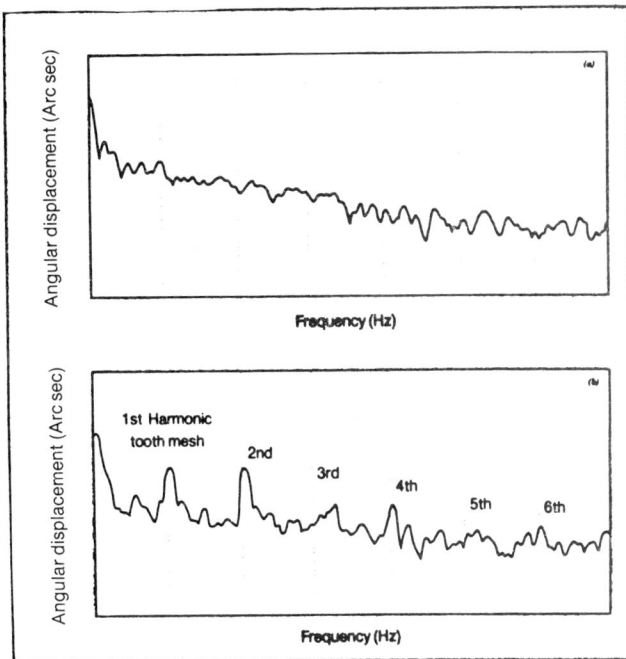

Figure 8 Spectral analysis represents single-flank test of precision ground helical gears. (a) Very conjugate gears have barely discernible peaks. (b) High second harmonic resulting from involute shape deviations ground into teeth. Deviation in (b) cannot easily be identified through analogue data or involute charts used in production control.

Pinion Runout — Bolt Hole

1st Harmonic of Mesh

2nd (Ghost)

2.6 (Ghost)

Figure 9 Harmonics of Mesh

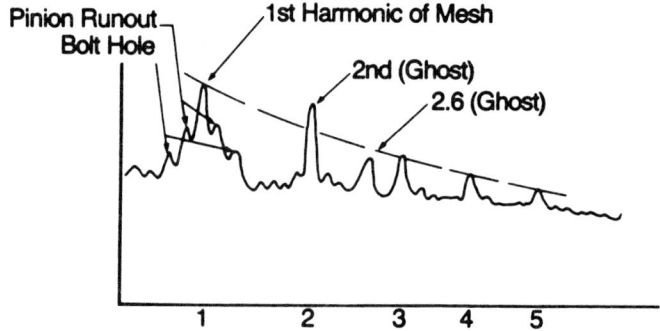

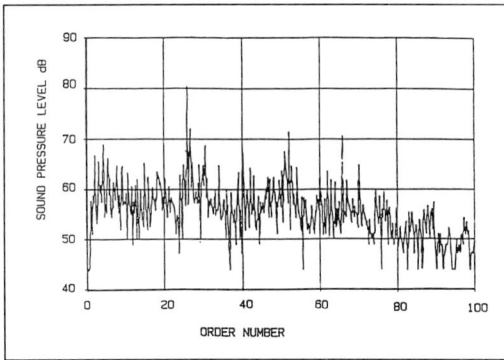

1   2   3   4   5

Figure 10 Order spectrum: note the strong tones at orders 26 and 66 (order 52 is a harmonic of order 26)

base is available by using the pulse train off the input shaft. This is referred to as an order analysis system. It is similar to conventional frequency analysis, but instead of using a real time clock for data sampling, a pulse train derived from the input shaft is used as the sampling clock. Assume, for example, a simple gearbox with a 26 tooth gear on the input shaft. With reference to the rotation of the input shaft, the tooth meshing noise will always be of order 26 (26 times per revolution). If a second pair of gears is in the train, the order can be derived with reference to the input shaft. If the ratio of the first pair was 1:1.5 and there were 36 teeth on the input gear of the second pair, the order of the second pair would be $1.5 \times 36 = 54$.

A pulse train off the input shaft is used to perform an FFT on the measurement data to give order spectra, as in Figure 10. The input data can be either acoustic noise from a microphone or accelerometer readings. The pulse train can be generated by shaft encoders or photo-gratings or by the use of a once per revolution signal combined with an electronic multiplier. The resultant spectrum makes it much easier to identify which gear pair is causing trouble. It has the further advantage that it is not necessary to run the system at a speed corresponding to the peak response; an offpeak measurement will still exhibit the typical spectrum of Figure 10. This particular system is known as PLATO (Phase-Locked Acquisition and Transfer to Orders).

Tooth Profiles

The vast majority of gears in use today employ the involute profile with a 20° pressure angle, for good engineering reasons. It happens also that the involute has certain advantageous properties when considering noise and vibration. In the absence of friction, the component of force in the transverse plane is constant in position and direction, and furthermore errors in the distance between the shafts of the meshing gears do not affect the constant velocity ratio. The presence of friction, which is usually low in a well designed pair, has a minor effect in that it causes a change in direction of the force at the pitch point. It has been demonstrated that a gear inaccuracy of 1 micron gives a greater change in force than that produced by friction. Friction reversal can be eliminated by using corrected gears, and by mating an all addendum gear with an all dedendum gear. The disadvantage is that the sliding velocity is increased and with it the risk of scuffing.

Other profiles have been suggested such as a cycloid or the more modern, conformal, Novikov-Wildharber gear, Figure 11. Whatever advantages they may possess (and it is claimed that the latter gears can transmit more power for the same size), they have not proved to be useful for noise reduction. The force between the gears varies either in position or direction, which gives rise to vibratory forces. These new gears are more sensitive to errors in centre distances and manufacturing variations than the traditional involute.

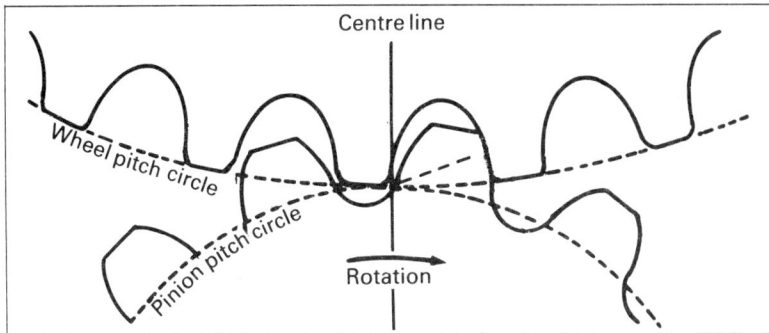

Figure 11 Novikov Wildhaber male and female gear meshing

Spur or Helical Gears

When spur gears rotate in mesh, the number of mating teeth in contact alternates between one and two, except for high contact ratio gears with long thin teeth, which may have three teeth in contact. The contact ratio, q, of a meshing pair is defined as the time average of the number of tooth pairs in contact. In standard gears this varies between 1.0 to 2.0. The higher the value of q, the lower the vibration. This is because, when more than one tooth is in contact, any deformations resulting from manufacturing errors are averaged out. It has been demonstrated that the attenuating effect of multiple contact is a function of q.

It will be apparent that normal spur gears have a maximum value of q equal to 2.0. Higher values are possible with slender tooth shapes, which brings in the probability of increased deformation and the consequent vibration amplitudes. Helical gears can be designed with q values of 10 or more.

A semi-empirical theory has suggested that the variation of q leads to the attenuation values of Table 2, which indicates the tremendous potential for the use of helical gears in noise suppression. The table indicates a 6 dB reduction in excitation for every doubling of the contact ratio.

TABLE 2 – VIBRATION ATTENUATION THROUGH USE OF HELICAL GEARS

Contact ratio q	1	2	5	10
Amplitude reduction (dB)	13	19	27	33

Note that when calculating the contact ratio for double helical gears, the appropriate value is that for a single helix.

Methods of Noise Reduction in Gear Trains

Some of the factors which affect the magnitude of the gear noise are summarised in Table 3, with indications as to the possible treatment. Deciding on the appropriate silencing treatment out of the options available is a matter of balancing the economic factors, and that is why it is important to find where the faults lie. Mass produced gearboxes can exhibit variations in TE due to manufacturing errors, and it may be considered uneconomical to improve the quality of manufacture or to have tighter quality control. It should be possible, by employing superlative manufacturing techniques, to reduce the TE to any desired value, but this would prove to be totally uneconomic.

The question arises as to what value of TE one might aim for. This is largely a matter of experience: ground spur gears are likely to have a per tooth error of about 10 micron, while helical gears with a large face width could be of the order of 5 micron. Well machined spur gears are likely to have errors of about 20 micron. Although helical gears are superior from this point of view, a warning should be given about the use of high helix angles. Any tooth errors apparent in the driving direction will also be present in the axial direction, and create forces which are directly transmitted through the thrust bearings to the casing. Double helical gears can balance the direct thrust forces but may not completely balance the varying forces.

Gear noise is a system problem, rather than just a gear problem. Vibrations caused by small defects in the tooth form are amplified by the resonances in the housing, producing the noise; there may be little audible direct noise from the gears. Some structures are such good amplifiers of noise that it is practically impossible to make gears good enough to run quietly in them. In attempting to reduce gearbox noise, the relative cost of improving the gears or the housing has to be assessed and the most economical method adopted.

Gears Manufactured of Materials Other than Steel

For maximum load carrying capacity, hardened steel is the preferred material, but alternatives may be used either for reasons of economy or to assist in the reduction of vibration. Non-ferrous alloys and cast iron provide higher damping and may offer a marginal assistance to vibration reduction, but the effect is small and is unlikely to be significant.

TABLE 3 – GEAR NOISE PARAMETERS

Parameter	Requirements or Treatment	Remarks

Gear type	Suitable mechanical performance, but see 'Remarks' for inherent noise level	Inherent level:
		  Spur – moderate to high
		  Helical – low
		  Herringbone – low
		  Hypoid – low to moderate
		  Bevel – moderate to high
		  Worm – low

Gear geometry parameter	(i) large number of small teeth preferred (ii) adequate gear stiffness (iii) accurate manufacture	Slim gears may be subject to disc vibration. Manufacture very important
Gear material	(i) optimum material selection (but may have to be based on mechanical requirements) see also Damping	Non-metallic gears may provide quiet operation in gear trains
Bearing	Good bearing design	Must give a rigid system
Lubrication	EP lubricants may be desirable, or necessary	Excess lubricant can lead to noisy operation
Damping	(i) gear material with high internal damping preferred, if practical (ii) damping rings or damping plates on gears (iii) isolation of gear rim	Resilient materials (*eg* nylon) provide excellent damping Most effective when applied to rim of gear Not necessarily effective and has practical limitations
Isolation	(i) resilient mounting can be used to reduce structural-borne sound (ii) enclosure can be used to reduce airborne sound	Treatment applicable in specialised or very bad cases of gear noise
Housing	(i) rigid housing essential (ii) good hydrodynamic internal shape (iii) large mass prefered	Housing material should have high damping if possible To provide efficient circulation and retention of lubricant Natural frequency should be well above rotational frequency range

Non-metallic gears are better able to provide a cure for a noisy transmission; either one gear wheel or the whole train may be replaced. The materials in use are nylon, polyacetal plastics or laminated and reinforced resins. They are weaker than steel gears and have a lower surface hardness. Their main advantage lies in the low modulus of elasticity and higher natural damping. They readily deform under load, greatly reducing the contact stresses, although the root stresses are unaltered. Providing that the gears are strong enough to resist the bending load imposed by the maximum torque, they can offer a solution to a noisy drive. Simply replacing a metallic gear with a plastic one of the same dimensions will also alter the system dynamics, moving the resonant peaks to another part of the spectrum.

Some precautions have to be taken when using these gears. Most plastic materials absorb moisture and increase in size by about 1%, so the designed backlash has to be sufficient to allow for this. Temperature is critical, so precautions need to be taken to remove the heat generated. Lubrication may not be necessary but may improve the load carrying capacity and assist in the cooling.

Summary of Techniques for Noise Suppression

After a noisy gear drive has been studied and the problem areas defined as in Table 3, some solutions need to be sought. In order of importance these are:

> Casing improvements
> Better manufacturing of the gears
> Design changes to the gears
> Internal changes
> Damping
> Isolation

If the noise is radiated from one face of the gearbox, which can be identified by sound intensity measurements or modal analysis, a change to the structure by adding stiffening ribs, Figure 12, can prevent the vibration reaching the environment in the form of acoustic noise.

Improved manufacturing may well be an expensive solution, if it entails ground gears, but is fundamentally better because it treats the noise at source.

If the noise is associated with a particular resonant peak, it may be possible to change just one gear by increasing or decreasing its weight or stiffening the support shafting. Moving to helical gears may also be considered, but this is likely to prove an expensive solution if vibration is the only problem.

Changing the ratios of some of the pairs while maintaining the same overall ratio may also be effective in altering the excitation frequency, but this solution may introduce a completely different problem, and should only be done if the dynamics of the system is fully understood, so that the proposed changes can be analysed at the design stage.

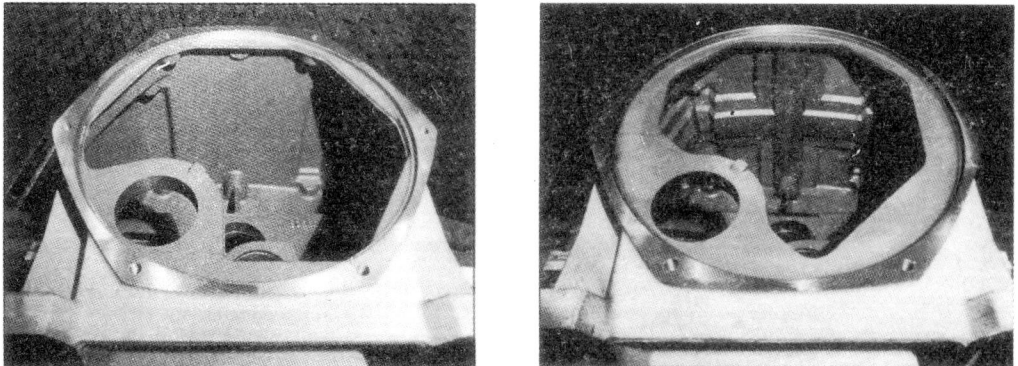

Figure 12 Left: Interior view of gearmotor housing before modification
Right: Interior view of gearmotor housing after modification

Damping may be introduced either into the gears themselves or into the casing. As mentioned above, a change to the material of the gear can be effective but only by the introduction of non-metallic gears. Some experimental work has been done on gears coated with a simple elastomeric layer or a sandwich of steel and elastomer. Early results are promising but it is too early to predict success for this technique.

Damping in the bearings can assist in the suppression of torsional vibrations. Rolling bearings have low friction and damping and do not help. Hydrodynamic bearings on the other hand give much higher damping at a once-per-revolution frequency.

The application of damping material to the outside of the casing might be thought to be a solution, but results do not support the idea. The application of a damping layer to a thin sheet such as a body panel can be effective, because the magnitude of the movement is significant, but this is not the case for a comparatively rigid body such as a gearbox; a layer which would be thick enough to suppress vibrations would be too thick to be practical. Splitting the casing and placing a gasket between the two halves is more likely to work.

Mounting the gearbox on a resilient support will be effective only if the noise source is radiation from the structure to which it is attached. See the chapter on *Resilient Mounting*. It can be difficult completely to isolate a gear drive. If the vibration is transmitted by the drive shaft, the whole train has to be isolated. If there are lubrication pipes or other rigid links between the gearbox and the structure, these can act as effective transmitters of vibration and should be well secured.

Chapter 3
Bearings

The main cause of friction noise generated by sliding or rubbing surfaces is lack of lubrication. It is essentially a form of high frequency vibration resulting from rapid intermittent contacting of the surfaces. Where this is transferred to a resonant structure this is heard as a 'screech'. In the case of plane sliding surfaces the frequency of the noise is related to the surface finish of the surfaces, and its intensity to both surface finish and surface loading.

In addition to plane sliding surfaces, all mechanical bearings are subject to friction. Vibration can also be a source of bearing noise – either separately or by virtue of increased friction.

Some types of bearing will generate noise if excited by vibration. Noise levels in such cases will increase with wear, accelerated by vibration. Machine balance is therefore of considerable importance, both for reducing wear and controlling noise levels which may be inherent.

Plain Bearings

Parameters affecting bearing noise are summarised in Table 1. Lubrication plays a significant part in controlling friction and thus noise. Under complete hydrodynamic lubrication conditions friction is low and the noise generated negligible. In the case of high speed bearings, however, there is the possibility of oil film whirl developing which can lead to vibration and an increase in noise level, particularly if any resonant effects are present. The speed of whirl, and thus the frequency of vibration produced, is generally low and never more than half the rotational speed of the shaft. Such noise is thus seldom objectionable unless amplified by resonance. It can, in any case, be eliminated if necessary by employing a discontinuous bearing surface – e.g. a segmental or pad type of bearing – or even a modification of the clearance in the original bearing. Other 'direct' cures may be produced by a change in lubricant viscosity or a change in bearing pressure (although this will usually necessitate replacement of the bearing by one of a different surface area).

In the absence of complete hydrodynamic lubrication, intermittent local seizure may occur through metal-to-metal contact.

The general case is where smooth rotation is inhibited by a series of relatively small drag forces due to localised friction or localised welding. Whilst this may be damaging to the bearing surfaces there may be no apparent effect on the running of the shaft in the majority of cases, although on occasions marked vibration may occur. This is because any form of stick-slip motion tends to excite the moving (in this case the rotating) system at its natural frequency. The extent to which this is noticeable as vibration and increased noise is dependent on the elasticity of the mass involved. Thus an increase in noise and vibration, as distinct from 'chatter', from a plain bearing is a likely indication that stick-slip motion is present, i.e. lubrication is inadequate. The fact that the bearing is tight will eliminate 'chatter' due to excessive clearance (e.g. in a badly worn bearing). Thus a distinction can be drawn between these two types of vibration.

TABLE 1 – PLAIN BEARING NOISE CONTROL

Parameter	Frictional Excitation	Oil Film Whirl	Use of Parameter for Quiet Design
Shaft speed	Likelihood increases with increasing speed.† Also increases at low speeds	Frequency less than half shaft speed. Generally not significant at high speeds.	Not usually an adjustable parameter. See Table 2
Bearing pressure	Possibility increases with increasing pressure.	Tends to increase with bearing pressure.	Reduce–*ie* use larger bearing area and lower bearing load.
Clearance	Small clearances increase possibility.†	Small clearances reduce oil film whirl.	1 micron per mm shaft diameter recommended to eliminate oil film whirl.
Bearing diameter	Increasing diameter increases rubbing speed for same shaft speed.†		Larger diameter decreases bearing unit load.
Lubricant viscosity	Generated if viscosity too low.	Varies with viscosity.	Change lubricant if necessary to maintain full hydrodynamic lubrication.
Lubricant film strength	Positive if film strength inadequate.		EP lubricants may be required with high bearing loads.
Lack of lubricant	Minimised with non-metallic bearings.		Select low friction materials.

† in absence of full hydrodynamic lubrication.

Bearing 'squeal' is the same form of vibration generated by friction, although a distinction can be drawn between seizure and localised overload conditions. In the case of metal shafts running in metal bearings, stick-slip motion will tend to generate high frequency vibrations in the skin of the shaft, the frequency increasing with the degree of seizure. Thus a bearing which is tending to seize will commonly generate a high frequency 'squeal'. On the other hand, in the case of a non-metallic bearing, squeal may be generated by localised overloading of the bearing without necessarily being an indication of potential seizure, particularly in the case of elastic materials. The bearing may be perfectly capable of adjusting itself to such conditions because of the relatively large bearing area available (i.e. because of the lower design unit load with such materials).

In the absence of non-metallic bearings, a number of materials are suitable for running dry. The addition of lubricant may have little or no effect on friction developed, although it can be effective as a coolant. This can be important in controlling the expansion of the bearing and the running clearances involved.

The load-carrying capacity of plastic bearings is relatively low, calling for large bearing surface areas; the material itself is usually resilient. These two factors tend to provide inherent damping of vibrations. Such bearings, however, are not necessarily completely free from 'squeal', particularly if excessively tight or subject to a degree of misalignment relative to the shaft. In the former case, the coefficient of expansion of plastic materials is always considerably higher than that of metals, so that tightness can readily arise from differential expansion. It is usual to allow far more generous clearances to take account of such possibilities. Also the dimensional stability of many plastics is relatively poor, so that changes in clearance can occur with differences in humidity.

Lubrication can help where excessive tightness has developed, although this is usually a frictional loss rather than a noise problem (for tightness in a plastic bearing is not necessarily accompanied by noise). A 'solid' type lubricant such as molybdenum disulphide is usually the most effective in such cases. For general lubrication, water is usually quite effective, the requirement being more for a coolant than an actual lubricant.

Certain non-metallic bearings must be operated only in a fully hydrodynamically lubricated condition. The segmental rubber bearing is a typical example. It is normally lubricated by having water continuously flushed through cooling channels in the bearing surface itself. If run dry the bearing will rapidly become very noisy (and also be irreparably damaged in a very short time). Rubber bearings are usually of fairly long length relative to their diameter and are prone to develop 'squeal' if badly misaligned. They will normally accept a small amount of misalignment without becoming locally overloaded and have excellent damping characteristics against shaft vibration. They are, therefore, basically quiet bearings.

In general, noise or vibration should never be a problem with any suitably designed plain bearing provided the bearing operates under conditions of full hydrodynamic lubrication. Plastic bearings can operate dry, but also act in the manner of isolating elements as regards vibration and noise.

Hydrodynamic Bearings

In high speed rotating plant, such as turbines and compressors, hydrodynamic bearings are the preferred choice, but can lead to a particular type of vibration, dependent upon the shaft speed; only journal bearings are likely to experience troubles. Vibration in these bearings can be caused by an out of balance condition (the amplitude limited by the clearance) or a misalignment, the cure for which is purely mechanical. Oil whirl vibration is caused by a complex interaction between the type of lubrication, oil film thickness and speed. The vibration characteristics can often be modified by a change in the oil supply as shown in Table 2. This can be achieved by a change in temperature, pressure or viscosity. Relieving the bearing with a stabilising groove, three lands or by using an elliptical bore can help in certain types of whirl.

Rolling Bearings

Although having much lower friction, rolling bearings tend to be noisier than plain bearings. Full hydrodynamic lubrication is seldom present because of the higher unit loadings and rolling and skidding metal-to-metal contact is usually present to some extent. The amount of vibration generated is largely dependent on the geometry of the bearing and geometric inaccuracies. Thus, for example, the running surfaces of a rolling bearing always have a certain roughness, which causes a succession of small impacts, the magnitude of which depends on the surface condition and the peripheral speed of the bearing. Ball bearings are more critical than roller bearings in this respect.

In a good bearing the level of vibration due to normal surface roughness will be very low, but will be aggravated by improper installation, overloading, lack of lubrication, contamination of the lubricant, cavitation, progressive wear or actual damage. Initially, however, the vibration level, and thus bearing noise, will be mainly dependent on race geometry, surface finish, waviness, eccentricity and parallelism, ball geometry and groove wobble.

The amplitudes of all excited frequencies will, in general, tend to be very small, unless serious irregularities are present or there is resonance. Resonance is the primary cause of noisy rolling

TABLE 2 – TYPES OF BEARING VIBRATION

Source of Vibration	Character and Frequency Relative to Shaft Speed n	Conditions of Occurrence	Suggested Remedy	Remarks
Out of balance (permanent)	Steady, n	Rotor out of balance, or journals misaligned (with 3 or more bearings)	Rebalance and check journal alignment	Some vibration usually present
Out of balance (thermal wander)	Varying amplitude, n	Thermal distortion of rotor	Improve starting and operating technique	Mainly on rotors with high temperature inlet
Bearing (light load instability)	Irregular, less than n	Light bearing load, e.g. turbine at 50 rev/s with bearing loading less than 0.4 MN/m^2	(a) Vary oil supply condition (b) Stabilised bearing	Mainly on small turbines
Bearing (half-speed whirl)	Whirl at or close to $n/2$	Within narrow speed range close to twice critical speed	Change critical speed of rotor	
Bearing (low-frequency whirl)	Whirl at lowest critical speed, below $n/2$	Over wide speed range	(a) Vary oil-supply conditions (b) Elliptical bearing or three-land bearing	Greatest risk when critical speed is below 0.4 n
Steam force	Whirl at lowest critical speed, below 0.65 n	Instability above certain load on turbine	(a) oil-supply condition (b) Elliptical bearing or three-land bearing	Mainly on H.P. turbine of set of high rating
Synchronous whirl	Very slow build-up of amplitude, n	May occur during starting or on change of load condition	(a) Vary oil-supply condition (b) Shorten bearing, or (if elliptical) increase vertical clearance	Intermittent on certain sets. Sometimes mistaken for thermal wander

bearing operation, and can be reduced or eliminated by employing a housing material with good damping characteristics. It can also be advantageous if the outer ring material has good damping properties.

The fundamental vibration frequency which will appear when any unbalance or eccentricity is present is:

$$f = N/60 \quad Hz$$

where N = vibrational speed in rev/min

If the outer ring is stationary and the inner ring is rotary, then a second fundamental frequency will be developed due to the rotation of the train of rolling elements:

$$f_i = \frac{f}{2} \left(1 - \frac{d}{E} \cos \beta \right) \quad Hz$$

where   d = diameter of rolling elements
        E = pitch diameter   } in same units
        β = contact angle, degrees

The suffix i denotes the inner ring rotation
If the inner ring is stationary and the outer ring is rotating:

$$f_o = \frac{f}{2} \left(1 + \frac{d}{E} \cos \beta \right) \quad Hz$$

The suffix i denotes the inner ring rotating.

Excessive noise at either of these frequencies would indicate an irregularity of a rolling element or a cage.

$$f_s = \frac{E}{2d} f \left\{ 1 - \left(\frac{d}{E}\right)^2 \cos^2 \beta \right\} \quad Hz$$

There would also be a spin frequency (f_s) for the rolling elements.

An irregularity (indentation or a rough spot) on a rolling element would generate a frequency of $2f_s$ because the spot contacts the inner and outer rings alternately, once per revolution.

If there is any irregularity on the stationary rollway, the resulting fundamental vibration frequencies generated are:

irregularity on rotating way : $f = N (f - f_r)$ Hz
irregularity on stationary way : $f = N f_r$ Hz
where $f_r = f_i$ or f_o as appropriate

All fundamental frequencies generated by irregularities can give rise to harmonics. If there are several such irregularities then the harmonic frequencies will be more pronounced.

It is not easy to identify a vibration caused by a defect in a rolling bearing, because such defects occur in a random manner. They may occur on one ball or several, and ring distortion may be symmetrical or unsymmetrical. A vibration spectrum resulting from a bearing fault may easily be confused with some other cause, see Figure 1 for example. A vibration which occurs at or near a fundamental frequency may point to a bearing fault, but other causes should not be ignored.

Some precautions can be taken at the design stage. A high precision bearing can be chosen, Class 4 or better (BS 6107) for critical machines, and the dynamic load rating, according to BS 5512, should be generously calculated. For vibration limits for rolling bearings refer to Table 3 of the chapter on *Machinery Health Monitoring*. Probably the majority of cases of vibration originating in the bearings are not due to faults in the manufacture of the bearings themselves, but rather to inaccurate machining of the shafts and housings or to assembly errors. A vibration signal brought about by an oval outer ring, for example, is indistinguishable from that caused by an oval bearing housing.

Elastomeric Bearings

The elastomeric bearing is a structure comprising alternating layers of rubber and metal laminates, with a configuration designed to accommodate various modes of load and motion. The

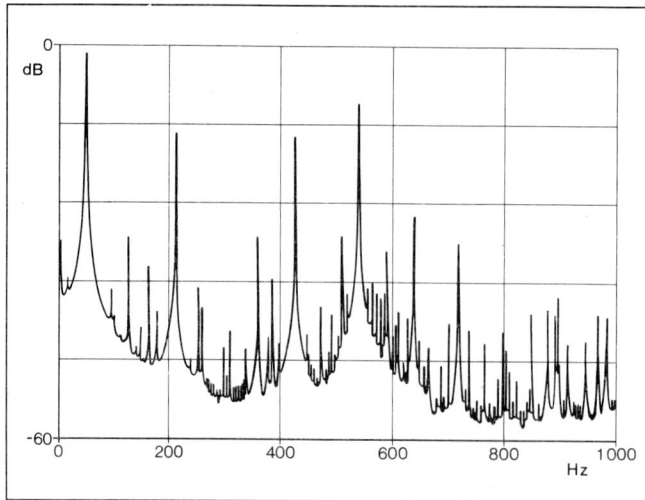

*Figure 1 Frequency spectrum for noise caused by deformed
inner rings*

elastomer is vulcanized and bonded to the metal laminates as well as to the attachment metal components (inner and outer 'races'). The attachment metal component configuration is optional and designed to attach the 'bearing' to the structure. The elastomer may be bonded to virtually any material including steel, aluminium, stainless steel, titanium, and many non-ferrous materials. Vulcanization of the elastomer to the metal laminates and to the attachment metal components eliminates potential brinelling problems, assures structural stability due to the integral construction, and provides a permanent, indestructible seal to keep out contaminants such as dirt.

The particular advantage of elastomeric bearings is that they can be made insensitive to vibration, shock or impact loading and can provide a noise barrier by proper selection of spring rates. On the mechanical side, they are self-lubricating, cannot seize and are free from pitting, galling or brinelling as well as offering very low friction and long life.

Configurations

Basic configurations for elastomeric bearings are:

(i)   *Sandwich*: The sandwich type bearing as illustrated in Figure 2 is designed to react to high axial loads while accommodating torsional or lateral motions through compression and shear of the elastomer respectively. This type of bearing is very stiff axial and soft in shear mode.

(ii) *Spherical sandwich*: The spherical bearing shown in Figure 3 is designed to permit motions around three axes through shear of the elastomer while carrying high axial loads in compression. The stiffness of the bearing is high in compression and low in the shear mode. This type of bearing can replace three or more separate conventional bearings.

(iii)   *Cylindrical*: The cylindrical type bearing reacts to high loads radially through elastomer compression while accommodating torsional or axial motions in shear. High stiffness in the radial compression direction while maintaining low stiffness in the axial and torsional modes is characteristic of this bearing, shown

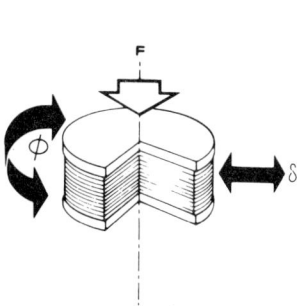

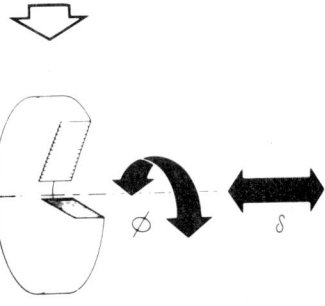

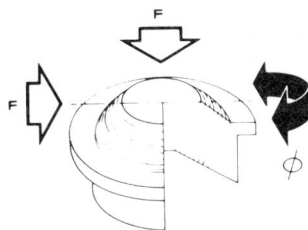

Figure 2 *Typical axial bearing*   Figure 3 *Spherical sandwich bearing*   Figure 4 *Typical radial bearing*

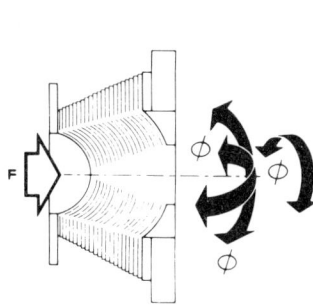

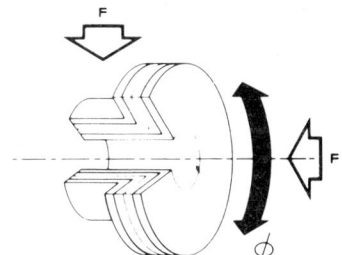

in Figure 4.

(iv)   *Spherical*: Figure 5 illustrates the spherical (ball-joint) type bearing which accommodates torsional or cocking motions about three axes, in addition to some axial motion, while reacting to high radial loads. The high compressive stiffness and low shear are effectively utilized to accommodate the radial loads and torsional, cocking and axial motions respectively.

(v)   *Conical*: High axial and radial loading capability in compression-shear loading of the elastomer and a low torsional shear stiffness are characteristic of the conical type bearing. Some cocking load or motion is possible with this configuration, as illustrated in Figure 6

The types illustrated provide the basic foundation for elastomeric bearings. Various combinations of the basic types can be integrated for virtually unlimited usage. For example, the sandwich and cylindrical types can be combined in order to react to axial and radial loads while permitting

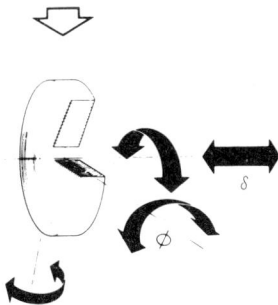

Figure 5 *Spherical tubular bearing*   Figure 6 *Conical bearing*   Figure 7 *Radial/axial bearing*

torsional motion as illustrated in Figure 7.

Bearing Mounts

The chief source of noise from metallic bearings is resonant vibration. The noise level developed by the secondary generator may be further amplified by virtue of being directly coupled to a resonant panel or vibrating system, or the unit containing the bearing may simply act as a rigid coupling to another resonator. Logically, therefore, damping should be provided at the source as far as possible, i.e. in the bearing support. This calls for a satisfactory design of housing or mounting, particularly where there is a degree of flexibility in how the bearing may be mounted, and in consideration of material specification.

Where the mounting is integral with a casing or end cover, sufficient rigidity is usually ensured by a generous thickness of metal around the bearing. Consideration can also be given to the selection of a material with good damping qualities, although cost and strength may be equally important. Cast iron is a low cost material with good damping qualities.

In a design for quiet operation where a material with high internal damping cannot be utilized for the bearing support because of other considerations, the same effect can be realized by restricting the use of such a material to damping rings or damping plates in contact with the bearing or, if necessary, even going to the extreme of isolating the main body of the machine or unit with a resilient material. Some bearings, in fact, are produced on this principle – an example being the elastomeric coating of a plain journal sleeve or the outer face of a rolling bearing. When fitted, this elastomeric layer provides isolation of the metal bearing elements from the rest of the unit. Thermo-plastic bearings provide similar isolation but are necessarily limited to lighter duty applications. However, the use of non-metallic rolling elements in rolling bearings cannot be overlooked as a method of providing isolation and resilience in the bearing itself.

Flexural rigidity of the bearing housing is less important in the case of rolling bearings than with plain bearings. A common design fault in the latter case is failure to provide adequate support for the length of bearing concerned (particularly in the case of bearings intended to support long shafts). As a consequence the bearing has a certain amount of freedom to flex under the shaft, when it will readily develop a barrel shaped longitudinal section allowing the amplitude of the shaft vibration to increase. This can only be avoided by stiffening the bearing support. If the initial vibration is high, bearing life and performance may both be improved by resilient mounting, but this can again destroy the original requirement of flexural rigidity if it permits the bearing to rock, as it will again wear to a barrel shape. A better solution may be a resilient type bearing surface such as a segmental rubber bearing, provided it can be lubricated satisfactorily, but much depends on the particular application involved.

Wineglassing of plain bearings can result from excessive clearances, especially where the shaft may be subject to a radial load (e.g. in a single cylinder i/c engine crankshaft). The freedom to rock provided by excessive clearance can rapidly lead to bellmouthing of the ends of the bearings with an accompanying increase in vibration and noise, although in a practical application

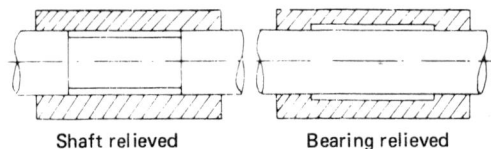

Shaft relieved   Bearing relieved

Figure 8

this will be masked by the noise of the prime mover. This is likely to be further aggravated by typical production geometry, such as light initial 'waisting' of the bearing as finished, and slight initial 'barrelling' of the shaft. A better design for a long plain bearing subject to radial loading is shown in Figure 8, where the shaft diameter is deliberately relieved, the effective bearing areas then being restricted to two relatively short lengths at each end of the bearing. Such a bearing is less likely to develop vibration noise due to wear.

See also chapter *Machinery Health Monitoring*.

Chapter 4
Gas Turbines

Sources of Turbine Noise

The three main sources of gas turbine noise are the inlet, exhaust and the turbine casing (with which are associated various ancillaries such as gears, pumps, coolers and ventilators). Intake noise levels are typical high frequency and sound power levels may be as high as 145 dB for larger units. Exhaust power levels may also reach 130 dB, although here the noise spectrum is generally shifted downwards to a lower frequency range. Casing levels associated with gas turbines are generally lower than intake or exhaust levels, but are nevertheless of sufficient magnitude to require consideration as a major source of noise. Figure 1 shows the three basic noise contents of a typical unsilenced gas turbine, with noise level measured at a distance of 400 feet (122 m).

There is also the fact that in general applications the gas turbine drives another machine, which itself will contribute noise to the environment. Figure 2, for example, analyses the various sound power levels generated in a typical 10 000 horsepower gas turbine drive compressor.

For the three sources of noise inlet, exhaust and casing the overall noise measured for a bare turbine with no silencing, can can be estimated in terms of the sound power level as:

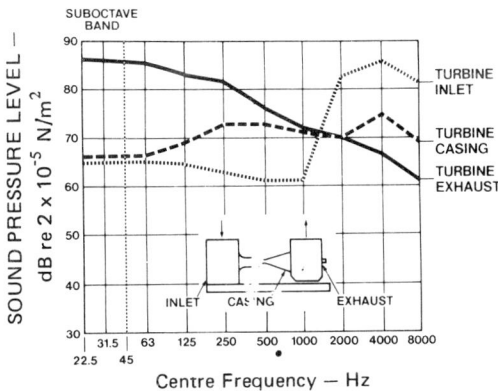

Figure 1 Typical unsilenced gas turbine noise levels at 400 ft (122 m)

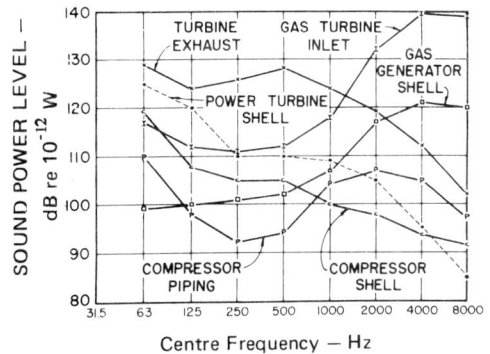

Figure 2 Typical 10 000 hp (7457 kW) aircraft gas turbine driven compressor noises.

inlet $L_w = 127 + 15 \log_{10} W$ dB
exhaust $L_w = 133 + 10 \log_{10} W$ dB
casing $L_w = 120 + 5 \log_{10} W$ dB

where W is the power of the machine in MW

For steam turbines, the following formula applies

$$L_w = 93 + 4 \log_{10} W \qquad dB$$

where, in this case, W is the power in kW.

In order to estimate the frequency spectrum for both gas and steam turbines, the following table can be used, which represents the correction to be subtracted from the overall value of L_w to obtain octave and A-weighted levels.

TABLE 1 – ADJUSTMENTS TO BE DEDUCTED FROM OVERALL L_W (DB)

Octave Band Centre frequency Hz	Gas Turbine			Steam Turbine
	casing	inlet	exhaust	
32	10	19	12	11
63	7	18	8	7
125	5	17	6	6
250	4	17	6	9
500	4	14	7	10
1k	4	8	9	10
2k	4	3	11	12
4k	4	3	15	13
8k	4	6	21	17
A weighted	2	0	4	5

These values also apply to an unsilenced engine. The degree to which silencing is possible depends on the extent to which sound insulation is applied to the casing and to the attention paid to exhaust and inlet casing. The techniques for these can be found in other chapters.

Gas turbines usually have high frequency pure tone components related to the fan frequency and the blades of the compressor; typically these tones are in the region of 4 kHz to 8 kHz. Such pure tones need to have rather larger attenuation than for a broad band spectrum. These pure tones are often very directional and some considerable environmental effect can be achieved by relocation of the turbine intake filter.

In addition to the main noise sources in the turbine, the other elements in the installation may also radiate noise. These can include the coolers for the lubrication circuit, electricity generating unit, gas compressor, gear boxes, waste heat boilers and other ancillary plant. As the installations become more complex, complete noise control packages are increasingly becoming standard. The total package concept has many advantages, since the responsibility for the overall noise specification is now with the noise control engineer.

Silencing Criteria

Two silencing criteria have to be considered: the noise in the turbine hall, affecting the exposure of the employees and the external noise affecting the outside environment. The former will be satisfied by a sound pressure level of 85 dB(A), the latter will require an analysis performed using the methods of BS 4142 as described in the chapter on *Environmental Noise Monitoring*; the sound level to be reached will depend on the needs of the nearby residents. Reduction in the turbine hall noise level can be easily satisfied by enclosure, but it should not be assumed that satisfying the internal criterion will necessarily be adequate for the external one; such sources as combustion exhaust, intake fan and cooling tower noise are primarily external, and have to be individually treated.

Intake Silencing

Intake noise is at maximum intensity at a frequency equal to the product of the axial compressor rotating speed and number of blades (blade passing frequency). Noise is reduced by the use of silencers and elbows, and sometimes even by pointing the intake upwards, to improve the sound contours.

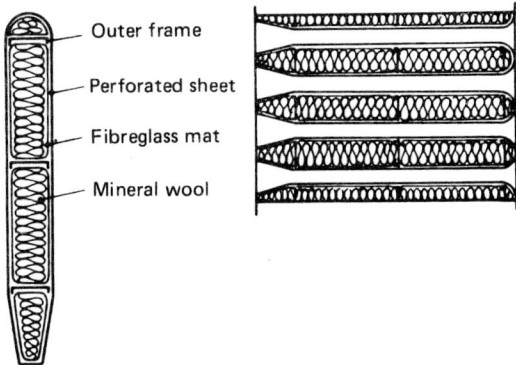

Outer frame

Perforated sheet

Fibreglass mat

Mineral wool

EXHAUST PLENUM

SILENCER

TRANSITION SECTION

Figure 3 Inlet dissipative silencer   Figure 4 Lateral exhaust silencer

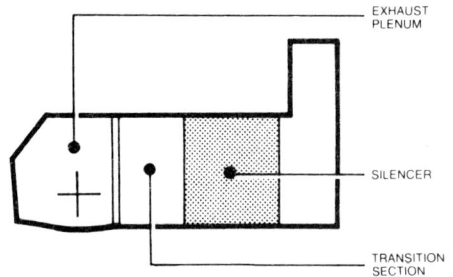

Dissipative (absorbent) type silencers are used, normally consisting of superimposed parallel baffles (Figure 3), the ideal baffle thickness being equal to one half of the wavelength to be treated. In practice they are generally made slightly thicker. Overall geometry is important to avoid self-generated noise, and the silencer is often split by vibration-isolating joints to minimise noise spread through the shell. The silencer is also normally coated to limit noise radiated from its walls.

Exhaust Silencing

The noise spectrum at exhaust is characterised by high intensities at low frequencies. This, plus the fact that gas temperatures reach up to 500–550°C (resulting in greater wavelengths) means that much thicker baffles (up to 800 mm) must be used.

Since the noise at higher frequencies is also considerable, it is often necessary to have to use two separate silencers of different thicknesses (one for high and one for low frequencies). Whenever very low residual noise levels are required, it is more important to limit self-generated noise at exhaust rather than at inlet, due to the fact that exhaust flow rates are higher and much more irregular.

The problem of self-generated noise can be overcome by exhaust systems such as in Figure 4. The silencer in this solution was located before the elbow so that the incident waves are planar,

One of eight IAC acoustic enclosures housing gas turbine sets at Greenwich Power Station of the L.T.E. The intake silencer and lagged exhaust silencer are also visible.

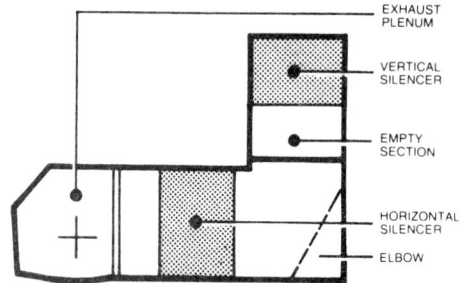

EXHAUST PLENUM

VERTICAL SILENCER

EMPTY SECTION

HORIZONTAL SILENCER

ELBOW

Figure 5 Lateral exhaust system

and thus the elbow provides maximum attenuation and cuts self-generated noise from silencer flow.

The configuration shown in Figure 5 where there is one silencer before and another after the elbow, may be used when even greater attenuation is desired. It allows maximising of both elbow efficiency and the efficiency of the vertical silencer which, coming after the elbow, is hit by random-incident waves. Evidently, in the second silencer, velocity must be reduced and distribution regularised as much as possible. This requires the use of large passage areas, turning vanes (made of perforated sheets so that acoustical effectiveness is not impaired), and a hollow duct right after the elbow to even out the velocity distribution.

Mechanical factors are much more important in exhaust silencers than in inlet silencers due to their great temperature variations (ranging around 500°C) and, above all, to the different temperature constants of the exhaust components. To overcome this, the silencer baffles are designed so that they can freely dilate with respect to the ducts they are housed in. Also, the panels themselves are assembled so that they can move internally to some extent, thereby preventing stresses that could result in damage.

A cutaway view of a typical silencer panel is shown in Figure 6. The layers, consisting of a screen, high temperature blanket, and fibreglass mat, serve to provide greater protection to the rock wool and guarantee silencer efficiency throughout the life of the turbine.

Recovery boilers used for gas heat recovery systems are also effective as exhaust silencers and can reduce the requirements for adequate exhaust silencing.

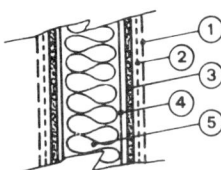

1. Perforated sheet
2. Stainless steel screen
3. High temperature mat
4. Fibreglass mat
5. Mineral wool

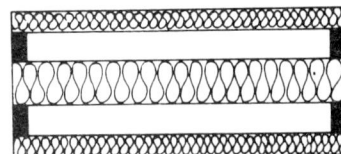

Section of typical turbine exhaust silencer

Figure 6 Exhaust silencer panel

Enclosures

It will usually be found that the overall noise level is best reduced by a well designed enclosure. For small units, prefabricated partitions made of wood or aluminium lined with glass fibre or mineral wool may be adequate. For permanent heavy duty installations, concrete or steel enclosures preferably of double skin construction are to be preferred. As with all enclosures, it is usually well worthwhile lining the inner surface of the enclosure with a sound absorbent material (see chapter on *Sound Insulation*). For all enclosures, careful design dictates well sealed access doors. Passages for pipes and ducts and electricity must also be sealed. Forced internal ventilation can itself be a noise source and wherever possible, ducting should be flexibly connected to the enclosure to minimise the transmission of noise and vibration outside the enclosure.

Usually the vibration level of a well designed gas turbine is low. If any vibration is present it probably originates from the driven machine and gearbox, and in that case the enclosure may require to be mounted on flexible supports, in addition to independent mounting of the turbine and generator itself if close coupled. The use of an isolating sandwich material is likely to be a satisfactory material for enclosure mounting.

Some examples of sound control measures, which have proved to be effective in specific installations are shown in Figures 7, 8 and 9.

Figure 7 shows how the various noise control measures can be applied to a combined cycle gas turbine plant. Figure 8 is gas turbine generator package for a chemical plant. In this example, the noise control package comprises:

A turbine intake attenuator, a parallel baffle type attenuator 2.5 m long, with transformation to fit directly to the filter housing.

Acoustic lined bend for gas turbine intake

Acoustic lined ductwork

An acoustic enclosure with ventilation system, housing the gas turbine and generator

Turbine exhaust attenuator, fabricated in Chromweld, 1.8 m long, with flexible bellows and transformation.

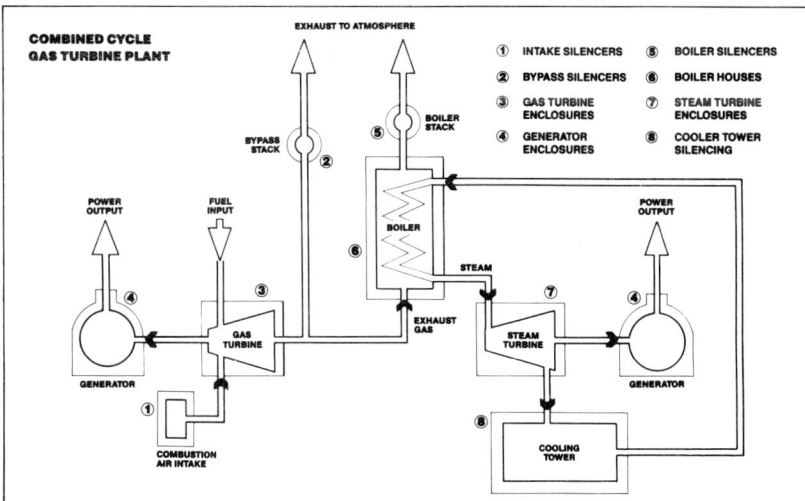

Figure 7 Combined cycle gas turbine plant, showing the various noise control measures required

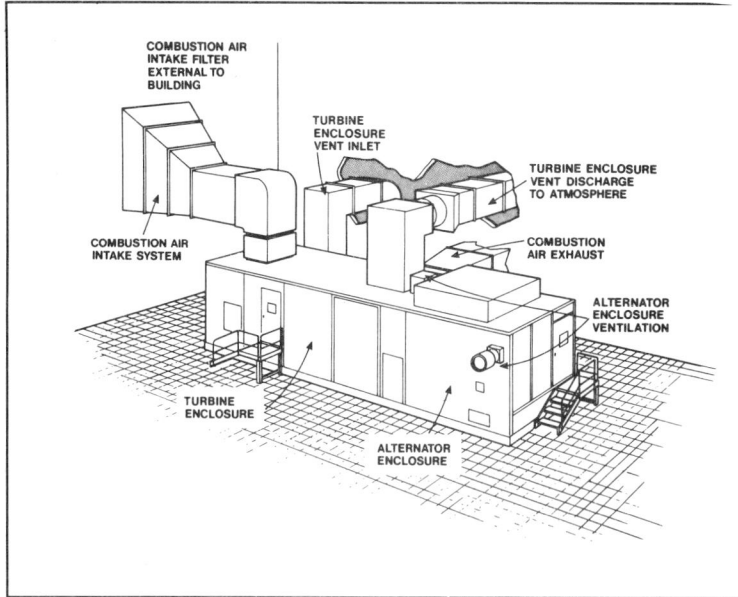

Figure 8

Figure 9 A typical noise control package installed on a gas turbine

Chapter 5
Noise on Construction Sites

Construction Site Equipment

Under this heading is included all equipment used for construction and demolition: compressors, pneumatic tools, hydraulic tools for concrete breaking, electric tools for drilling and dressing stone and concrete, pumps, concrete mixers, etc. Heavy plant includes bulldozers, dumpers, excavators, pile drivers, rollers and concrete mixers. Some of these are found outside the confines of a construction site. In the street or in residential areas they can generate many complaints, and have the been the subject of legislative controls.

Standards and Directives

British Standard BS 5228 is a helpful code of practice for noise control on construction and open sites. It is now in several parts, dealing with different aspects of noise control on building sites and open cast coal sites. The 1984/1986 edition should be used in preference to the earlier versions. European Community Directives have been issued, defining the limits of permissible sound levels for a number of items of equipment used on building sites: tractors, compressors, tower cranes, welding generators, electric generators, hand-held breakers, excavators and loaders. The limits quoted apply only to trade within the European Community, but they prescribe sound power levels which are reasonably easy to achieve with a modest extra cost over that of the unsilenced equipment. Refer to the chapter on *Standards* for details of British Standards and Directives.

Estimating the Noise on a Construction Site

BS 5228 should be used as a comprehensive guide to the estimation of the noise likely to be experienced. Not only does it give methods of estimation, but has also one of the most comprehensive lists of typical noise levels for items of construction equipment. More information about typical noise levels from equipment may be found in the Construction Industry Research Association (CIRIA) Report 64 'Noise from construction and demolition sites measured levels and their prediction'. Two other CIRIA reports may be of interest: No 120 A guide to reducing the exposure of construction workers to noise, and No 138 Planning to reduce noise exposure in construction.

There are the two aspects of construction site noise which have to be considered: the protection of the employed workers from noise-induced hearing loss, and the neighbourhood nuisance that may result from the operations. Both these matters are discussed in other chapters.

The fundamental parameter for construction sites is the activity L_{Aeq}, the equivalent continuous A-weighted sound pressure level determined at a distance of 10 m from and over the given activity. Sound levels of construction equipment have been extensively determined, and the values of this parameter for the various machines in use are available from the reports quoted above. Although 10 m is a typical distance in a construction site between equipment and bystander, the actual value of the sound pressure level at the site boundary (or at some other critical point) can be calculated using one of the methods discussed elsewhere in this text. The site workers will be much closer to the source and different conditions apply to them. The calculation has to take account of the duration of the operation and the actual distance. The various operations then have to be added together to determine the combined L_{Aeq} for the assessment period. The flow chart describing the methods of estimation in BS 5228 is given in Figure 1. A superior method to the activity L_{Aeq} method is applicable when the sound power level of the equipment is known, which it will be for example for the items in Table 3 and any other equipment tested under controlled conditions; this is known as the plant sound power method. As a general rule, it is always preferable to use the best sound level information available; the best will normally be when measurements have been taken under the strict test conditions specified in the directives, corrected for a difference in the environment. (note that for road breakers, the sound power method of the Directive underestimates the noise generated by the steel ring). A composite method, using the sound power values when known and L_{Aeq} (10 m) for the rest will be appropriate. Whichever method is used, the end result should be the same.

The time over which L_{Aeq} is calculated will depend upon the circumstances and the point at which the prediction is made. If the prediction point is an office building, the time during which it is occupied will be the calculation time. For domestic annoyance, the period may be chosen as 12 hours or longer, with a different calculation for day and night time activities. The Local Authority in the UK, following the procedure outlined in Part 2 of BS 5228, may prescribe a time and a level, and those will be the conditions used in the calculations. Reference should be made to the local legal framework for other countries.

General Methods of Noise Control on Construction Sites

A number of the methods discussed throughout this text can be adopted for reducing the noise levels of equipment used on construction sites. Primarily for the reduction of external noise, the following checklist may be helpful, see also Table 4.

Ensure that the equipment used is at least as good as the requirements of the EEC Directive. Note that only equipment placed on the market after the commencement date is covered. If the equipment is older it may not meet the requirements; approved equipment bears the CE mark, which quotes the sound power level.

Where Community Directives do not apply, search for those manufacturers which supply sound reduced equipment. Do not apply direct silencing techniques at source without the approval of the manufacturer. Only purchase equipment which has been noise tested by a manufacturer who is willing to quote the values. A warning should be given about the use of silenced equipment which has been so modified as to reduce its efficiency, and so requires a longer time to complete its intended function.

Choose the time of operation so as not to interfere with the nearby residents. As far as possible activities should be limited to the hours of 7 a.m. to 7 p.m.

Use enclosures where other methods fail. Some suggested designs are to be found in BS 5228: Part 1.

Use rubber linings for chutes.

Stationary plant		Mobile plant	
Activity $L_{A\,eq}$ method	Plant sound power method	On site (limited area)	On haul roads

Select sound power for stationary items of plant, obtain the percentage on-time noise at maximum level

Select sound power for mobile items of plant

Select $L_{A\,eq}$ (10 m) from tables of stationary or quasi-stationary activities

Calculate sound level at point of interest from sound power level and the actual distance. Then allow for screening and reflections

Calculate sound level at point of interest from sound power level and the minimum distance. Then allow for screening and reflections

Calculate $L_{A\,eq}$ from sound power level for mobile plant on haul roads

Calculate $L_{A\,eq}$ for actual distance and allow for screening and reflections

Combine actual levels with the percentage on-time to give $L_{A\,eq}$

Estimate the distance ratio and obtain the equivalent on-times

Estimate the percentage of the assessment period for which the activity takes place and calculate the assessment period $L_{A\,eq}$ for that activity

Estimate the percentage of the assessment period for which the activity takes place and calculate the assessment period $L_{A\,eq}$ for that activity

Estimate the percentage of the assessment period for which the activity takes place, correct for equivalent on-time and calculate the assessment period $L_{A\,eq}$ for that activity

Estimate the percentage of the assessment period for which the activity takes place and calculate the assessment period $L_{A\,eq}$ for that activity

Calculate combined $L_{A\,eq}$ for assessment period

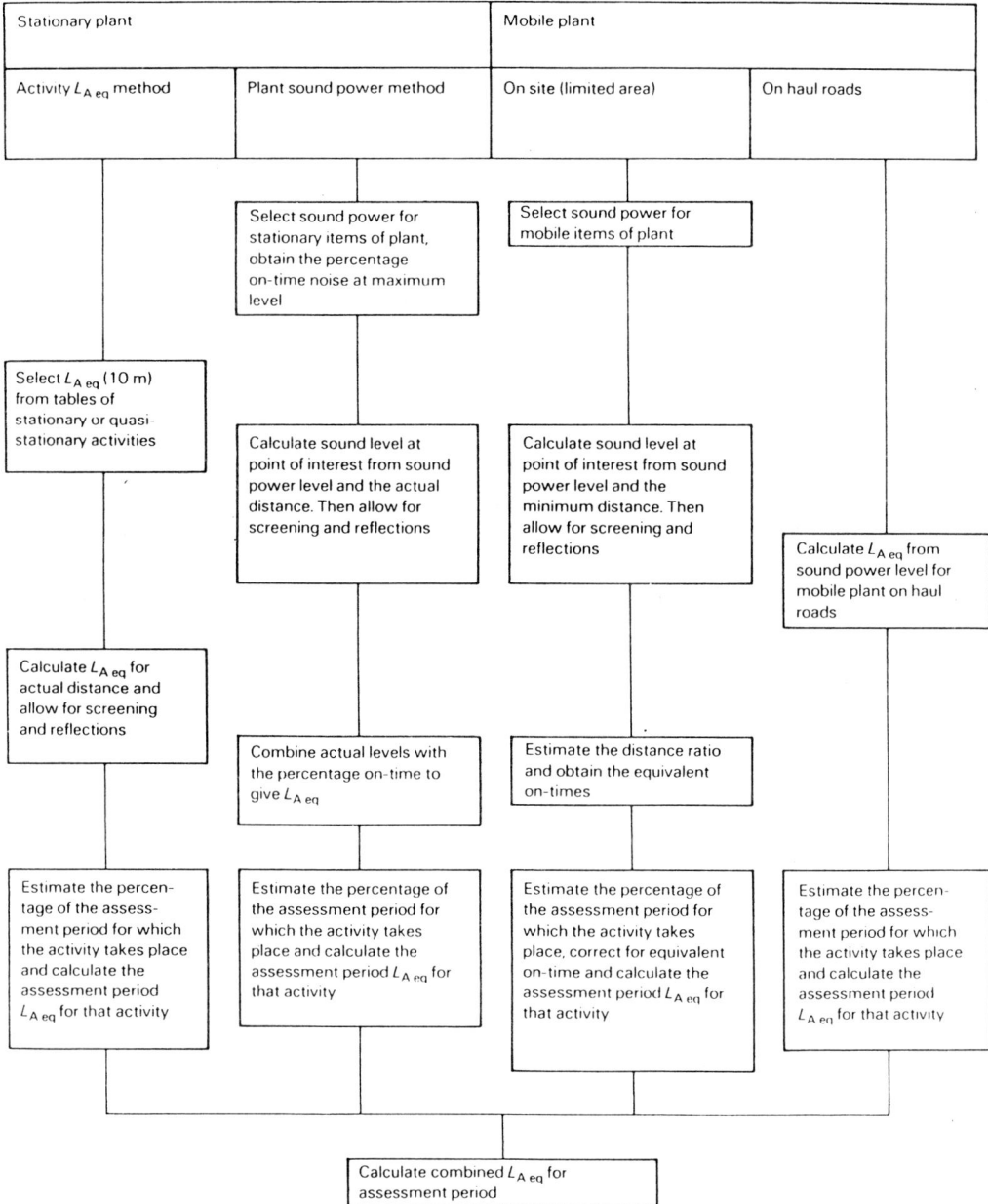

Figure 1 Flow chart for the prediction of site noise

Site the equipment remote from the boundaries of the site or behind natural screens.

Consider using the least noisy of the available working alternatives, as for example in piling (see section below).

Observe the approved maintenance instructions and ensure correct and regular servicing of the equipment.

Precautions to reduce the exposure of the employed workers to noise are:

Mark all areas where the sound pressure level exceeds 90 dB(A). Refer to Noise at Work Regulations 1989.

Supply ear defenders, individually selected, and insist on their being used. For workers in critical areas, a dosemeter should be worn.

Ensure that sound treatment is applied to the inside of cabs.

Make available the use of quiet rest rooms and offices.

Consider the rotation of staff to limit individual exposure to high levels

Noise from Pneumatic Equipment

All pneumatic machines and tools tend to produce noise, caused in the main by the discharge of high pressure air from the exhaust port. In addition to the exhaust noise, there may be mechanical noise, particularly from percussive tools.

There is a growing body of legislation limiting the noise from all types of equipment, and this places a duty on both the manufacturer and equipment user. The manufacturer is required to produce a tool which is within the prescribed legal limits, and the employer is obliged to pay attention to the noise exposure of the tool operator by limiting the working time or by providing suitable protection.

In considering the noise generated on a building site, it should be remembered that it is the total noise environment that is important when considering the exposure of the operator or the other site workers. To demonstrate the comparison between tools and compressors, Table 1 has been prepared.

TABLE 1

Machine	Noise Level dB(A)
3.5 m/min compressor at 7 m	
unsilenced	88
silenced	75
supersilenced	70
Breaker at 7m	
unsilenced	94
muffled	89
muffled with steel damping	88
Breaker at 1m	
unsilenced	106
muffled	101
muffled with steel damping	100

Unsilenced compressors and road breakers are not available today (at least in the developed world), but the figures in Table 1 are useful in indicating the progress of silencing techniques. It can be seen that the breaker is by far the noisier piece of equipment and swamps the noise from the compressor. It makes very little sense to pay for an expensively silenced compressor and use it to operate a noisy breaker. As a general statement, the noise of a compressor can be reduced to any prescribed level, merely by adding progressively better treatment to its enclosure. A compressor can be as bulky as necessary without affecting its functioning; the only drawback is the expense. The same is not true of a breaker, which has to be both efficient and convenient to use. If a breaker cannot be handled, it will not find favour with the operator; so, however well silenced it may be, ergonomic considerations will predominate in the choice of a tool.

When trying to reduce the total noise on a site it is clearly desirable to tackle first the most noisy equipment but there may be a practical limit to the noise reduction possible for certain kinds of equipment. The limit appears to have been nearly reached by a modern pneumatic breaker working on a conventional percussion system. Even the use of a hydraulic or electric breaker will not result in a quieter tool a hydraulic breaker is just as noisy as a pneumatic one, because most of the residual noise comes from the vibration of the drill bit, as explained below.

Noise Reduction in Pneumatic Tools

Most tool manufacturers apply some sort of noise reduction treatment to their tools. Figure 2 shows the proportion of noise from the various sources in a road breaker. It can be seen that the first and largest noise source to tackle is the exhaust noise, followed by the ringing noise from the steel, and then the internal clatter of the working parts. Fortunately exhaust noise is reasonably easy to suppress, at least in theory.

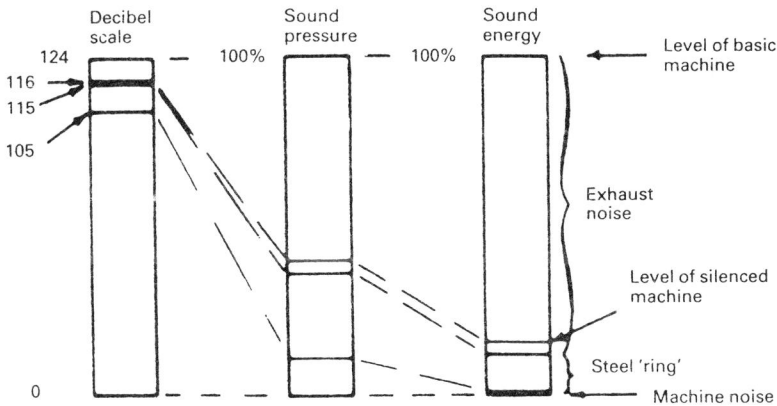

Figure 2 Comparison of the decibel scale with sound pressure and energy for an experimental silenced rock drill

The principle to follow is to reduce the velocity of the jet noise from the exhaust port. The pressure ratio at exhaust is as high as 2 or 3 to 1, which implies that the exhaust velocity is sonic and the noise is produced by turbulent mixing of the high velocity jet. The principles that have been developed for silencing engine exhausts are based on the assumption that the pressure variations are small. Such an assumption is not applicable to pneumatic tools. The technique that has proved most successful is diffusion of the exhaust stream by a gradually increasing cross section of the air passages. It has to be admitted that much of the design of exhaust mufflers is empirical,

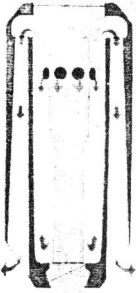

and most mufflers have been developed by trial and error. Figure 3 shows how a typical silencer functions. If the only problem were to reduce the exhaust noise, the solution would be easy it would consist of a succession of expansion volumes joined by restricted passages. However the noise reduction must be achieved without changing the performance of the tool; so there must be no back pressure to impede the motion of the piston and no restrictions in the flow passages which could be clogged by ice formation. Balancing these factors is not easy.

Most manufacturers have found that a flexible plastic such as polyurethane is the most suitable material for manufacture of a muffler. Solid polyurethane is a a very robust material which is capable of withstanding hard usage, and ideally it should form an integral part of the tool construction so that it cannot be removed without the the tool ceasing to operate. The chapter on *Contractors Tools* describes the construction.

The next source of noise, particularly in a road breaker, is the steel 'ring'. As the impact stress wave passes down the tool stem, part of the energy will be absorbed by the road surface, but a proportion of it is reflected back and forth along the length of the tool. The stem is made of a high quality steel and so has a low internal damping, which ensures that it acts as a tuning fork, radiating noise. Although the energy from this source is small, it occurs at a single frequency and is subjectively very annoying.

The one technique that has proved to be of some value in reducing noise from this source is the addition of a damping ring to the hexagon of the tool stem, as in Figure 4. An extra collar is forged on the stem to hold a steel damping ring which is swaged down over a lead-filled rubber sandwich. The rings have a limited life, although it is possible to renew a worn ring with special equipment. These tools are more expensive and are normally only available for chisel stems, so their use is limited to those places that are particularly sensitive to noise. Attempts have been made to attach a damping ring with adhesive only, but so far unsuccessfully.

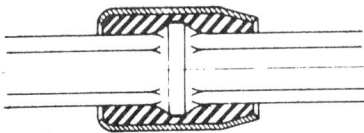

Figure 4 Diagram of 'muffle steel'

The clatter from the internal working of the tool is reduced by the presence of the muffler itself, particularly if made of a flexible plastic. A recent road breaker development by CompAir has its cylinder component constructed as a composite of steel and plastic which has natural damping and appears to be effective at reducing internal clatter.

The one source of noise which is very difficult to suppress is that produced by vibration of the material being worked. In the case of a road breaker, noise is produced by the shock waves in the road surface. In the case of a chipping hammer or a riveter, noise is generated by vibration of the casting or other component being worked.

Legislation on Construction Equipment Noise

Manufacturers in the Common Market are now bound by the Treaty of Rome (in particular Article 100). Legislation has recently been introduced which limits the sale of certain construction equipment to those which meet the prescribed noise levels. Reference should be made to the appropriate EEC Commission Directive, and its national implementation document which can be found in the chapter on *Standards*. Each directive quotes the maximum levels of noise and the proper test methods to be used. It should be noted that since 26 May 1986 the marketing or use of specified construction equipment manufactured after that date which emit noise in excess of the permitted levels is a criminal offence. Equipment which satisfies the regulations has to bear an approved mark of the same form as shown in Figure 5. The value 104 shown on that plate represents the sound power in dB(A) as measured by the testing station. It may be less than that required in the directive; it must, of course, not be greater. In order to establish conformity with the directives, the measurements have to be determined at a test station approved by the appropriate authority inside the member country of the EEC. Equipment approved in one country can be freely imported into another without further inspection. In the UK, the authority is the Department of Trade and Industry. The laboratories which are able to perform the tests are commercial bodies (rather than government laboratories, as in some countries of the EEC) and are inspected by NAMAS (National Measurement Accreditation Service), a service of the National Physical Laboratory, see Table 2. They are in commercial competition, so the charge for performing the tests will vary.

Figure 5

The regulations have no force outside the EEC although many other Western countries have similar, if not stricter, rules.

It should be understood that for road breakers the noise level measured according to the directive is not necessarily typical of the actual noise that is likely to be experienced by an operator of a tool or by a passer-by. The test is performed for type approval of the breaker alone, and so it is devised in such a way that the steel ring and the radiated noise from the concrete block are suppressed. The quoted value is the sound power level emitted by the breaker, in the rather

TABLE 2– APPROVED BODIES FOR CONSTRUCTION EQUIPMENT NOISE TESTING

Organisation	Road Breakers	Compressors	Welding/Power Generators	Earth-moving Equipment	Tower Cranes
A V Technology Avtech House Cheadle Heath Stockport SK3 0XU	x	x	x	x	
Acoustical Investigation and Research Org Ltd Dixons Turn Maylands Avenue Hemel Hempstead			x		
BSI Testing Services Maylands Avenue Hemel Hempstead H2 4SB	x	x	x		x
Ricardo Consulting Engineers Bridge Works Shoreham-by-Sea BN4 5FG		x	x	x	x
Sound Research Labs Saxon House Downside Sunbury-on-Thames Middlesex UB4 2QX			x		
Wimpey Laboratories Beaconsfield Rd Hayes UB1 0LS	x	x	x		x
Taywood Engineering 345 Ruislip Road Southall UB1 2QX	x		x		
Lloyds Register of Shipping 71 Fenchurch Street London EC3M 4BS	x	x	x		x

artificial test arrangement shown in Figure 6. In order to assess the actual noise exposure in practice, sound pressure measurements should be taken. The type approval test is useful in comparing one breaker with another, but should not be used to determine the noise environment on a particular site.

Measurement of Noise from Tools other than Road Breakers

To measure the noise emitted by other items of site use, particularly other pneumatically operated equipment, the only alternative test procedure is the CAGI-Pneurop Code, which should be studied for further details. This code defines the measurement procedure for all kinds of pneumatic equipment. The readings have to be reported in terms of the sound pressure level. The

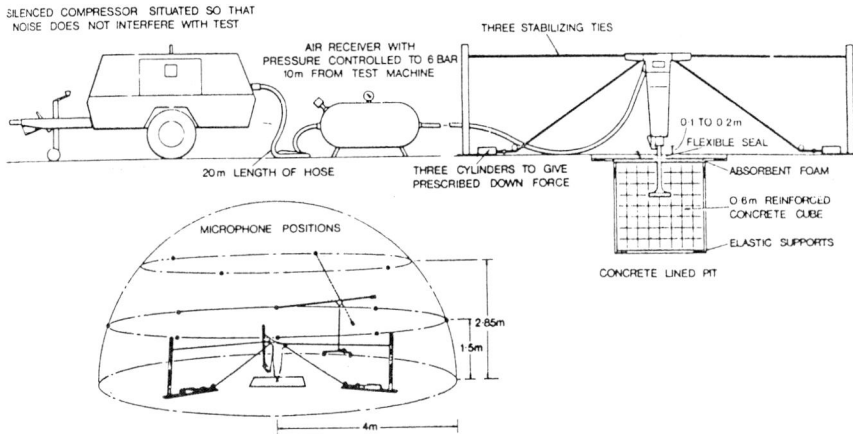

Figure 6   Test arrangement to measure sound power from road breakers according to EEC Test Code

measurement distance from the noise source is 1 m for tools and 7 m for compressors and other large equipment, and the measurement points are situated on the sides of an (assumed) enclosing parallelopiped, see for example Figure 7. This code is useful for assessment of the actual noise experienced by the operator or by the public, since it measures the noise from all sources. It has been shown that when this code is applied to compressors it gives results (when the sound pressure readings are converted to sound power) that are as accurate as the EEC method. For the reasons given above, this is not the case for road breakers. For more information on pneumatic equipment refer to the Pneumatic Handbook published by Elsevier Science Publishers.

Mobile Compressors

The mobile compressor is frequently blamed as one of the noisiest items of construction plant which affect the environment. In fact the public are probably unable to differentiate between noise from the compressor and noise from the tool connected to it. The tool is by far the largest producer of noise, and in a typical roadside site, noise from the tool swamps that from the compressor. Despite this, there has been much pressure on the legislators to introduce regulations to limit the permissible noise level of compressors.

There may be some confusion about the designation of the sound emitted by a compressor (or any equipment for that matter). The approved method now is to specify the sound in terms of sound power against a base of 1 pW 10^{-12}W. This takes into account that a piece of machinery emits a different amount of noise in different directions, and therefore the only true assessment of its noisiness is to measure the total noise emitted over a surface enclosing it. However large the surface is, the total noise is the same. This is the principle behind the EEC Directives on the Approximation of Laws of the Member States on Noise. An alternative way of specifying the sound emitted by a compressor, which is frequently quoted in manufacturers' literature, is by the use of the Pneurop/CAGI Test Code. This is a simple method of noise measurement and consists of taking an average of four readings of sound pressure level at a distance of 7 m from the side of the compressor (not from its centre as in the EEC code). An approximate method of converting from sound power to sound pressure at 7 m using the Pneurop/CAGI Code is to subtract from the former 26 dB for a small compressor (up to 10 m^3/min) and 27 dB for larger ones.

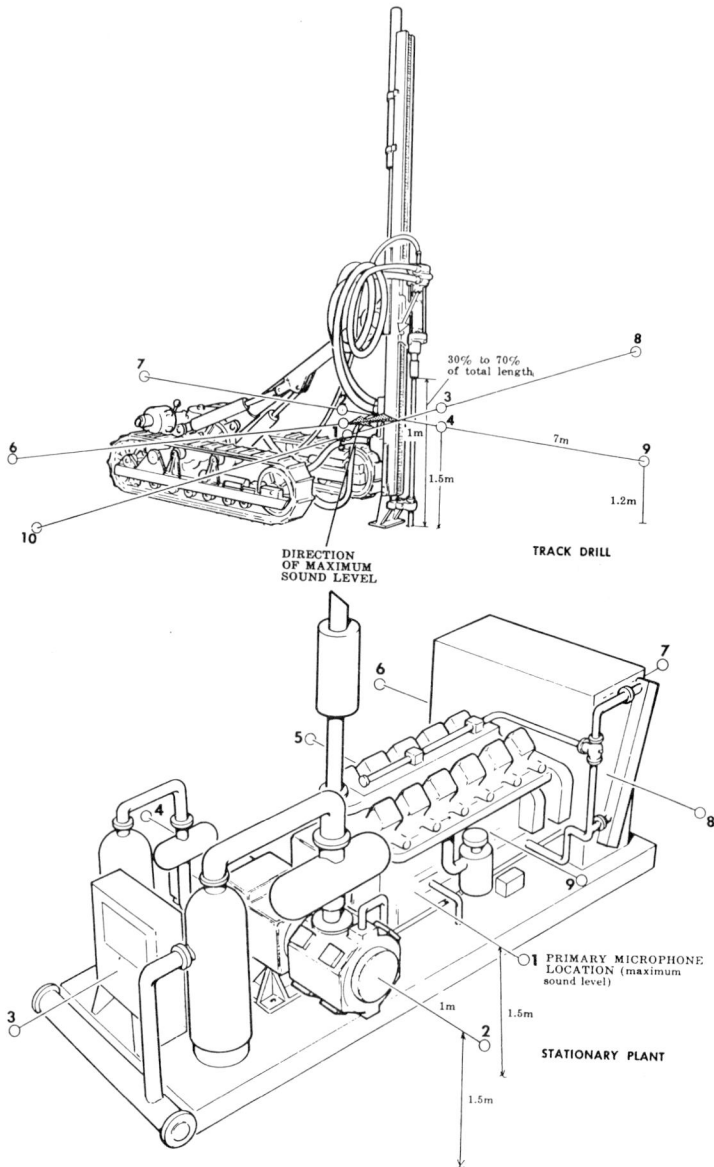

Figure 7 Measurement position according to Pneurop/CAGI Test Code

Out of the widespread use of the Pneurop/CAGI Code there arose the description of a compressor as standard, silenced or super-silenced. Standard means that no particular sound proofing has been applied, silenced means that the level of silencing is down to 75 dB(A) at 7 m, super-silenced means that a level of 70 dB(A) has been achieved.

If one wishes to determine how much noise an operator or a member of the public is exposed to, the local level of sound pressure has to be calculated from the sound power value and the distance from the centre of the compressor. The required relationship, ignoring such factors as the presence of reflecting objects and the directivity of the noise source, is:

$$L_p = L_w - 8.0 - 20 \log_{10} r$$

where L_p is the sound pressure, expressed in dB, at a distance r metres from the centre of the compressor

L_p is the sound power level, expressed in dB

Methods of Reducing the Noise Level of Portable Compressors

Most of the noise is generated by the engine, and since the engines used are standard industrial units with no particular attention paid to silencing for the compressor market, the burden is on the compressor manufacturer to apply sound reduction techniques to the canopy enclosure. The following techniques have been found useful:

Line the canopy with sound-deadening material. The use of such materials as absorbent foam in the engine/compressor section where there could be contamination by fuel and oil leading to a fire hazard should be avoided. Double skinning could be used instead.

Use flexible engine mounts between engine and chassis

Ensure the enclosure is complete. Use undertrays. Any opening doors should be well sealed. Instruments must be capable of being read from outside without opening the doors.

The canopy should be soundly made, reinforced where necessary to prevent panel drumming and rattle.

Extra engine exhaust silencing.

Line the cooling air intake and exhaust passages with sound deadening material. This is an important point to observe much noise can escape through the cooling air passages. Use silenced exhaust valves.

Compressor intake from inside canopy.

Mount canopy flexibly on chassis.

Some or all of these methods are capable of reducing noise level down to the required level.

Noise Reduction in Piling

The noise experienced during a piling operation will depend on the type of piling method employed. In principle the three types of piling driving, jacking and boring can be used for either bearing piles or retaining piles as shown in Figure 8.

In driven piling, a hammer is used to strike the top of the pile through a dolly. The hammer can be a simple drop hammer, relying on gravity to produce the impact force, or can be driven by steam, compressed air, hydraulic power or diesel. For some types of soil, it is possible to employ a vibratory pile driver, for part at least of the operation. Very high impact noise can be generated by the hammer (of the order of 120 to 130 dB(A) or even higher sound power levels). A vibratory driver may prove to be a quieter alternative but the vibration created in the ground may give rise to a disturbance at least as troublesome as the noise of the impact. Noise can be reduced by the use of an aluminium alloy or non-metallic dolly interposed between the hammer and the driving helmet. A dolly can be made of wood, plastic or a composite material, although if this method is used, the pattern of the stress wave in the pile is likely to be changed, reducing the efficiency of the operation.

Jacked piling is a method of installing piles, by which the force used to push in the pile is reacted by a group of already inserted adjacent piles. This is basically a much quieter method, but is only suitable in certain kinds of cohesive soils and silt.

Bored piles employ a rotary or rotary-impact boring machine. If the ground is cohesive, the piles can be inserted in the bored hole, otherwise a casing may be needed to support the ground. Another method is continuous flight augur, with the concrete formed by injection through the augur. The noise level is lower unless impact is used for the insertion or removal of the casing, in which case the same conditions as a driven pile apply. Reference should be made to BS 5228: Part 4.

Bearing piles

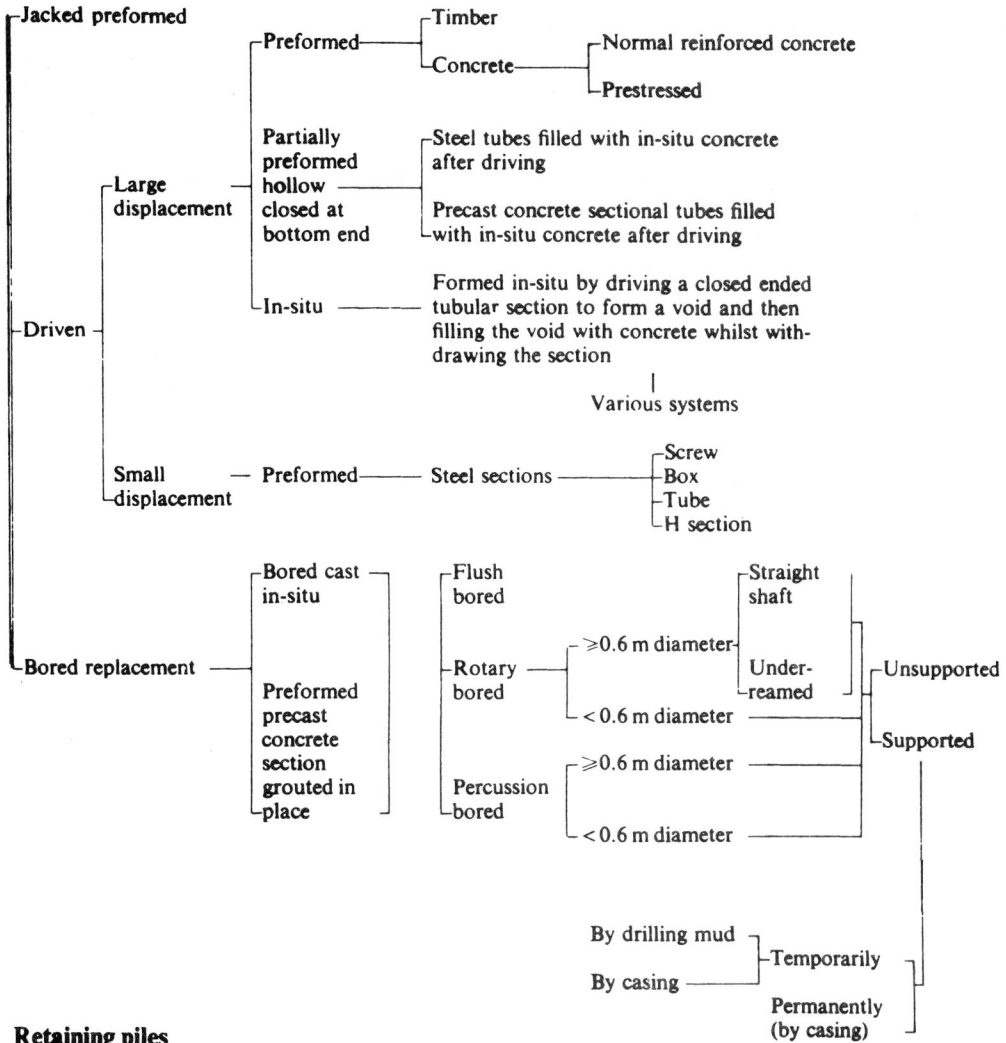

- Jacked preformed
- Driven
  - Large displacement
    - Preformed
      - Timber
      - Concrete
        - Normal reinforced concrete
        - Prestressed
    - Partially preformed hollow closed at bottom end
      - Steel tubes filled with in-situ concrete after driving
      - Precast concrete sectional tubes filled with in-situ concrete after driving
    - In-situ
      - Formed in-situ by driving a closed ended tubular section to form a void and then filling the void with concrete whilst withdrawing the section
        - Various systems
  - Small displacement
    - Preformed — Steel sections
      - Screw
      - Box
      - Tube
      - H section
- Bored replacement
  - Bored cast in-situ
    - Flush bored
    - Rotary bored
      - ≥0.6 m diameter
        - Straight shaft
        - Under-reamed
          - Unsupported
          - Supported
      - < 0.6 m diameter
    - Percussion bored
      - ≥0.6 m diameter
      - < 0.6 m diameter
  - Preformed precast concrete section grouted in place

- By drilling mud
- By casing
  - Temporarily
  - Permanently (by casing)

Retaining piles

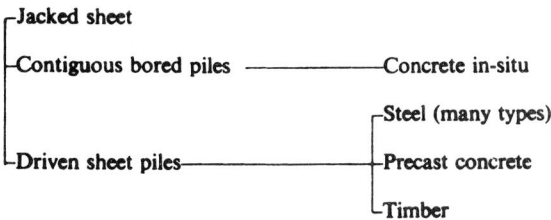

- Jacked sheet
- Contiguous bored piles ——————— Concrete in-situ
- Driven sheet piles
  - Steel (many types)
  - Precast concrete
  - Timber

Figure 8 Piling alternatives (BS5228)

TABLE 3 – PERMISSIBLE SOUND LEVELS FOR CONSTRUCTION EQUIPMENT

	Permissible sound power level in dB(A)/1 pW	
	Before Sept 1989	Current
Portable Tools (Concrete Breakers and Picks)		
Mass of appliance m in kg		
$m \leqslant 20$	110	108
$20 < m \leqslant 35$	113	111
$m > 35$ (and devices with an internal combustion engine)	116	114
Portable Compressors		
Nominal Air Flow q in m³/min		
$q \leqslant 5$	101	100
$5 < q \leqslant 10$	102	100
$10 < q \leqslant 30$	104	102
$q > 30$	106	104
Power Generators		
Power P in kVA		
$P \leqslant 2$	104	102
$2 < P \leqslant 8$	104	100
$8 < P \leqslant 240$	103	100
$P > 240$	105	100
Welding Generators		
Current A in amp		
$A \leqslant 200$	104	101
$A > 200$	101	100
Tower Cranes		
Lifting mechanism	102	100
Energy generator	As for power generator above	
Combined lifting mechanism and energy generator	Higher value of two components	
Note that the sound pressure level at the operator's position is limited to 80 dB(A)		
Hydraulic Excavators, Rope Operated Excavators, Dozers, Loaders and Excavator Loaders		
Net installed power P in kW		
$P \leqslant 70$	106	
$70 < P \leqslant 160$	108	
$160 < P \leqslant 350$		
Hydraulic and rope-operated excavators	112	
Other earth-moving machines	113	
$P > 350$	118	

TABLE 4 – CONSTRUCTION SITE NOISE SOURCES AND POSSIBLE REMEDIES BS5228

Machine	Source of noise	Possible remedies (to be discussed with machine manufacturer)		Possible alternatives
Piling equipment	Pneumatic/diesel hammer or steam winch vibrator driver	Enclose hammer head and top of pile in acoustic screen, acoustically dampen sheet steel piles to reduce vibration and resonance		(1) Alternative method of piling. (2) Alternative methods of soil retention and ground improvement e.g. diaphragm walls, ground anchors, shafts formed of pre-cast concrete segments sunk into the ground under Kentledge, use of pre-treatment prior to excavation such as dewatering freezing soil injection etc.
	Impact on pile	Use resilient pad (dolly) between pile and hammer head e.g. 2 layers of asbestos cloth stuffed with glass fibre or mineral wool and protected by plywood. Packing should be kept in good condition.		
	Crane cables, pile guides and attachments	Careful alignment of the pile and rig		
	Power units or base machine	Fit more efficient silencer or exhaust. Acoustically dampen panels and covers. When intended by the manufacturers engine panels should be kept closed. Use acoustic screens where possible		
Bulldozer Compactor Crane Dumptruck Dumper Excavator Grader Loader Scraper Shovel	Engine	Fit more efficient silencer or exhaust Enclosure panels, when fitted, should be kept closed		
Compressor Generator	Engine	Fit more efficient exhaust silencer	Screen the compressor or generator	Use electric motor in preference to diesel or petrol engine for compressors. If there is no mains supply, a sound reduced compressor or generator can be used to supply several pieces of plant. Use centralized generator system
	Compressor or generator	Acoustically dampen metal casing. Enclosure panels should be kept closed		
Pneumatic concrete breaker and tools	Tool	Fit a muffler or silencer, this will reduce the noise without impairing efficiency	Use the breaker inside a portable acoustic enclosure	Use rotary drill and burster. Hydraulic and electric tools are also available. A thermic lance can be used to burn holes in concrete and to cut through large sections of concrete; any rein-forcement helps the burning process. For breaking large areas of concrete, equipment which breaks concrete in bending could be used
	Bit	Use dampened bit to eliminate 'ringing'. Little noise once surface is broken		
	Air line	Leaks in air line should be sealed		
	Motor	Fit muffler to pneumatic saws		
Power saws	Vibration of blade or material being cut	Keep saw sharp. Use a damped blade. Clamp material while cutting with packing if necessary		
Rotary drills diamond drilling and boring	Drive motor and bit	Use machine inside an acoustic cover		Thermic lance

TABLE 4 – CONSTRUCTION SITE NOISE SOURCES AND POSSIBLE REMEDIES BS5228 (contd.)

Machine	Source of noise	Possible remedies (to be discussed with machine manufacturer)		Possible alternatives
Riveters	Impact on rivet	Enclose working area in acoustic screen		Design for high tensile steel bolts instead of rivets
Cartridge gun	Explosion of cartridge	Use a sound reduced gun		Drilled fixings
Pupm	Engine pulsing	Enclosure in acoustic screen (allow for engine cooling and exhaust)		
Batching plant Concrete mixer	Engine	Fit more efficient silencer on diesel or petrol engine. Enclose engine	Locate static mixing plant as far as possible from those likely to be inconvenienced by the noise	Use electric motor in preference to diesel or petrol engine
	Filling	Do not let aggregates fall from an excessive height		
	Cleaning	Do not hammer the drum. See Department of the Environment Advisory Leaflet 'Making concrete' Appendix J		
Hammer	Impact on nail			Screws
Electric impact chisel	Impact			Rotary hand milling machine
Materials handling	Impact of material	Do not drop materials from a height. Screen dropping zones especially on conveyor systems		Cover surface with resilient material or unload elsewhere
Steam cleaning	Escaping jet of steam	Pass escaping steam through silencer or screen the outlet zone		

Care should be taken to ensure that when selecting a quiet process the ancillary equipment such as cranes and compressors do not become the major source of noise.

Figure 9 The new, quieter Hydrovane HV5L portable air compressor, fitted with the Lister Petter LPW4 diesel engine. The EC plate shows $L_{WA} = 100\,dB$

Chapter 6
Noise Cancellation Techniques

This technique, which is also known as active noise control, has had much publicity recently because it seems to offer an elegant solution to acoustic noise problems. The technique was proposed in the 1930s, but it has only now been possible to put the ideas into practice through the advent of fast acting digital electronics and computing technology. The basic idea is that noise and vibration can be reduced by generating a pressure wave which is exactly 180° out-of-phase with the original signal and of the same magnitude. The opposing signal may be called *contranoise*, and for simple situations the original and the generated signal cancel each other out. Contranoise is generated electronically and passed to a loudspeaker system, or in the case of vibration cancellation through to a transducer or active ant-vibration mount.

Figures 1 and 2 show the system applied to the exhaust of a diesel engine.

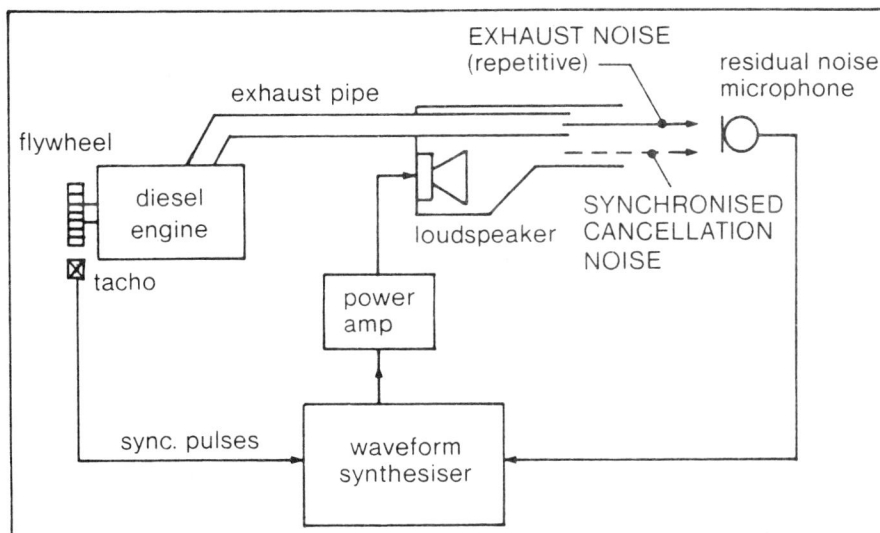

Figure 1 Synchronised, synthesised diesel exhaust cancellation –
Essex system

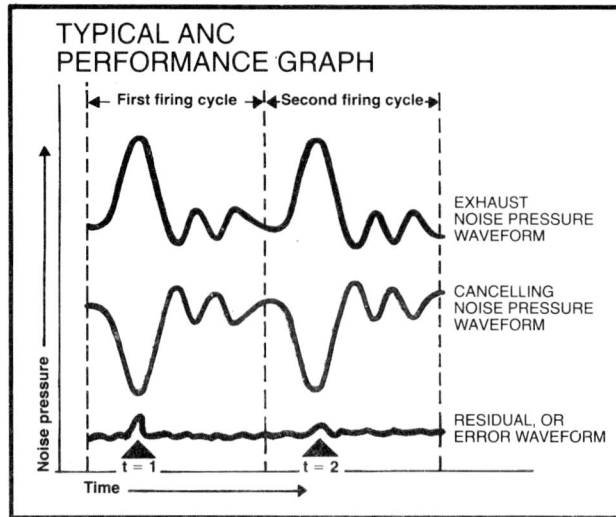

Figure 2 Principle of Active Noise Reduction

Some of the applications that have proved successful are

Diesel engine and exhaust noise

Active engine mountings

In-car noise reduction through multiple speaker systems

In-duct fan noise

Jet engines

It will be apparent that there are a number of inherent features of this system which limit its application. The noise signal needs to be be stationary or near stationary, the source should be located in a small area and the propagation should be one dimensional (in ducts or pipes, for example). It is most effective at low to medium frequencies, up to about 250 Hz, which is the region where passive noise control is often least effective. The ability of electronic signal processing to synthesise an out-of-phase signal is necessarily time limited, so it becomes progressively less effective at higher frequencies. The more repetitive or coherent the signal is, the easier it is to predict the magnitude and phase.

There are two basic techniques for active noise suppression: the conventional, which measures the sound pressure, changes its phase and then passes it to a loudspeaker; and the synchronous or periodic which is suitable for repetitive (deterministic) signals. The latter makes use of information from the previous cycle and synthesises a waveform to match it at exactly the right time; it takes a synchronisation signal from the machine (which may be an engine or a fan, for example) and is required only to adapt that signal to the variation from one cycle to the next. Provided that the signal is reasonably repetitive, it is more effective than the conventional method, because there is no phase error in the contranoise signal. It is also known as the Essex method, because the principles were developed by a research group at Essex University.

The two techniques are compared in Figure 3.

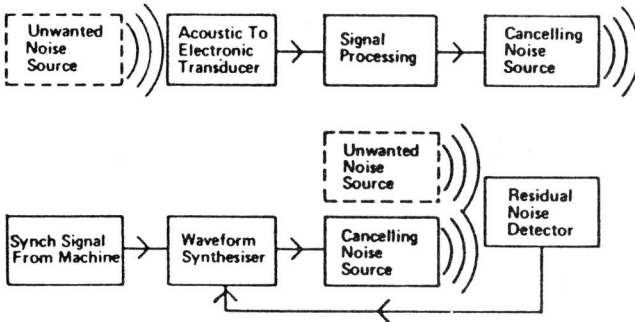

Figure 3 a) The traditional active system. Delay, inherent in all three functions, separates the unwanted and cancelling noise sources.
b) The Essex (periodic) system. The two sources are now coincident. The residual detector is an inherent part of the system and enables it to adapt to a changing acoustic environment

To have an effective noise cancellation system in a room, the anti-noise generating installation must be at least as complex in a spatial sense as the originating noise source. For automobile applications, it may be satisfactory to produce a local 'zone of silence' around the head of the driver. Improving the acoustic quality at one location may well make things worse elsewhere.

For such applications as noise control of an exhaust, the expense of installing a loudspeaker and the accompanying electronics may not be acceptable. The equipment may have to be located in an area vulnerable to accidental damage.

The Quiet Road Vehicle

The tractor or lorry driver in a modern vehicle can be isolated from the noisy environment by an acoustically insulated cab and from the vibration by a soft suspension system. The danger inherent in this approach is that he is unable to hear warnings or detect changes in the behaviour of the machinery. The problem can be dealt with in another way by the use of an active system which is able to remove the repetitive engine sourced noise, leaving the warning noises unaffected. In theory the driver could have open windows yet still work in a comparatively quiet environment.

Lotus Engineering have done much research into the use of the technique in high performance cars. The system using four microphones and two speakers is illustrated in Figure 4. The speakers

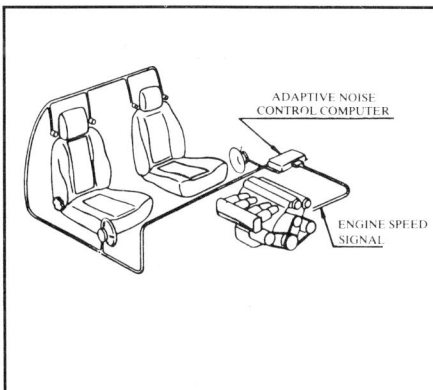

Figure 4 Schematic representation of microprocessor control system

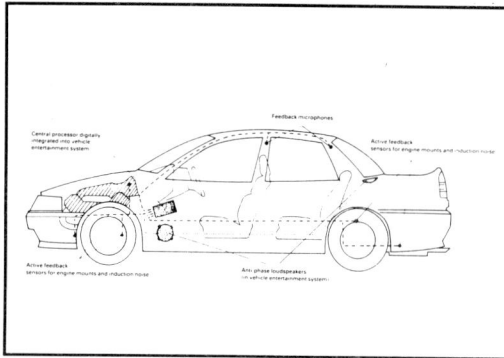

Figure 5 Representation of adaptive noise system in a car

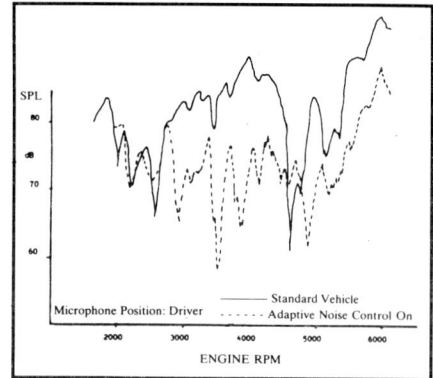

Figure 6 The reduction in second order sound pressure level using adaptive noise control in A

are good quality 6 inch diameter with 40 W capacity. the complete layout is shown in Figure 5. A typical improvement is shown in Figure 6, for a 2.2 litre two-seater sports car.

Application in Air Ducting

Most of the noise in a ventilating and heating system is generated by the fan. Typically the frequency is in the region of less than 300 Hz. The conventional way of suppressing the noise produced by the air pulsations in the fan is by the use of absorptive silencers made of perforated metal and filled with a material such as mineral wool. The amount of attenuation depends on the dimensions of the duct and the silencer. The lower the frequency, the larger the dimensions of the silencer have to be, at least the same order as the wavelength of the sound. The dynamic insertion loss (DIL) of an absorptive silencer falls off dramatically at frequencies below 250 Hz, which is the region where active noise suppression is most effective. This points to the use of a dual silencing regime – active and passive – for optimum performance. When considering active attenuation, the economics are of crucial significance. An absorptive silencer necessarily introduces a pressure drop of the order of 100 Pa, which has to be overcome by the fan, whereas a properly designed active system has a negligible pressure drop, less than about 10 Pa. The extra capital and running costs of the higher pressure fan have to be set against the cost of installing the electronic circuit and loudspeaker. This kind of noise is very repetitive so a synthesis method is applicable. The wave propagation is one-dimensional along the duct, and so a conventional system can also be used, with the time delay of the electronics compensated for by a spatial separation of microphone and loudspeaker along the length of the duct, Figure 7. The separation distance is equivalent to the time taken for the electronic processing to be completed, multiplied by the sound of wave propagation in the medium.

The real situation in a duct turns out to be rather more complex than the simple description above indicates. If several frequencies are present and are in phase at the microphone, they will probably not be in phase at the speaker position, since the phase relationship varies with frequency. The propagation characteristics of the duct vary from day to day with temperature, pressure and humidity, so the situation is not stable. There is a further problem in that a speaker radiates upstream as well as downstream; the downstream component attenuates the sound level, but the upstream component is reflected back to the source and is sensed by the microphone as

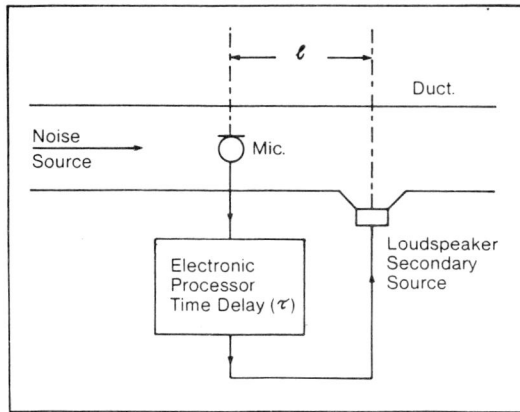

Figure 7 Geometric arrangement of monopole active
attenuator in a duct

another source, producing acoustic feedback instability. This problem can be overcome through the use of directional microphones and speakers and by more complex speaker arrays, see Figure 8. Advances in digital signal processing make it possible to overcome some of these drawbacks, but as the processing gets more complex, the longer time it takes and the greater the spatial separation needed to compensate.

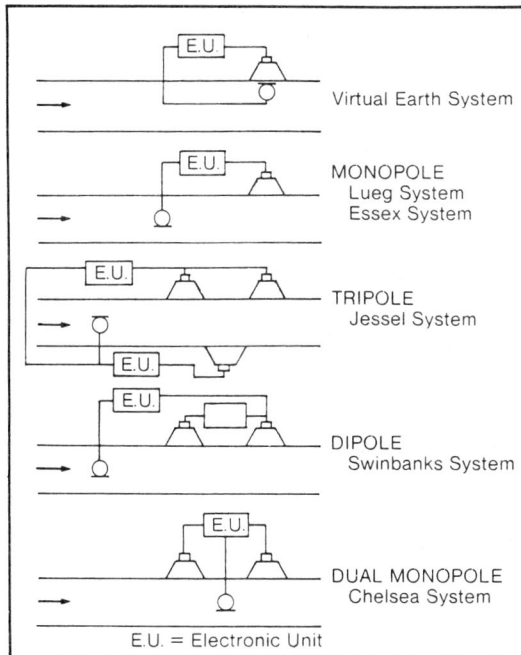

Figure 8 Types of active noise attenuators for ducts

Active Ear Defenders

Ear defenders can also be made using the principle. Either conventional or active techniques are possible, depending upon the application. One advantage of the synchronous type, Figure 9 is that only the repetitive signal is cancelled, leaving the non-repetitive signal, which can be a warning or a speech message, unaffected. This has particular applications for engine test bays. See chapter on *Hearing Conservation*.

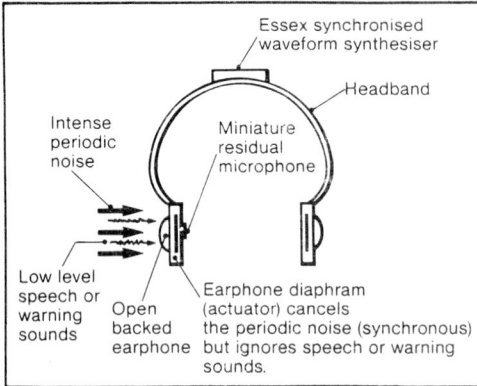

Figure 9 Selective ear defender. Standard open-backed (acoustically transparent) hi-fi earphones with Essex system fitted to cancel unwanted engine-induced noise, but to let through speech and warning sounds

Figure 10 Active Noise Control at a refrigerated food distribution depot

Chapter 7
Noise From Fluid Power Systems

Noise From Fluid Power Systems

One of the sources of noise in process plants and in hydraulic systems is pressure pulsations in the pipework. These pulsations are transmitted internally throughout the system, where they can set up vibrations in the pipes, valves, pumps and motors and may result in premature mechanical failure. Or they may be transmitted externally, where the internal pulsations set into vibration the surfaces of the components and so radiate airborne noise. The main sources of noise are pumps, valves and motors. It is usual to distinguish between fluid-borne noise (FBN), structure borne-noise (SBN) and airborne noise (ABN). FBN is transmitted throughout the system and can cause noise to be generated at a point remote from the origin; thus, where the noise starts at the pump, it may set the pipes into vibration and cause airborne sound to be produced by them or by the structure to which they are attached. SBN is generated internally, transmitted to the pump casing and from there to its mounting and attached pipework. ABN is radiated directly from the pump casing to the air.

A simple analysis of a hydraulic system assumes an ideal hydraulic fluid (i.e. an incompressible one) and inextensible piping. In practice, slight compression of the liquid always occurs, and pressure waves are generated by alternating compression and relaxation of the fluid. These waves are readily transmitted throughout the whole system, unless precautions are taken to intercept them.

Two standards give useful assistance in determining noise from fluid power systems. BS 5944 deals with the measurement of airborne noise from various components of a hydraulic system (pumps, motors and valves), and BS 6335 (Part 1 for pumps is the only one at present available) deals with the determination of pressure ripples in hydraulic systems. The latter standard describes the experimental procedure for rating a positive displacement pump and also describes a procedure for the calculation of the bulk modulus of a fluid via a measurement of the speed of sound. The relationship between the speed of sound and the bulk modulus is given by

$$c_o = \sqrt{\left(\frac{B}{\rho}\right)}$$

c_o is the speed of sound in the fluid, B is the bulk modulus and ρ is the density. The bulk modulus in this case is the effective tangential isentropic modulus, which is the value to be used in pulsation calculations.

Pump and Motor Pulsations

Positive displacement hydraulic pumps can take a variety of forms: gear, radial piston, axial piston and vanes, each of which can generate pulsations in the fluid on its pressure side . The worst offender is probably the piston pump, which is widely used because of its high efficiency and ability to generate high pressures. Pumps can also generate noise in the suction line by cavitation and by drawing in aerated fluid.

A piston pump produces its main noise at a frequency given by the product of the rotational speed and the number of cylinders. The pump delivery is not sinusoidal, so there are harmonics of the fundamental, whose magnitude depends on the type of valving employed and how smoothly the hydraulic pressure changes from suction to pressure. One common method of silencing piston pumps is through the use of silencing grooves, machined into the portplate prior to the pump openings. Silencing by this method is in the hands of the pump designer.

Gear pumps present less of a problem than piston pumps, but they can generate noise through the trapping of oil in the meshing space of the gears. Relief grooves can be milled into the end plates of the gear casing, or the design can rely on leakage across the tips of the gears. Vane pumps can be dealt with in a similar way.

Computer analysis is available to help in pump design. The aim of the design is a compromise between low noise and efficiency.

Motors present less of a problem, but since they have some of the features of a pump, they may give problems for much the same reasons. Downstream cavitation is the most likely FBN. ABN can be checked by the method of BS 5944, and dealt with in the same way as the other airborne sources.

If the pump generates excessive ABN, there is little that can be done other than to change the pump, isolate it well from the support structure or enclose it. Damping material applied to the pump casing is unlikely to be effective, unless very thick layers are applied.

Valve Noise

There are several ways in which valves can generate noise. If relief valves are set very close to the working pressure of the system, they can frequently operate to relieve the pressure. They are prone to instability if there are pressure fluctuations in the pipe. Probably the most common form of noise generated by a valve is downstream turbulence. When there is a restriction in a pipe, the fluid accelerates through it with a consequent loss of pressure; if the pressure falls below the vapour pressure of the liquid, cavities will form which generate noise as they collapse, see Figure 1. Valve noise can easily be distinguished from pump noise. The latter is strongly periodic, with the noise occurring at pump speed and its harmonics, whereas the former is a broad-band

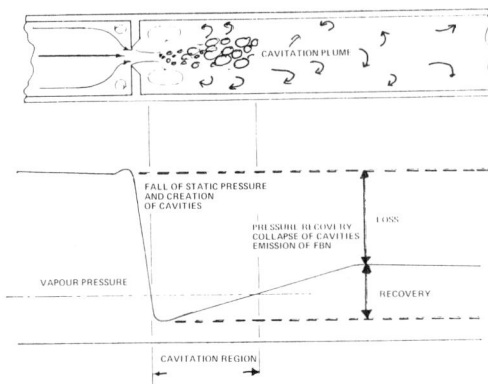

CAVITATION PLUME

FALL OF STATIC PRESSURE
AND CREATION
OF CAVITIES

PRESSURE RECOVERY
COLLAPSE OF CAVITIES
EMISSION OF FBN

LOSS

VAPOUR PRESSURE

RECOVERY

CAVITATION REGION

Figure 1 Cavitation at a restriction in a
pipe (above)
Plot of static head downstream (below)

random noise, from 2 kHz upwards. One way of preventing this kind of turbulent noise is to make sure that the downstream pressure is kept at a level which does not permit cavitation. Some modern hydraulic systems are pressurised on the return side, which helps to suppress both valve cavitation and the pump cavitation mentioned above. This kind of system carries with it the complication of having a pressurised reservoir. Suppressing valve noise is a matter of good design, with the avoidance both of sharp edges in the flow path and unnecessary changes of direction. Pilot operated valves tend to be superior to direct operated valves.

Noise Suppression by the Use of Gas Accumulators

Gas accumulators are widely used for a variety of purposes in hydraulic circuits. They are made from a steel shell, open to the hydraulic circuit, containing a rubber bladder which is internally pressurised with an inert gas, usually nitrogen. The nitrogen is precharged to a little less than the system pressure (about 90%). The accumulator is connected to the circuit by a port which is closed either by a poppet valve, or a button moulded into the bladder or sometimes by a perforated plate, Figure 2. The purpose of each of these is to prevent the extrusion of the bladder through the port when the system is unpressurised.

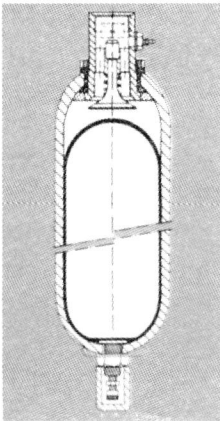

Figure 2 Standard bladder accumulator for fitting into a branch pipe in a hydraulic circuit

Accumulators are used:

to even out the flow demands from a pump, by charging up during periods of low demand and supplying fluid during periods of high demand,

to maintain pressure between set levels and so avoid continuous pump running,

to cater for thermal expansion in a closed circuit,

for shock absorbing and

for dampening of pressure pulsations.

The last two are of relevance to noise suppression. The accumulator is usually connected into the main pressure line close to the pump, as shown in Figure 3.

For low frequency pulsation, the size of the accumulator can be determined as follows:

$$P_{max} = P_o + \Delta P$$
$$P_{min} = P_o - \Delta P$$

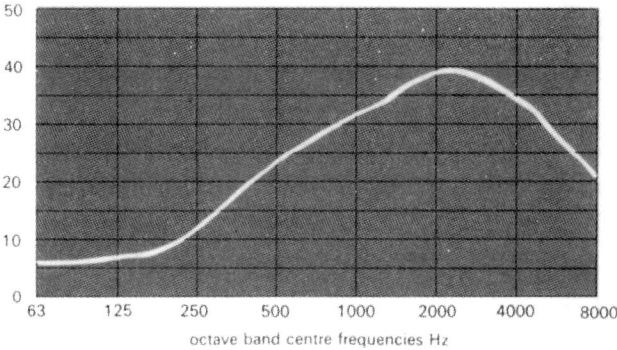

Figure 3 Installation of a gas accumulator in a hydraulic circuit

P_o is the system pressure and ΔP is the maximum permissible amplitude of the pressure pulsations.

The volume of the accumulator can be determined from:

$$V = \frac{\pi D^2 L K}{4} \cdot \frac{(P_o/P_{min})^{1/n}}{1 - (P_o/P_{max})^{1/n}}$$

D and L are the diameter and stroke of the piston, respectively. K is a constant obtained from Table 1 and n is the adiabatic index, equal to 1.4 for nitrogen.

TABLE 1 – TYPICAL PUMP PULSATION FACTORS

Pump Type	K
Simplex single-acting	0.6
Simplex double-acting	0.25
Duplex single-acting	0.25
Duplex double-acting	0.15
Triplex single-acting	0.13
Triplex double-acting	0.06

Accumulators behave in much the same way as a Helmholtz resonator as discussed in the chapter on *Sound Absorption*. They have a characteristic frequency where the absorption is most effective. Well above that frequency, the absorption may be inadequate. An estimate of the frequency of a gas accumulator can be made by first calculating the effective spring stiffness of the accumulator.

$$k = \frac{n P_o S^2}{V}$$

where S is the cross-sectional area of the neck of the branch connection to the accumulator and k is the effective spring stiffness. The effective resonator mass, m, is found from:

$$m = m_n + \left(\frac{S}{S_a}\right)^2 m_a$$

m_n is the mass of the liquid in the neck, m_a is the mass of liquid in the accumulator, S is the area of the neck and S_a is the area of liquid in the accumulator. In many cases, the second term in the equation can be ignored.

The frequency of the resonator is now found from

$$f = \frac{1}{2\pi} \quad (k/m)^{1/2}$$

A special design of accumulator incorporates into the design a baffle plate which directs the fluid flow into the bladder chamber, as in Figure 4. This adds a degree of high frequency attenuation to that obtainable with a standard accumulator.

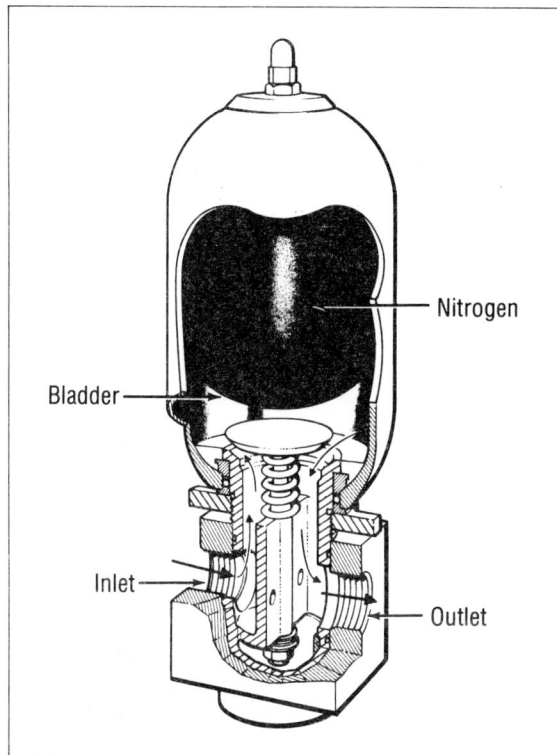

Figure 4 *Bladder accumulator showing a baffle to direct fluid flow into the bladder chamber*

Reactive Fluid Silencers

Although the fluid in an hydraulic pipe is much less compressible than that in the exhaust pipe of an engine, the same principles of silencing can be adopted, taking into account the differences in density and compressibility. A fluid-filled Helmholtz resonator behaves in exactly the same way as described above, except that the flexibility comes from the compression of the oil (and to a lesser extent, the expansion of the walls of the chamber). This has given rise to a number of proprietary designs of reactive silencers, on the principle of Figure 5.

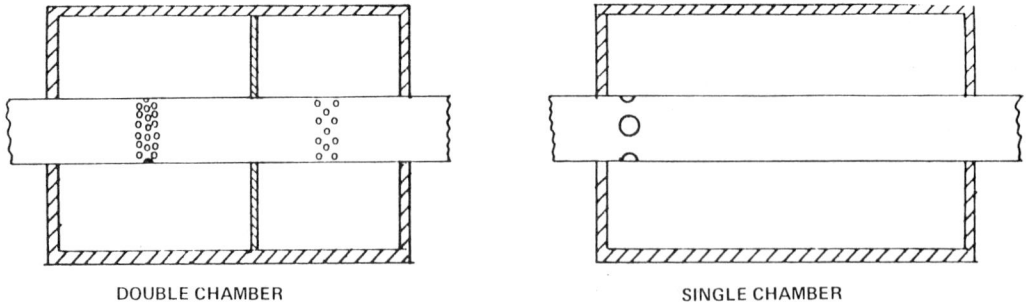

DOUBLE CHAMBER   SINGLE CHAMBER

Figure 5 Diagram of a hydraulic resonator in two forms

Piping

The damping properties of flexible hose are superior to those of rigid piping, so where possible flexible hose should be used. It is more expensive, but it may be a cheaper solution than the incorporation of a silencer in the system. Where rigid piping is used, it should be supported by well damped pipe hangers, two examples of which are illustrated in Figures 6 and 7. Rigid piping can also be enclosed in an acoustic sheath.

When designing the circuit, it is best to keep the piping size as generous as possible, using the connections to the valves and pumps as a guide to the correct sizing.

The length of the pipe between the pump and the first valve should be chosen so that it does not coincide with the wavelength of the pump pulsations.

The wavelength of sound in the fluid is

$$\lambda = \frac{1}{f} \left(\frac{B}{\rho_f} \right)^{1/2}$$

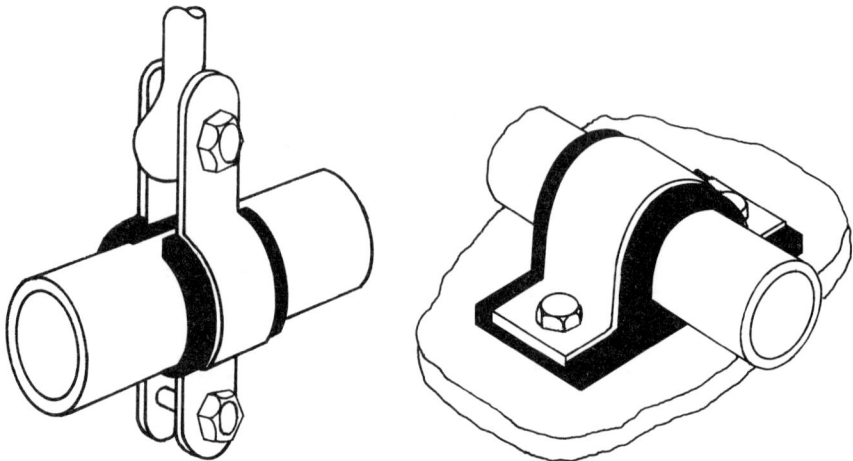

Figure 6 Pipe clips using Tico strip

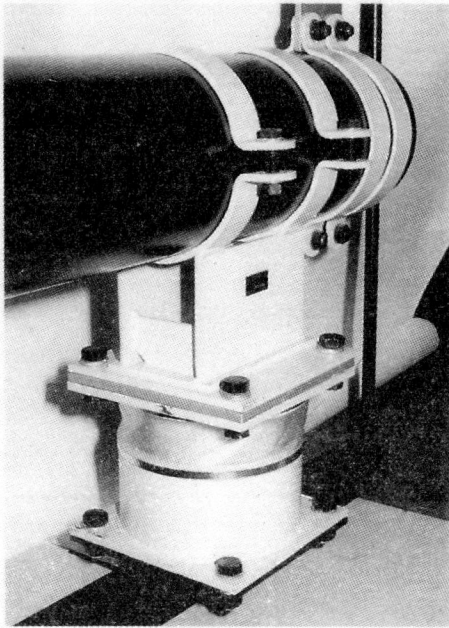

Figure 7 Pipework Damper available in sizes up to 450 mm diameter This is purely a viscous damper and will not support static loads

B is the bulk modulus of the combined fluid and pipe, λ is the critical wavelength, p_f is the frequency of the pump and f is the fluid density.

$$\frac{1}{B} = \frac{1}{B_f} + \frac{d}{tE}$$

B_f is the fluid modulus, t the pipe thickness and E the elastic modulus of the material of the pipe. This relationship is for thin wall pipes only. The calculation of B for rubber hose is more problematical. It depends on the manufacture of the hose, the type of reinforcement and the temperature. Not all suppliers are able to quote values.

An effective silencer can be made from a length of flexible hose inserted into the pipe, as in Figure 8.

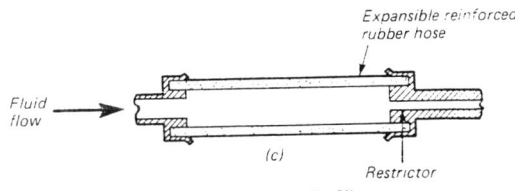

Figure 8 Pipe damper using a length of standard hose

Ways to Minimise System Noise

Use generously sized pipes on the pump inlet, and ensure either a good gravity head from the reservoir or a boost pump.

Use a large reservoir with baffles to encourage the release of air from the return fluid.

Fit flexible pipes where possible, and particularly on the suction side of the pump.

Consider a pressurised return system.

Mount pumps and motors on suitable isolators and avoid attaching them to resonant panels.

Minimise flow restrictions and sudden changes in pipe diameters.

Fit a pulsation damper or a silencer close to the pump.

Avoid lengths of pipe which are close to the wavelength of the pump pulsations or its multiples.

Choose valves, motors and pumps that have been designed to minimise pressure pulsations.

Consider the design of an enclosure around the noisy pumps and valves.

Use isolating pipe clamps.

Use servo valves which give a controlled opening and closing, thus avoiding the pressure transients.

SECTION 6

Chapter 1
Sound Insulation

Distinction Between Sound Insulation and Absorption

There is a clear distinction between sound insulation and sound absorption. A sound insulating material or structure attenuates the sound waves passing through it and thus acts as a sound barrier to the passage of sound from one room to another. A sound absorbent material, on the other hand, absorbs a proportion of the sound incident upon it so that the level of the sound reflected from the surface is reduced. Sound absorption is intended to reduce the loudness of reflected sound in a room or enclosure and decrease reverberation. At the same time a high proportion of sound energy may be transmitted through the absorbent surface, as sound absorbers are generally poor sound insulators. Sound insulation is effected by the use of heavy panels, the heavier the more effective, while sound absorption materials are made of foamed plastic, building boards with air cavities and other lightweight materials.

Sound absorbing materials applied to the internal room surfaces can contribute to sound insulation by reducing the general sound level in the room, so that the reduction to be achieved by the insulating treatment is smaller than it would be without the absorbent treatment.

Sound Insulation Principles

Figure 1 shows diagrammatically the principles of sound insulation.

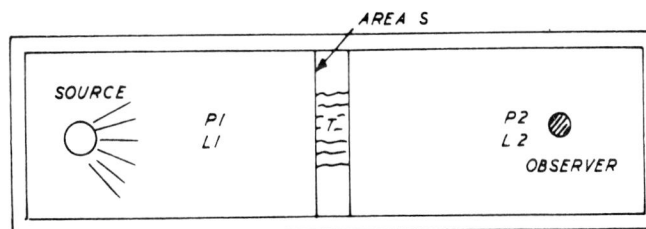

Figure 1

A number of terms have been introduced to define the material properties used for insulation.

Transmission Coefficient, t, is the ratio of the sound power transmitted through a partition to the sound power incident upon it.

$$t = \frac{W_2}{W_1}$$

Sound Reduction Index (R) is the ratio of the sound power transmitted through a partition to the sound power incident upon it *expressed in dB*.

$$R = 10 \log_{10} \frac{W_1}{W_2} = 10 \log_{10} \frac{1}{t} \quad dB$$

so $t = 10^{-0.1R}$

where W_1 is the sound power on the source side and W_2 is the sound power on the observed side. The Sound Reduction Index is also known as the Sound Transmission Loss. When values of these parameters are given it is usually understood to be for a unit area (per square metre). When one is calculating the value of t for a partition comprising n elements, the overall value is given by

$$A_{tot} t_{tot} = \sum_{k=1}^{k=n} A_k t_k$$

where A_k and t_k represent the individual areas and transmission coefficients of the elements of the panel, A_{tot} and t_{tot} represent the total area and coefficient for the complete panel.

Where the coefficient is that of a complete element, e.g. a window which includes the glass and the frame, it replaces one of the individual terms $A_k t_k$ in the summation.

If R_e, the effective value of R for the panel, is required, it is found from

$$R_e = 10 \log_{10} \frac{1}{t_{tot}}$$

The Apparent Sound Reduction Index, R', is expressed as

$$R' = 10 \log_{10} \frac{W_1}{W_3} \quad dB$$

W_3 differs from W_2 in that it represents the total sound power transmitted into the receiving room. In any practical situation, there will always be paths for the sound additional to that through the partition. These paths include flanking transmissions (round the sides of a panel), leaks, ventilation ducts and the like, with the result that the measured noise in the receiving room is greater than pure calculation would indicate and consequently the effective sound reduction is smaller. R' takes care of this; where known, this is value to be used for calculations.

R depends upon the angle of incidence of the sound wave, but since in many cases it is convenient to deal with diffuse fields, the value of R (and R') is then given by

$$R = L_1 - L_2 + 10 \log_{10} \frac{S}{A} \quad dB$$

alternatively

$$R = L_1 - L_2 + 10 \log_{10} \frac{ST}{0.163 V}$$

L_1 and L_2 are the sound pressure levels, S is the area of the partition, A is the equivalent absorption area in the receiving room, T is the reverberation time and V the volume of the receiving room. In accordance with Sabine's Law,

$T = 0.163 \, V/A$

The Level Difference, D, between the two rooms is defined as

$D = L_1 - L_2$

The Standardised Level Difference, D_{nT}, is defined as

$$D_{nT} = D + 10 \log_{10} \frac{T}{T_0}$$

T is the reverberation time in the receiving room and T_0 is the reference reverberation time 0.5 s, assumed to be correct for a furnished domestic room.

When evaluating R or D, the correct procedure has to be followed as explained in BS 2750. The method requires the use of a special room as described in the chapter on *Room Acoustics*.

R varies with frequency, and when a laboratory test to evaluate it is carried out, the full frequency range from 100 Hz to 3150 Hz has to be covered either in octave or third octave bands. Sound transmission loss is poor at low frequencies but improves at the higher frequencies, and the variation with frequency follows a similar pattern for all building materials in common use. For the practical purpose of evaluating the effectiveness of a partition, an average is required, which has to be calculated from the individual octave band measurements.

There are several ways of finding the average. The easiest, and for many purposes a perfectly satisfactory way, is to find a numerical average of all the values in the full frequency range. If octave band measurements are taken, there are six readings to be averaged, and for third octave measurements there are 16 readings.

The Weighted Sound Reduction Index, R_w, is determined by a procedure given in BS 5821 and requires the use of the reference curve of Figure 2. This curve can be used to find averages for R, R', D or D_{nT}. The measured values are first plotted on a grid similar to Figure 2. The reference curve is shifted in steps of 1 dB towards the measured curve, until the mean unfavourable deviation, calculated by dividing the sum of the unfavourable deviations by the total number of measurement frequencies (6 or 16, according to whether octave or third octaves values are available), is as large as possible , but not more than 2 dB. An unfavourable deviation at a particular frequency is when the measured value is less than the reference value. Only unfavourable deviations are taken into account. R_w is then the value of the reference contour at 500 Hz.

A parameter used in USA and similar to R_w is known as the Sound Transmission Class, STC, which is calculated in accordance with ASTM E413. It is practically identical to R_w, except that the frequency range is from 125 Hz to 4 kHZ in third octave bands only, and there is a further requirement that no measured value of R at any frequency shall be more than 8 dB below the reference curve.

When making an initial estimation of the likely effect of a particular panel, it is often adequate to take a average value, calculated as above, providing that the full frequency information is known. An example of the several ways of determining R is shown in Figure 2. If measured information is unavailable the following empirical relation may be used.

$R = 15 \log_{10} m + 12 \, dB$

Using this equation, the value for R in the example of the brick partition of Figure 2 (weight 220 kg/m^2) is 47 dB.

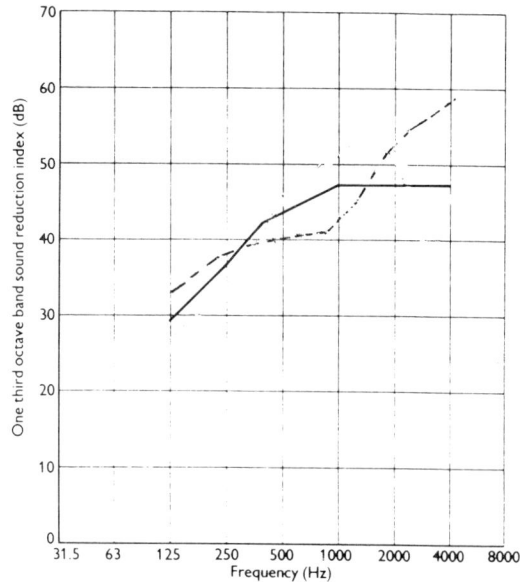

Figure 2 Sound reduction index for a brick partition
with the weighting curve of BS5821 superimposed
$R_W = 43\,dB$, $R_{average} = 44\,dB$, $R_{500} = 40\,dB$

It will be seen from this that the mean value of R increases by 4.5 dB per doubling of the weight ($15\,\log_{10} 2 = 4.51$). Some authorities quote 5 dB. It should be emphasised that the formula is empirical and should only be used where measured data are not available.

In summary the actual performance of a sound insulation panel is subject to modification in a number of ways:

The sound reduction index, R, is frequency dependent; low frequencies are more readily transmitted than high frequencies. The performance of a sound insulating material is only fully described by knowledge of its R value throughout the frequency range. Average values, based on measurements at specific frequencies, are normally used, except that where particular frequencies predominate it may be necessary to calculate the sound reduction throughout the spectrum.

R depends on the angle of incidence of the sound wave received from the source. In most cases, practical applications involve random incidence, so R can be defined by sound waves of random incidence, equivalent to hemispherical distribution from the source.

At certain frequencies of the incident sound, coinciding with the natural frequencies of the panel, resonant vibrations will occur with a marked reduction in sound insulation. This is because the panel is acting as a generator of sound energy. Loss of performance as a sound insulating material will be most marked at the lowest frequency.

At certain frequencies, the phases of incident sound will tend to coincide with the phase vibrations of the panel. This can induce flexural vibrations in the panel, substantially reducing the sound insulation. This is known as the coincidence effect and occurs at frequencies greater than a critical frequency, defined as that frequency at which the flexural wave velocity equals the velocity of sound in air. For a homogeneous panel, the critical frequency is proportional to the thickness of the material. The coincidence problem therefore is most likely to occur in the case of thin panels with low damping. Increasing the damping or the mass can easily render the coincidence effects negligible.

Sound transmission through the panel may be supplemented by secondary transmissions through the adjacent structure, Figure 3. These are known as flanking transmissions. Where the insulation is of the order of 35 dB or less, the flanking transmissions are usually negligible. There is also the possibility of sound being transmitted through vibration of the building structure or through any gaps or openings. A reduction of 60 dB for example is equivalent to a ratio of one million, so a small opening can vitiate the effect of an otherwise well designed panel.

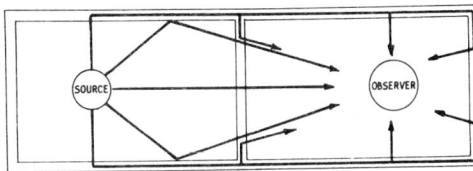

Figure 3 Flanking Transmission

Panels may have porosity or cracks in them which reduce the mass effect. Lightweight concrete or brickwork with cracks in the mortar can have sound reduction properties worse than calculations would indicate because of the unseen passages through which the sound can travel. For this reason, it is recommended that brickwork should be faced, at least on one side by a dense rendering.

Sound Insulation of a Panel

Figure 4 shows the characteristic shape of the transmission loss curve for a single leaf partition. The three regions shown are dominated by stiffness, mass and mass/coincidence.

Sound transmission is brought about by the partition being forced into vibration by the pressure variations in the incident wave. The vibration of the panel on the receiving side excites the air into motion and creates the sound waves experienced by an observer in the receiving room. The

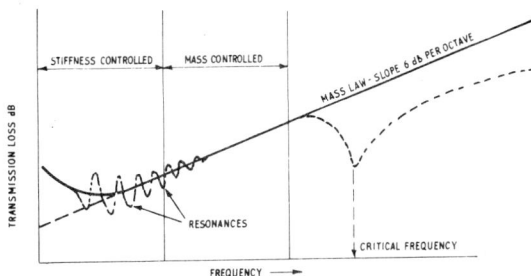

Figure 4 Transmission loss for a single leaf panel

panel vibrates mainly as a panel supported by the floor, walls and ceiling. The heavier the panel, the more difficult it is to excite, and so the better it is as a sound insulator. The general law relating mass and frequency to the Sound Reduction Index is known as the Mass Law. It is an empirical expression and so different authorities have different expressions for it. One common form is:

$$R = 20 \log_{10} (m\,f) - 48\,dB$$

m is the surface density in kg/m^2 and f is the frequency in Hz.

As pointed out above, the value of R is approximated by its individual value at 500 Hz. Inserting 500 Hz into the above equation gives a mean value of

$$R = 20 \log_{10} m + 6\,dB$$

which can be compared with the previously quoted formula

$$R = 15 \log_{10} m + 12\,dB$$

Discrepancies of this kind are unavoidable when empirical formula are used, and point to the desirability of always using the best available information obtained from experimental measurements.

The relationship in this form applies to sound transmission at random incidence. For sound waves of normal incidence, R is about 5 dB greater.

The important point to emphasise is that, whatever formula is used, R is determined primarily by the mass of the panel. There is no such thing as a lightweight acoustic panel made from some magic material. Apart from the minor factors discussed below most of which reduce R, for a homogeneous panel one cannot do much better than accept the predictions of the Mass Law; doubling the thickness of a brick wall will only increase its R value by 6 dB at best. However, a double panel partition separated by an air cavity is considerably more effective than a double thickness wall without the cavity, provided it is well designed.

Deviations from the Mass Law

The Mass Law is normally presented as 6 dB per octave and this is the way it is shown in Figure 4. The mass can be increased either by choosing a denser material or increasing the thickness. Figure 4 shows the 6 dB/octave line modified by two other factors, the stiffness effect at low frequency and the coincidence effect at high frequency. Both of these are directly related to vibrations induced in the panel.

The first kind of vibration, dominated by the stiffness, occurs, as indicated above, when the panel acts as a plate supported on its edges; it can be studied by the methods described in the chapter on *Modal Analysis.* Plate vibration modes depend upon the shape and the method of support, as well as the stiffness and mass, so it is not possible to predict the sound reduction capability of a panel in this region without doing a complete study of the modal behaviour. The frequency of a panel of conventional construction is usually of the order of 100 Hz or less, and in this region lightweight panels of sandwich construction made from thin skins with foam, plastic or honeycomb cores can be effective, provided that there are no other complex resonances to limit their performance.

The second kind of vibration is of flexural waves, produced by the incident sound wave at a grazing incidence to the panel. It is customary to deal with the sound wave in the source room as diffuse, where the waves have a random incidence to the panel. The components tangential to the surface will excite flexural motion. The frequency of the wave depends upon the material thickness and its stiffness. When the speed of the flexural waves equals the speed of acoustic wave propagation in the air, a phenomenon known as coincidence occurs at a frequency known as the critical frequency. The change it can make to the sound reduction index, i.e. the magnitude of the coincidence dip, is determined by the amount of damping in the material. Some panels have very good damping properties. Those containing limp materials such as lead, or ones made of studding held together by screws or nails, which allow relative internal movement, are good. Windows in which the glass panels are made of two thin panes with a plastic layer between them also have good damping. Composite constructions with rigid glues are less good.

The value of the critical frequency may be found by reference to Table 1. If it is necessary to calculate it for materials not in the table, a formula developed by Cremer may be used.

The bending stiffness of the panel per unit width is first calculated:

$$B = E I (1 - \nu^2)$$

where B is the bending stiffness in Pa m^3, E is the elastic modulus in Pa, I is the second moment of area per unit width in m^3, ν is Poisson's Ratio.

The critical frequency is when the speed of propagation of bending waves equals the speed of acoustic wave propagation in the surrounding medium. This leads to:

$$f_c = \frac{c^2}{2\pi} \sqrt{m/B} \quad Hz$$

f_c is the critical frequency, c is the velocity of sound in air (330 m/s) and m is the surface density in kg/m^2.

This calculation is based on the assumption that the panel is isotropic (equal stiffness in all directions). In the case of orthotropic panels, for example with stiffening ribs in one direction, there will be two values of the critical frequency. For a panel made from a *homogeneous* material, the term $\sqrt{(m/B)}$ is inversely proportional to the thickness, so this leads to the simple relationship $f_c = A/t$, where t is the thickness in mm. Values of A for many materials in common use are in Table 1. As an example, a brick wall 100 mm thick, has a critical frequency of 18,000/100 = 180 Hz.

If one of the dominant frequencies of the source is close to the critical frequency, it may be necessary to pay particular attention to this effect, by changing the construction or by adding damping.

Double Surface Panels and Cavity Walls

In this type of construction, much improved sound reduction properties are achieved. If the cavity is large the result is almost as if there were an intermediate room between the source and the observer. One might expect in this case that the value of R would be doubled (in dB) for a double panel. For example, a wall with an R value of 25 dB, would only increase to 31 dB if it were doubled in thickness, but if it were made with a cavity between, one might expect that it would increase to 25 + 25 dB = 50 dB. There are several reasons why this does not happen in practice.

In a cavity the two panels are usually tied together with some form of bridging construction which acts as a sound path. Even if the two panels are separate, flanking transmission can take place round the junctions with the floor and walls. The ideal design for a cavity construction is where the two panels are mounted independently.

The sound level in the cavity can build up through internal reflections, so that, although the

**TABLE 1 – CRITICAL FREQUENCIES OF COMMON
MATERIALS IN CONSTRUCTION**

Material	Density (kg/m² per mm)	Constant A (Hz mm)
Aluminium	2.7	12,900
Brick	1.9	22,000
Concrete		
dense poured or		
reinforced	2.3	18,700
hollow block	1.1	20,900
Flax board	0.39	33,800
Glass	2.5	15,200
Hardboard	0.81	37,800
Lead	11.0	55,000
Partition board	1.6	77,500
Plasterboard	0.75	40,000
Plywood	0.6	21,700
Perspex	1.15	30,600
Steel	7.7	12,700
Timber (fir)	0.55	8,900

Note: Properties of construction materials show considerable variation within a range and the above values
should only be taken as typical.

value of R for the individual panels may be correctly determined, the sound level on which it has
to operate may be higher. This can be overcome, to some extent, by lining the cavity with a sound
absorbent material, but the most effective use of a filling is in acting as a resistance to the air flow
in the cavity, by preventing standing waves. For this purpose a fibrous material such as glass fibre
or rockwool is preferable to a closed cell foam.

The air in the cavity acts as a spring, transmitting energy from panel to panel, and introducing
the possibility of a phenomenon known as mass-air-mass resonance, when the surfaces vibrate
on the spring of the air in the cavity. This is particularly noticeable in double glazed panels which
are hermetically sealed. Many double skin constructions suffer from this kind of resonance at low
frequencies. The lowest order resonant frequency is found from:

$$f_m = K \sqrt{\frac{m_1 + m_2}{d\, m_1\, m_2}}$$

Theory would indicate that the value of K should equal 60, however to take account of experience of typical panels used in construction, K is often taken as equal to 80 for an air-filled cavity
and to about 60 for one filled with an absorptive material. m_1 is the surface mass of the first layer,
m_2 is the surface density for the second layer and d the gap width in m.

The theory on which this is based assumes that there are no standing waves in the gap and so
there is air no coupling between the panels. Coupling can be prevented by filling the gap with a
sound absorbent material such as mineral fibre. The filling will not alter the mass-air-mass resonance frequency, although if dense enough it may suppress the peak response.

As an example of the use of this formula, take a double-glazed panel made of 3 mm glass with
a 6 mm gap. From Table 1, the surface mass is $2.5 \times 3 = 7.5$ kg/m²

$$f_m = 80 \times \sqrt{\frac{7.5 + 7.5}{0.006 \times 7.5 \times 7.5}} = 530 \, \text{Hz}$$

Increasing the gap to 18 mm changes the value of f_m to 305 Hz.

If it is possible, the two panes should have different thicknesses so that there is a separation between the individual resonant frequencies.

To calculate the sound reduction index for a double skin panel, there are two regimes to consider: below and above the resonant frequency f_m. Below it, the composite panel behaves as if the separate leaves were connected and acting as a single panel; above it, the composite index is calculated by adding about 6 dB to that for the separate panels added together. At frequencies close to f_m, the curve of R shows a characteristic dip, whose magnitude is dependent on the internal absorption of the cavity. There is some evidence to show that R increases at about 9 dB per octave above f_m. If there is coupling in the form of rigid studding which connects the two panels together, they effectively make the panels behave as one and the mass law applies.

While the above analysis gives an outline of the theoretical background to the calculation of R, it is emphasised that it is always preferable to use measured values wherever possible. Table 2 gives some typical values for construction materials in common use.

TABLE 2 – SOUND REDUCTION INDICES VALUES IN dB

Type of partition (figures in brackets are thicknesses in mm)	m kg m^{-2}	Frequency/Hz						
		100	200	400	800	1600	3150	Mean
Brick, unplastered (115)	220	34	37	40	41	51	57	43
Brick (225) plastered both faces	480	48	40	45	50	61	67	52
Cavity wall, brick (50 × 115) airspace (50) plastered both faces, butterfly ties	480	30	36	45	58	65	82	53
Clinker concrete wall (75) unplastered	100	16	15	20	23	29	38	23
Clinker concrete wall (75) plastered both faces	150	22	33	36	47	55	56	42
Concrete floor (100) reinforced	230	36	40	38	47	55	65	47
Wood joist floor, boarding (22) nailed to joists (225 × 50) plasterboard ceiling (9.5) skim coat plaster	80	16	26	34	35	42	41	32
Fibreboard (12)	4	11	15	18	22	27	30	21
Plasterboard (9.5)	10	13	19	22	28	34	35	25
Plasterboard (9.5) each side of (100 × 50) studs (airspace 100)	25	12	23	32	36	44	46	32
Plywood (6)	4	8	10	15	20	25	28	18
Steel, 16G	12	14	20	25	31	37	43	28
Plate glass (6)	17	18	23	24	30	30	35	27
Window glass (3)	7	18	12	20	26	30	28	22
Window glass, double, airspace (50)	14	19	19	27	36	47	52	33
Window glass, double, airspace (180)	14	28	25	36	43	50	53	39

Insulation Properties of Concrete Blockwork

An example of noise insulation typical of cast concrete blocks is presented in Tables 3 and 4. Information from Boral Edenhall

TABLE 3 – VALUES OF SOUND REDUCTION INDEX FOR DENSE BLOCKS

Thickness (mm)	Form	Density (kg/m³)	Average SRI (dB)
75	S	2000	43
90	S	2000	43
100	S	2000	43
	C	1800	43
140	S	2000	45
	C	1800	45
	H	1575	44
150	S	2000	46
	H	1575	44
190	S	2000	47
	H	1575	46
200	S	2000	48
	C	1800	47
	H	1575	46
215	S	2000	48
	H	1575	47

TABLE 4 – VALUES OF SOUND REDUCTION INDEX FOR LIGHTWEIGHT BLOCKS

Thickness (mm)	Form	Density (kg/m³)	Average SRI (dB)
75	S	1400	39
90	S	1400	40
100	S	1400	41
	C	1100	40
140	S	1400	43
	C	1100	42
	H	1000	41
150	S	1400	44
	H	1000	42
190	S	1400	45
	H	1000	43
200	S	1400	45
	C	1100	44
	H	1000	43
215	S	1400	46
	H	1000	44

Notes: Values of SRI are given for single leaf plastered work in the range 100 – 3150 Hz
S = solid, C = cellular, H = hollow
Surface density can be found as product of density and thickness.
Special acoustic blocks are available, which in addition to their insulation properties, have some sound absorbent properties, achieved through the form of the block and by incorporating an acoustic filler, Vitafoam. They act as damped Helmholtz resonators. Sound absorbing coefficients of this type of block are given in Table 5. Their insulation properties are similar to those of a standard block.

TABLE 5 – PROPERTIES OF ACOUSTIC BLOCKS

SOUND ABSORPTION COEFFICIENT‡				SRI COMPARISON	
Frequency (Hz)	100 mm Type 'A'	100 mm Type 'R'	140 mm Type 'R'	140 mm Type 'F'	140mm Noisemaster Blocks	140mm Standard Cellular Blocks†
100 125 160	0.19	0.20	0.38	0.62		
200 250 315	0.83	0.72	0.80	0.63		
400 500 630	0.41	0.90	0.43	0.39		
800 1000 1250	0.38	0.55	0.40	0.40		
1600 2000 2500	0.42	0.52	0.42	0.53		
3150 4000 5000	0.40	0.50	0.40	0.62		

‡Based on either A S T M C 423-60 C423-66 and C423-77 or BS3638 : 1963.

†See relevant Data Chart

140mm Noisemaster

STL (dB)

Frequency (Hz)

140mm Ordinary Two-core blocks

STL (dB)

Frequency (Hz)

Chapter 2
Sound Absorption Materials

General Principles of Sound Absorption

The principle of sound absorption is shown in Figure 1. When a source of sound is in a room or enclosure, sound waves strike the surface of the room at random incidences and will be reflected. If the surface is absorbent, a proportion of the sound will be absorbed and the remaining proportion reflected.

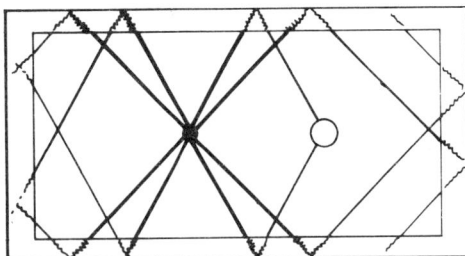

Figure 1 Absorption in an enclosed space

The overall effect on the observer is more complicated. He will receive both direct and reflected sound waves. He has no means of distinguishing between the two and experiences only the resultant sound, the intensity of which will depend upon his position in the room; if close to the source, direct sound will predominate, with a magnitude governed by the inverse square law, and with a frequency spectrum equal to that of the source. At a more distant part of the room, the sound level will be more constant, consisting mainly of components of the sound reflected off the walls; the frequency spectrum may differ from that of the source, since the wall surfaces do not reflect equally throughout the frequency spectrum. The more absorbent the surfaces are, the lower will be the sound level well away from the source.

The value of a sound absorbent treatment applied to the surfaces lies in lessening the general noise level in the room. It can be particularly valuable in large workshops where noise is confined to certain parts of the room. Absorbent treatments can prevent noise being reflected into areas of quieter activity. It should be noted, however that additional sound insulating barriers may be necessary to reduce the noise radiated directly from the source. When a complaint about noise is received, one should first determine whether the complainant objects to direct sound or to general reflected sound, and choose the treatment accordingly.

All the generally used materials of construction have surfaces which absorb some sound energy, varying from about 5% for hard painted surfaces to 80 – 90% for acoustic materials. To achieve lower than 5% or higher than 95% requires special techniques, such as those employed in acoustic test rooms.

The level of reflected sound can, in theory, be reduced to any desired extent by installing sufficient absorbing material, but in practice this is not so easy. A doubling of the total absorption of a room results in a halving of the sound level (3 dB reduction). When there is little absorption already existing, it is easy to double it, but when the whole of the room surface has been covered, with acoustic tiles on the walls and ceiling, carpets on the floors and curtains at the windows, it becomes difficult to double it again. It becomes necessary to use suspended ceilings, thicker absorbent materials backed by an air gap and other such expensive techniques.

Several general conclusions can be drawn:

> The addition of sound absorbent material is particularly effective in reducing noise if the amount of existing absorption is small.

> If the room is well furnished, the application of further treatment will bring little benefit. In dwelling rooms containing soft furnishings, it is unlikely that a noise reduction more than 5 dB will be achieved with acoustic treatment. A improvement of up to 10 dB may well be achievable in workshops or classrooms.

> Because sound absorbent treatment is ineffective at reducing the direct sound, the operator of a noisy machine will not benefit significantly from wall treatment. This is particularly true for large workshops and offices, where most of the sound is experienced directly.

> When most of the sound is experienced indirectly by people remote from the source, a benefit can be expected by sound treatment applied to the surfaces.

> A benefit can be obtained by the use of moveable screens with absorbent treatment applied to the surfaces. This will add to the effective absorbency of the whole room, and may also act to insulate direct radiation from the source.

Types of Sound Absorber

Sound energy can be absorbed by the following mechanisms:

> Friction between the fibres of a porous material or in the voids of a non-fibrous material – a dissipative absorber.

> A membrane absorber which works by vibration of a highly damped panel.

> A tuned cavity absorber, working on the principle of a Helmholtz resonator.

The performance of an absorber is expressed as its absorption coefficient, which has a minimum value of 0 and a maximum of 1.0. A coefficient of 0 represents total reflection and 1.0 represents total absorption. The coefficient, α, is found from

$$\alpha = 1 - r$$

where r is the ratio of the sound energy reflected from a surface to that incident upon it. It is frequency dependent so, when considering the use of a material, it is important to know the spectrum of the noise with which one is dealing and the particular frequencies one wishes to suppress. Its value is not absolute but depends upon the technique used for measurement. For most purposes, the approved method is to be found in BS 3638, which gives a value when measured in a diffuse field and is known as the reverberation absorption coefficient; it is also sometimes known as the Sabine coefficient. Other methods of measurement use an acoustic tube and are not so representative of a practical situation. The frequencies of measurement are 125 Hz, 250 Hz, 500 Hz, 1 kHz, 2 kHz, 4 kHz.

Some methods give rise to an anomaly, in that the coefficient at certain frequencies apparently exceeds 1.0. The explanation for this is complex, and for most practical purposes it is wise to limit the value of the coefficient to 0.95, despite higher figures being quoted. Even a value of 0.95 is unlikely to be reached with normal materials available for domestic and industrial applications.

Noise Reduction Class (NRC) of a material is the average value of absorption coefficients measured at the frequencies 250 Hz, 500 Hz, 1 kHz and 2kHz, expressed to the nearest 0.05. NRC is used for material specification, but for analysis purposes the coefficients throughout the range are to be preferred. As an example, a typical 12 mm thick urethane foam has the characteristics:

Frequency (Hz)	Absorption Coefficient
125	0.04
250	0.07
500	0.18
1000	0.4
2000	0.65
4000	0.83
NRC	0.35

Types of Dissipative materials

The requirements of a sound absorbing material are that it should be sufficiently porous to allow sound waves to enter the material so that the maximum proportion of the sound energy can be absorbed in friction. In a porous material, the pores should be interconnected, rather than closed, and the material must have a high damping quality. A fibrous material by its nature allows air flow through the thickness. A material which is good for thermal insulation usually has unconnected pores which is not so good for sound absorption. The ability of a material to absorb sound is related to its resistance to a steady air flow through it. Flow resistance is the ratio of the pressure drop through a sample of material to the velocity of air passing through it. It is measured in rayls (Pa s/m) or acoustic ohms (Pa s/cm).

The specific flow resistance, R_f, of a material of uniform consistency is the flow resistance per unit thickness.

$$R_f = \frac{P}{dv} \quad \text{rayls/m}$$

P is the pressure drop in Pa, d the thickness in m and v the flow velocity in m/s.

The calculation relating flow resistance to acoustic absorption is complex, but it has been shown that optimum results are obtained when the value of the flow resistance lies in the range of 800 to 2000 rayls. It is not usually helpful to apply this theory to predict the value of the absorption coefficient. In practice, measured values are always used.

An increase in the thickness of fibrous or porous absorbers improves the sound absorption in the low and middle frequencies, but seems to have less effect at high frequencies. An example of this is shown in Figure 2. Usually 25 mm is the minimum thickness that should be used, unless absorption is required only at high frequencies, and this applies whether or not the material is mounted over a cavity.

The change of absorbency with density depends primarily on the increase in flow resistance

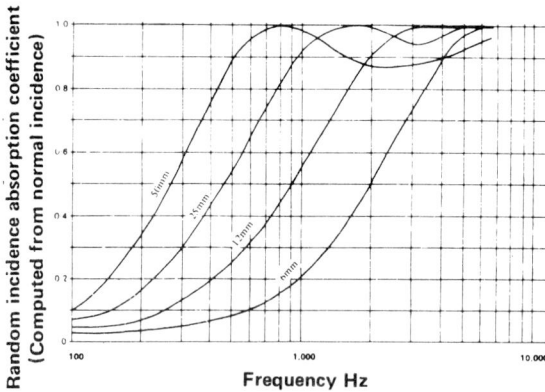

Figure 2 Absorption characteristics of
'Revertex' foam pads

rather than on a direct density effect. One might expect that as the material density increased, the sound absorbency would also increase until the optimum was reached, after which point, with increasing density, the absorbency would decrease again. There is some evidence for this.

A sound absorbent material which relies on absorption by dissipation through friction in the pores should preferably be backed by a cavity so as to allow air to flow through. If the material is backed by a solid wall, the velocity at the wall is zero and the material close to the wall is ineffective. The depth of the air cavity is governed by the frequency which it is required to suppress. Mounting the material off the wall has the same effect as increasing the thickness of the absorbent material. The optimum condition is when the air gap thickness equals the material thickness, in which case the absorbency is equal to the that of a material of twice the thickness with no air gap. The absorption is most effective when the sound propagation is normal to the surface. This can be assured if lateral propagation is prevented through the use of battens or ribs.

The frequency below which the absorption characteristics of porous materials start to deteriorate is determined by the wavelength in relation to the total depth of material plus cavity. This turnover frequency is approximately

$$f = 0.1 \, c/d$$

where c is the velocity of sound and d the total thickness. This implies that the turnover frequency occurs when the total depth is one tenth of the wavelength. At higher frequencies, the absorbency can be high (greater than 0.9). The only practical way of improving low frequency absorbency is by increasing the effective depth; thus for a high absorbency at 250 Hz, the overall thickness needs to be 100 mm.

The use of an exposed porous absorbent material is impractical when it is necessary to protect the surface from contamination by moisture, chemicals, dust, oil, etc. Provided that the material is faced with a limp membrane such as a plastic sheet, its presence makes little difference to the sound quality, except that the high frequency properties tend to deteriorate. A paint layer which seals off the surface is not recommended. The frequency at which a membrane causes a reduction in absorbency is given approximately by:

$$f = 100/m$$

where m is the surface density in kg/m^2

In many practical situations, it is necessary to protect the porous material against mechanical damage, which requires a cover made of perforated hardboard, plywood or some similar material. If the facing panel has an open area in the form of holes occupying greater than 20% of the total, the properties of the absorbent material are the same as if the panel were not there.

Panel Absorbers

These absorbers work by absorbing the sound energy in a vibration of a damped panel. This type of absorber is particularly advantageous for low frequency sound, in which dissipative absorbers are not effective as demonstrated in Figures 2 and 4. Usually the stiffness of the panel can be ignored, the effective frequency being governed by the panel mass and the volume of the air cavity behind it. The construction of these panels usually consists of flexible panels mounted on battens attached to the wall or they may be mounted on a suspended ceiling. If the panel is backed by porous material, some of the energy can be dissipated in friction of the material as well as in the air space. The most effective frequency of a panel absorber is found from

$$f = \frac{60}{\sqrt{m\,d}}$$

m is the mass per unit area in kg/m^2 and d is the depth of the cavity in m.

The prediction of the absorption coefficient of a panel absorber is not only theoretically suspect, but practically difficult. It is not possible to determine a value for the coefficient under laboratory conditions, which can be related to the actual coefficient when placed in a room. The acoustic behaviour of the panel depends on the nature of the coupling between it and the sound field, and on the design of the support structure.

From the above equation it can be seen that a panel absorber is most effective at low frequencies. A typical example of a panel absorber is a suspended ceiling, which can be made from plasterboard, hard board, plywood or some similar 'hard' material. One advantage that this kind of absorber has is that the surface of the panel can be decorated without affecting its performance. If the hard panel is replaced by a soft material such as acoustic tiles, the combination is an absorber effective over a wide frequency band.

A perforated panel is not effective as a panel absorber, because the air volume can no longer act as a cushion. It does, however, become a cavity absorber, with the characteristics of a Helmholtz resonator.

Cavity Absorbers

By this is usually meant a Helmholtz resonator. In its simplest form, it can be represented by Figure 3 and such isolated cavities can be used where it is necessary to absorb sound at a single frequency. The resonant frequency of this device is found from

$$f = \frac{c}{2\pi} \sqrt{\frac{A}{L\,V}}$$

c is the speed of sound , A the cross section area of the neck, L its length and V the volume of chamber.

At the resonant frequency a cavity absorber is capable of an equivalent absorption coefficient of 0.9, but the response curve is highly tuned and away from resonance, the effect is much less. If the cavity is lined with an absorbent material, the effect is to flatten the peak and broaden the

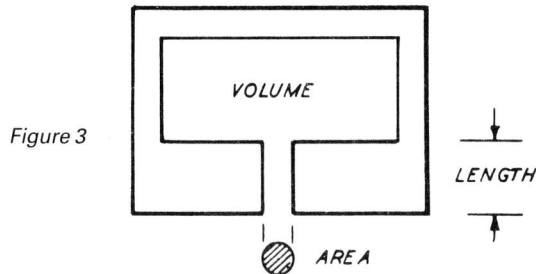

Figure 3

bandwidth. Acoustic panels with an array of Helmholtz cavities are available commercially.

An alternative form of resonant absorber is made from a perforated panel mounted over an air cavity. The perforations are usually round holes but can be slits or other decorative shapes. This type works best if the cavity behind the perforations is divided up into small volumes. The resonant frequency of this panel is given by

$$f = \frac{550\,P^{1/2}}{[L\{t + 0.85d\,(1 - 0.22d/a)\}]^{1/2}}$$

where P is the percentage open area, t the thickness of the panel, d the diameter of the perforations, a the spacing between the holes, all dimensions except P in mm.

A further development of the cavity resonator is the incorporation of an absorbent material placed behind the perforated panel. If this is done, care must be taken to ensure that the liner is not in contact with the panel, otherwise the effect of the cavity will be lost. A mesh should be placed between the panel and the liner.

A combined construction of this kind is a hybrid, which can be looked upon either as a porous absorber with a perforated cover or as a resonator with a lined cavity. To decide which mechanism predominates, the resonant frequency should be calculated, to assess whether it lies within the frequency range of interest. If it is below 125 Hz or above 4 kHz, the behaviour will approximate that of the porous absorber. If it lies between the two, it behaves as a flattened Helmholtz type.

As always, an experimental determination is to be preferred after the preliminary calculations have been done.

Absorption Coefficients of Acoustic Materials

While BS 3638 is the approved method of measuring absorption, it is not always possible to obtain values determined strictly in accordance with the requirements of the standard. Provided that reverberation coefficients are quoted, the differences are not likely to be great. Table 1 gives some widely quoted values from *Sound Absorbing Materials* by Evans and Bazley (HMSO), which gives values for many more materials.

Foam Plastic

Polyester Urethane Foam is a material combining the features of both closed and open cell formulations. As previously noted, if the cells are all closed, the air cannot flow through; conversely if the cells are all open, the sound waves are not easily absorbed. A careful combination of open and closed cells gives optimum conditions. Table 2 gives some examples.

TABLE 1 – REVERBERATION ABSORPTION COEFFICIENTS

Material	Thickness (mm)	Frequency/Hz					
		125	250	500	1000	2000	4000
Acoustic plaster	13	0.15	0.20	0.35	0.60	0.60	0.50
Acoustic tiles (perforated fibreboard)	18	0.10	0.35	0.70	0.75	0.65	0.50
Asbestos (sprayed)	25	0.10	0.30	0.65	0.85	0.85	0.80
Brickwork	–	0.02	0.02	0.03	0.04	0.05	0.07
Carpet (Axminster)	8	–	0.05	0.15	0.30	0.45	0.55
Carpet on underlay	–	0.14	0.35	0.55	0.72	0.70	0.65
Glass fibre (resin-bonded)	25	0.10	0.25	0.55	0.70	0.80	0.85
Glass wool (uncompressed)	25	0.10	0.25	0.45	0.60	0.70	0.70
Mineral wool	25	0.10	0.25	0.50	0.70	0.85	0.85
Polystyrene, expanded (rigid backing)	13	0.05	0.05	0.10	0.15	0.15	0.20
Polystyrene, expanded (on 50 mm battens)	13	0.05	0.15	0.40	0.35	0.20	0.20
Polyurethane foam (flexible)	50	0.25	0.50	0.85	0.95	0.90	0.90
Snow	25	0.15	0.40	0.65	0.75	0.80	0.85
Wood panelling (oak, on 25 mm battens)	13	0.20	0.10	0.05	0.05	0.05	0.05

TABLE 2 – ABSORPTION COEFFICIENTS OF POLYETHER URETHANE FOAM MEASURED IN ACCORDANCE WITH ASTM C423–84 (SIMILAR TO BS 3638). DATA FROM E.A.R.

Top Surface	0.04 mm aluminised polyester				0.04 mm reinforced aluminised polyester			0.04 mm clear polyester
Thickness (mm)	12	17	25	50	12	17	25	25
Density (kg/m^2)	0.4	0.63	0.83	1.7	0.4	0.63	0.83	0.83
R 125 Hz	0.11	0.12	0.2	0.5	0.05	0.07	0.21	0.04
250 Hz	0.08	0.3	0.81	0.62	0.2	0.39	0.52	0.07
500 Hz	0.6	0.67	0.61	0.91	0.75	0.84	0.71	0.18
1 kHz	0.67	0.81	0.73	0.86	0.49	0.46	0.87	0.4
2 kHz	0.87	0.77	0.71	0.75	0.65	0.75	0.75	0.65
4 kHz	0.73	0.78	0.69	0.73	0.48	0.55	0.66	0.83
NRC	0.55	0.65	0.70	0.80	0.5	0.6	0.7	0.35

Results are for material attached to a rigid wall. The low frequency behaviour only be improved by mounting over an air gap.
The facing material has an important influence on the properties of the foam, in most cases improving the absorbency.

Rockwool

Figure 4 shows the effect of changing the density of Rocksil (rockwool). As the density increases, the high frequency absorption improves, with little change to the low frequency properties. Table 3 shows the effect of rockwool mounted over a cavity.

Wood Fibre Boards

Insulation board is a processed wood fibre board with a density less than 160 kg/m^3 . Boards of this type are also produced with perforations as acoustic boards. Particle board or chip board is made from solid fragments of wood bonded with a resinous binder, with densities from 160 to

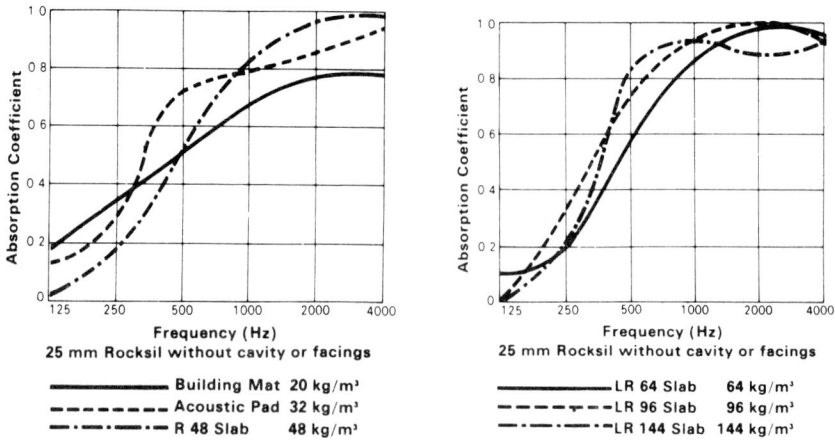

Figure 4

TABLE 3 – ABSORPTIVE PROPERTIES OF ROCKSIL PRODUCTS MOUNTED OVER 300 mm (12 in) CAVITY AND WITHOUT FACINGS

Rocksil Product (Rock Wool)	Density		Thickness		Absorption Coefficients (ISO)					
	kg/m³	lb/ft³	mm	in	125 Hz	250 Hz	500 Hz	1 000 Hz	2 000 Hz	4 000 Hz
Building Mat	20	1.3	25	1	0.45	0.65	0.55	0.65	0.75	0.85
Building Mat	20	1.3	50	2	0.65	0.85	0.75	0.85	0.95	0.95
Acoustic Pads	32	2.0	50	2	0.80	0.95	0.80	0.90	0.95	1.10
R 48 Slab	48	3.0	25	1	0.45	0.70	0.65	0.95	0.95	0.90
LR 64 slab	64	4.0	50	2	0.95	0.85	0.95	1.00	1.05	0.95

TABLE 4 – SOUND ABSORPTION COEFFICIENTS OF WOOD FIBRE PRODUCTS

Type of board	Thickness (mm)	Cavity behind board (mm)	Absorption Co-efficient Frequency (Hz)					
			125	250	500	1000	2000	4000
Plain insulating board	12.7	152.4	0.13	0.56	0.15	0.13	0.17	0.20
Acoustic boards								
Regular perforations	12.7	152.4	0.09	0.67	0.47	0.61	0.73	0.80
Regular perforations	12.7	25.4	0.20	0.30	0.35	0.55	0.70	0.70
Regular perforations	12.7	Solid	0.10	0.20	0.40	0.50	0.45	0.50
Micro-perforations	12.7	19.1	0.19	0.68	0.46	0.65	0.73	0.61
Micro-perforations	12.7	Solid	0.11	0.28	0.67	0.65	0.71	0.59
Regular perforations	19.1	25.4	0.20	0.50	0.70	0.85	0.75	0.65
Irregular perforations	19.1	25.4	0.20	0.40	0.49	0.51	0.80	0.86
Cross grooves	19.1	50.8	0.08	0.13	0.34	0.50	0.67	0.80

320 kg/m^3. The sound reduction properties depends upon the form of construction used. Table 4 gives some examples.

Several configurations of perforated chipboard are shown in Figure 5, with the corresponding sound absorption coefficients in Tables 5 and 6. The frequency dependent behaviour varies with the frequency of the slots.

For an attractive surface appearance, solid timber can also be manufactured to give good absorption properties. Figure 6 shows some examples. The absorption qualities can be improved with a mineral wool backing.

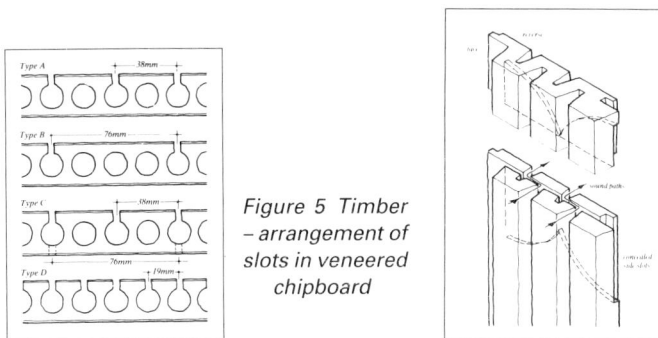

Figure 5 Timber – arrangement of slots in veneered chipboard

Figure 6 Arrangement of slots used as an example. Table below shows example solid timber acoustic wall lining sound absorption coefficients

TABLE 5 – EXAMPLE SOUND ABSORPTION COEFFICIENTS FOR TIMBER VENEERED CHIPBOARD (*DENOTES SLOTTING IN THE REAR AND MINERAL WOOL BACKING)

| | Frequency Hz: | | | | | | Test |
	125	250	500	1000	2000	4000	method
Type A	0.15	0.35	0.40	0.60	0.85	0.55	BS3638
Type B	0.18	0.33	0.31	0.32	0.57	0.36	DIN52212
Type C*	0.26	0.72	0.54	0.42	0.63	0.51	DIN52212
Type D	0.25	0.20	0.23	0.53	1.10	0.65	DIN52212

TABLE 6 – SOUND ABSORPTION COEFFICIENTS FOR PRE-FINISHED, LIGHT-DENSITY CHIPBOARD (MIKROPOR)

| | Frequency (Hz): | | | | | | Test |
	125	250	500	1000	2000	4000	method
Laquer finish	0.15	0.40	0.40	0.40	0.35	0.45	BS3638
Hessian finish	0.36	0.18	0.39	0.90	0.83	0.64	DIN52212

Chapter 3
Noise in Air Distribution Systems

The primary source of acoustic energy in an air distribution system is the fan, the energy it generates moving along the system with the air through the main transmission path. At the end of the system this is radiated to produce a sound pressure level in the receiver area (see Figure 1). Additionally, acoustic energy is generated directly as the result of airflow past the solid surfaces of the ducting system. This is classified as secondary noise. Such an input of secondary noise energy is likely to occur at any point along the air distribution system depending upon its design. It can be represented on the energy flow diagram by a number of separate inputs which have to be added to the fan noise energy at that point.

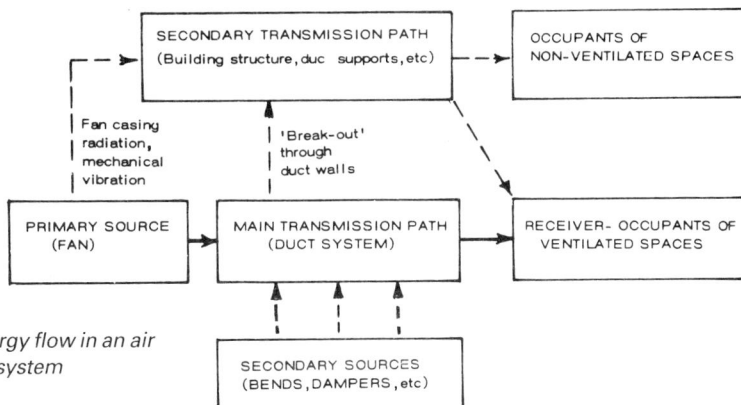

Figure 1 Acoustic energy flow in an air distribution system

Finally energy can reach occupants of the building by paths other than the distribution system, the most obvious source being mechanical energy transmitted from the fan casing into the plant room and subsequent transmission through plant room walls to adjacent areas, and break-out of duct borne energy through the walls at various points where the system passes through noise sensitive areas. These are labelled collectively 'flanking transmission'.

The control of noise in air distribution systems thus first means identification of all the possible noise sources and transmission paths, followed by analysis of the significance of and treatment required for each:

(i)   Determine sound power level of the fan;

(ii)   Calculate attenuation that will be provided by system as designed, both for system side and atmospheric side, and hence determine sound pressure level at receiver's position;

(iii)   Compare this sound pressure level with appropriate criteria and select suitable silencing treatment for main fan;

(iv)   Estimate secondary sound pressure level generated at various critical points along the system, compare this with the *silenced* fan noise at that point in the system, and prescribe suitable attenuation measures if the fan noise is significantly increased.

(v)   Ensure all flanking transmission paths are adequately treated to prevent excess energy reaching the occupied spaces by indirect means.

Transmission Path

Determination of fan sound power level is covered in detail in the chapter on *Fan Noise*. The figure used for any direct calculation must include all possible contributions arising from the installation design. That is the sound power level must be representative not only of the basic vortex noise produced by the fan, but also of any additions due to intake turbulence or interaction noise.

Starting with the fan sound power level, the attenuation that can be expected from the distribution ducting system as designed is then subtracted to give the sound power actually radiated into the room. Given this sound power, the sound pressure at any point in the room can be calculated from the standard expression:

$$L_p = L_w + 10 \log \left(\frac{Q}{4\pi r^2} + \frac{4}{R} \right) \quad dB$$

Q is the directivity factor, r the distance in m, R the room constant m^2 sabines.

It may not be immediately obvious to the system designer, but as soon as he had laid out the system of ducting to distribute the required air throughout the building, he has in fact designed a form of fan silencer. Although acoustic energy produced by the fan is, as was stated earlier, all constrained to pass along the duct system, the transmission of energy between the fan and the ventilated space is not entirely without loss. Most elements of conventional ducting systems do provide some attenuation of acoustic energy through them, as the following sections indicate.

Plain Duct Runs

Real duct walls are flexible to some extent. As sound energy passes through the duct in the form of a succession of pressure waves, its action is to vibrate the duct wall, The work required to overcome resistance of the walls to movement is provided by extraction of energy from the sound field itself, resulting in a progressive loss of sound power along the duct run. The order of magnitude that these losses may be expected to amount to is shown in Figure 2.

Bends

Energy is lost at bends by a process of diffraction and reflection rather than by pure absorption. The effect is most pronounced for plain 90° mitre bends, when the attenuation to be expected is as indicated in Figure 3. Note here the resonance effect which gives maximum attenuation at the frequency whose wavelength is twice the duct width. For long radius bends, or bends fitted with long chord turning vanes, the loss of energy will not be so pronounced, and it is normally the practice to ignore any attenuation that they might contribute.

Figure 2 Approximate attenuation of plain
ducts of minimum width 75–175 mm up to
150 mm

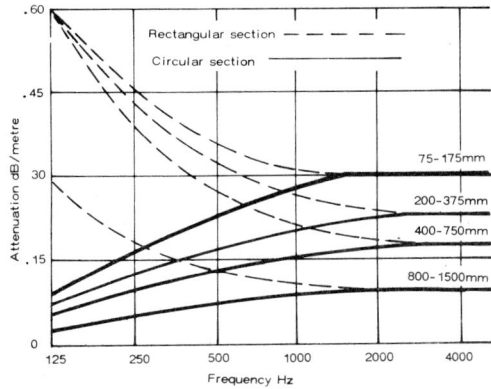

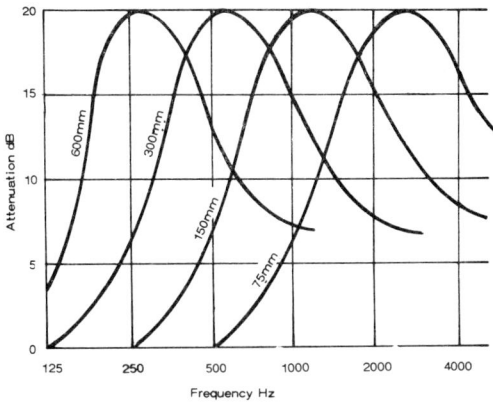

Figure 3 Approximate attenuation of
unlined 90° elbows of different duct
widths

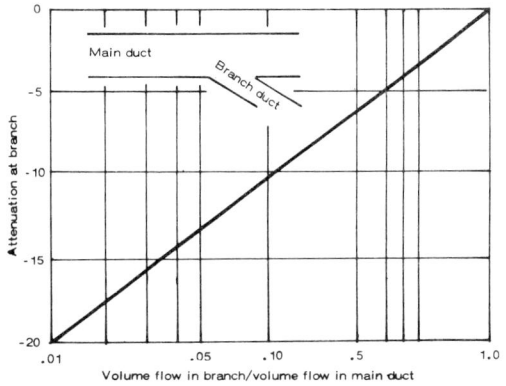

Figure 4 Attenuation at a branch

Branches

Strictly speaking, there is no nett loss of energy at a branch; rather it simply divides between the main and branch duct in much the same proportion as the airflow. Although the total energy in all branches immediately after a junction is the same as that in the main duct approaching it, the energy actually arriving at the grille at the end of a single branch is less than was present before the branch and must therefore be counted as attenuation. If, eventually, all the branches were to lead to the same room then all the individual branch energies are simply added back together in the course of the reverberant part of the room calculation. If not, we are concerned only with the energy reaching a particular grille. Attenuation to be expected by taking one single branch is determined only by the airflow in that branch as a proportion of the total flow approaching the junction. This is shown graphically in Figure 4.

Plenum Chambers

A large cavity in the system such as builders' ducting or even the fan chamber itself, with the fan either pressurising or extracting from the room, and the duct system starting at the walls, can be used to form an acoustic plenum. Energy fed into a plenum tends to regard the chamber as a small room, and as in the case of a room, the sound pressure level at the entrance to the ducting on the other side determines how much sound power is transmitted. It has two components.

Direct sound energy is radiated straight from the source and reverberant sound energy is the remaining energy fed into the chamber but reaching the outlet duct after reflection of the chamber's internal surfaces. Clearly the first component will then be more dependent upon the geometry than anything else, while reverberant energy will be determined largely by the absorption characteristics of the plenum surfaces. Figure 5 defines the geometric parameters in the calculation, and the expression for attenuation across the plenum is:

$$LW_1 - LW_2 = -10 \log S_2 \left[\frac{\cos \theta}{2\pi d^2} + \frac{1}{R} \right] \; dB$$

where $R = \dfrac{S_T \alpha}{1 - \alpha}$ the room constant

In fact, the above expression underestimates the attenuation in the lower frequency bands as no account is taken of reflection at the inlet to the plenum. A value for this to be added to the attenuation calculated above, can be taken from Figure 6.

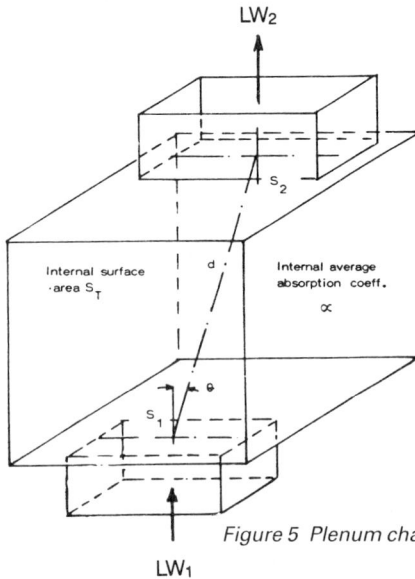

Figure 5 Plenum chamber geometry

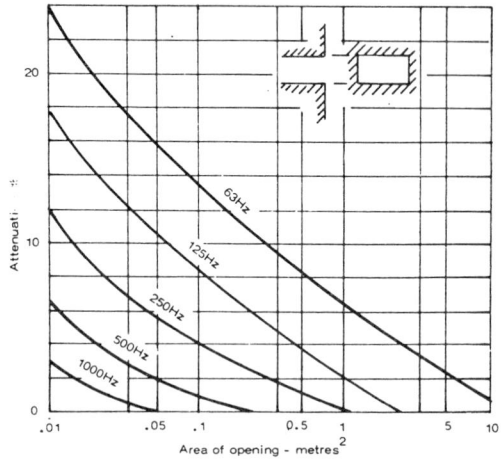

Figure 6 Attenuation at duct terminations

Proprietary Units

Not many in-duct elements supplied as a package serve to give any appreciable attenuation of sound energy. Such items as heaters, cooling coils or even filters are normally neglected, except with regard to their ability to produce sound as mentioned later. There are one or two however, where attenuation of sound may be significant, and notable among these are the constant volume or constant pressure controllers of the type employed in high velocity systems. It is not possible to give a general expression for the loss as this varies with individual manufacturers' designs, but a statement of any attenuation should be obtained from the manufacturer and included in the duct calculation.

Terminal Units

When the attenuation provided by all the various duct elements have been added to the fan input sound power level, the result is the sound power which reaches the grille. Not all of this energy is radiated into the ventilated space however because of the phenomenon of reflection which always occurs when waves travelling in a medium of given acoustic impedance, in this case the confines of the duct section, meet a sudden change in impedance, here represented by the emergency into the large volume of the ventilated space. The amount of energy reflected and hence 'lost' to the occupants of the room is dependent upon the magnitude of the change. In the case of a ductwork termination into a room, the smaller the duct the greater the change of impedance. More precisely the parameter is the duct size compared with the wavelength (frequency) being considered. A summary of the attenuation to be expected is shown graphically in Figure 6.

Noise Control

Referring to Figure 1, it can be seen that the noise control engineer has a number of options available to ensure that the correct amount of acoustic energy reaches the occupants of the ventilated spaces. They may be summarised as follows:

1. Reduce strength of the primary source (the fan).
2. Attenuate fan noise in the main transmission path (the duct system), between the fan and area affected.
3. Ensure all sound power generated by secondary sources is kept down to a level which is insignificant compared with the level of fan noise, silenced or unsilenced, at that part in the system.
4. Eliminate the transmission of all flanking and indirectly produced energy.

Reducing Noise in Transmission Along the Duct System

Once the minimum possible fan sound power has been achieved by careful selection, the next point of attack is the transmission path by which it reaches the ventilated space. The obvious way to reduce the amount of fan noise passing along the system is to employ a fan silencer.

It is often the case however that some savings in size and hence cost of a proprietary silencer, can be effected by careful design of the air distribution itself. Most duct elements, for example, offer a certain amount of attenuation. It should always be the initial aim to note by what means it may be maximised. For example:

(a) Plain duct runs provide more low frequency attenuation when rectangular in section (Figure 2). A disadvantage here is, of course, that much of the energy extracted from inside the duct often appears as airborne energy outside ('breakout') but providing this can be accepted, or dealt with as described later, the benefits to be gained by using rectangular section can be appreciable.

(b) Plain mitre bends can give quite significant lower frequency attenuation if correctly sized (see Figure 3). The additional pressure loss is of course an important consideration, but the addition of short chord turning vanes can alleviate this to some extent.

(c) Any plenum that can be incorporated in the system will be of considerable benefit, as a trial calculation using Figure 5 and the associated equation will reveal. Again, however, pressure loss at inlet and outlet, together with the volume required to form the chamber, must be balanced against the benefits.

(d)   Reflection of energy at the duct termination can be seen from Figure 6 to vary
       inversely with duct area, at least for the important low frequency energy. A
       number of small terminations is better than a few large ones.

When all has been done along these lines to ensure maximum 'natural' attenuation in the system, it may well be found that additional attenuation is required. For the main silencer, a packaged splitter type will probably be required, but for more moderate attenuation, particularly for terminal silencing of secondary noise (see later), a simple lined duct or bend may suffice. Performance of lined ducts and attenuation for some representative bend sizes and linings is given in Figure 7.

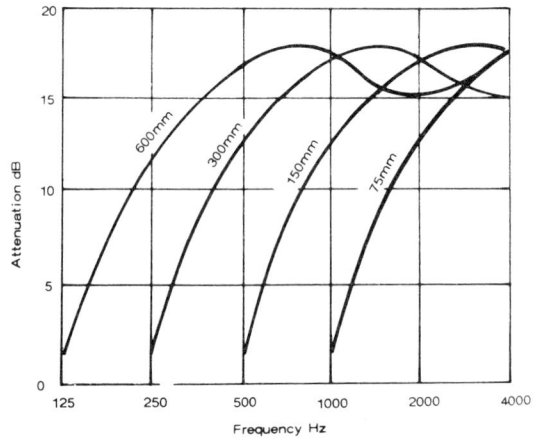

Figure 7 Typical attenuation of lined ducts – duct widths as shown. (Duct lining thickness approximately 10% of duct width)

Secondary Noise

It is important to compare the secondary sound levels generated in a system with the main fan noise in the same part of the system. If significantly lower – say more than 6 dB difference – then the secondary noise may be ignored. If on the other hand secondary noise is comparable with – or greater than – the fan noise at that point in the system, then the uplift due to it is bound to create an excessive sound pressure level in the room. In that case, some secondary silencing will be required.

System components which generate secondary noise are conveniently examined in two groups. The first comprises in in-duct elements which, whilst being exposed to higher airflow, have at least one advantage in that the energy they produce is confined within the duct where, if excessive, it is reasonably amenable to conventional attenuation techniques. The second comprises the various types of terminal units, which usually cause the greater problem because the energy they produce actually starts in the ventilated space. The possibilities of attenuation, therefore, before it reaches occupants of the same space, are limited.

In-Duct Elements

It is difficult to predict with any accuracy the noise generated by elements in the system which can themselves generate noise. These elements are: bends, branches (tees and crosses), contractions, dampers and proprietory fittings. The noise generated is heavily dependent on the flow velocity, so at speeds of the order of 1 m/s, there is a negligible noise generated, but at speeds of 10 m/s the noise becomes significant. The governing parameter is the Strouhal number fD/V, where f is the octave band centre frequency, D the duct diameter and V the flow velocity. The Strouhal number is dimensionless and so is independent of the units used, provided they are consistent (e.g. D in m and V in m/s). One equation which should be used with reserve is:

$$L_W = K + 10 \log S + 55 \log V - 45 \text{ dB}$$

> where   K is a constant for the particular duct element considered, and is given in Table 1.
>
> S is the cross sectional area in m^2 of the duct, across which velocity is V.
>
> V is the flow velocity in section S, m/sec.

TABLE 1 – AIRFLOW GENERATED NOISE IN DUCT FITTINGS (after Stewart)

Fitting				Overall sound power level dB (octave bands 250–8000 Hz) K		
90° bends radiused				48		
mitred with turning vanes				56		
mitred without turning vanes				57		
Duct with 90° tee (5% draw-off)				55		
Damper: Open				44		
15° closed				53		
45° closed				65		
Corrections to give octave band SWL						
Octave Band Centre Frequency	Hz	250	500	1000	2000	4000	8000
Spectrum correction to overall SWL	dB	−7	−7	−8	−10	−17	−29

Distribution of the overall sound power across the frequency spectrum is also indicated in Table 1.

Consideration must also be given to the noise generating propensities of a turbulent flow on to the duct element in question although relative velocities for duct elements are rather less than those for rotating blades, so the effect may not be so drastic. If, however, a turbulent inflow is likely, for example in the case of a damper located immediately after a bend, it would be prudent to add 5 dB to the estimated sound power level across the spectrum.

With some proprietary units, such as dampers, the estimates given previously should be sufficiently accurate for a design estimate, although it must be said that there is no substitute for laboratory based manufacturers' test figures. With others though, such as individual manufacturers' designs of mixing boxes or constant volume units, their test figures are at present about the only data available.

If the test entry conditions were substantially smoother than those which will occur in practice, a 5 dB addition should be made for safety.

Terminal Units

Secondary noise generation at terminal units, grilles, diffusers, induction units and the like, is of critical importance. Unlike the noise of in-duct elements, there is no opportunity to attenuate energy produced by the terminal unit before it reaches the occupant of the same room.

In plain grilles and diffusers, the mechanism of producing noise is essentially the same as the vortex noise mechanism set up by moving fan blades.

First an estimation has to be made of how much power is likely to be produced. As with so many other components of the system which are of proprietary supply, there is no substitute for the manufacturer's figures based upon test data, providing the background to the figures is examined to establish that test conditions were not too dissimilar to those of the installed product. In the absence of such information, one can for some units obtain a design estimate from one of the empirical expressions which are available. One estimate expresses the overall sound power level in terms of the minimum neck area in the case of diffusers, or free area in the case of grilles, $A(m^2)$ and the maximum flow velocity in neck or between vanes, $V(m/sec)$, as follows:

$$L_W = 13 \log A + 60 \log V + 33 \text{ dB}$$

For design purposes, the spectra corrections given in Table 1 may be assumed to apply.

Fitting any sort of control damper to a grille or diffuser will add two more noise sources. First there will be the vortex shedding component from the damper blades themselves, and second the action of the damper blade wake turbulence upon the grille or diffuser vanes as illustrated in Figure 8. The second is by far the more important.

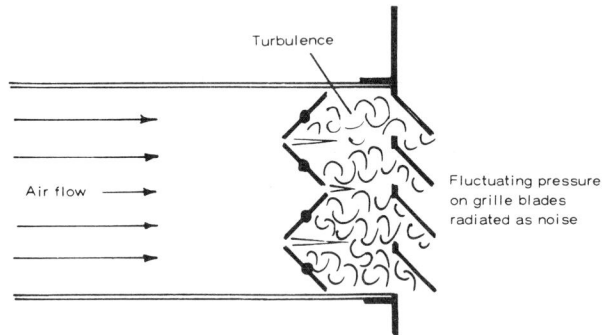

Figure 8 Noise generated by dampers

In current grille design, it is the practice to manufacture the damper as a multi-leaf opposed-blade pack which clips on to the rear of the grille. The system has a number of advantages including the ability to balance flow through a number of grille easily, adjusting the damper by turning a screw reached through the front face of the grille. Unfortunately, the system also pays a penalty, sometimes a heavy one, in additional noise generation.

Nearly all manufacturers of grilles and diffusers can supply quite detailed figures for the sound power level produced for their grille/damper assemblies, as a function of either face velocity or volume flow through, or, more usually, as a function of pressure drop across the unit.

In the absence of such information at the design stage, one can assume an increase in the aforementioned estimated sound power level, of 5 db in each octave band for the addition of an internal damper vane fully open.

For subsequent flow adjustment, i.e. increased drop across the damper, it can be assumed that the overall sound power level will increase approximately by the following:

$$L_w = 30 \log \frac{p}{p_0}$$

where

p is the new static pressure drop across the unit

p_0 is the pressure drop with damper pack added but vane fully open, with the same volume flow.

Induction units are rather less amenable to a general estimate of the noise they produce. The generating mechanism itself is basically one of free jet noise, but with the possibility of additional output from turbulent flow impingement on solid surfaces.

If free jet noise is the only source, its sound power output may be estimated from a design chart such as Figure 9. Note here that the power level estimated applies to one jet. For the total jet noise from one unit, an addition of 10 log N dB should be made, where N is the number of nozzles. Note also that the directivity of the free jet will not apply in general, being affected by the casing of the unit to a very marked degree.

In addition to the noise of the free jets, additional sources will be present on most induction units. Any damper into the unit plenum will of course produce noise as discussed earlier in this

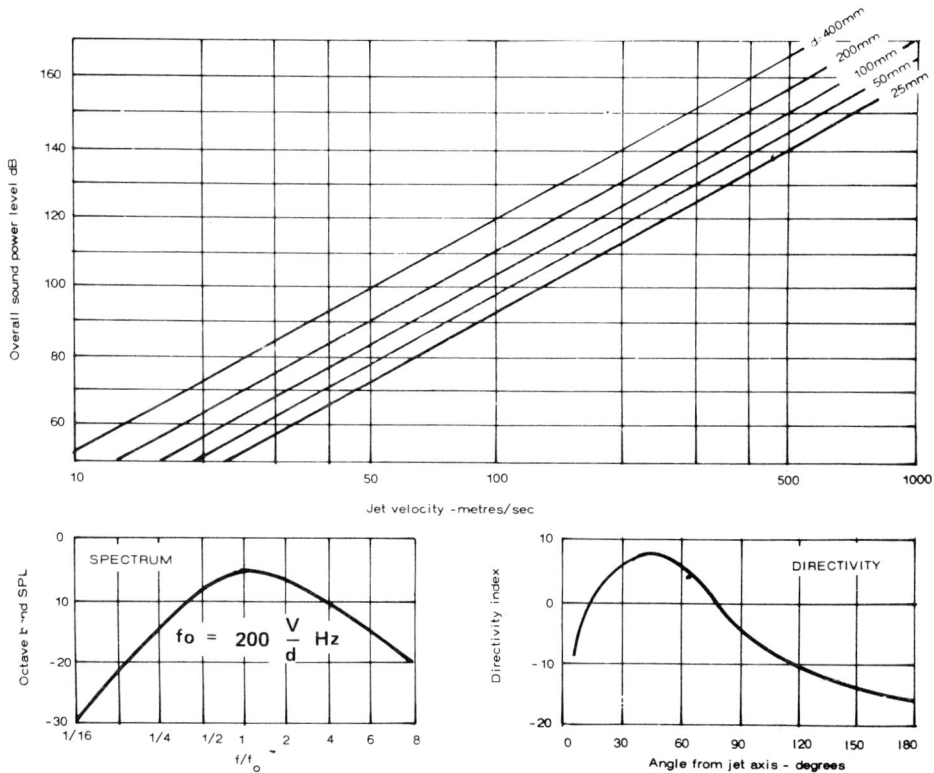

Figure 9 Design chart for estimating induction unit jet noise

chapter, and secondary generation at the air entry is to be expected. Sound power from both of these sources, however, may be expected to be reduced by some 15 dB across the spectrum in emerging from the unit to the room. Additional noise of the turbulence interaction type may be expected if the primary nozzle jets are allowed to impinge upon any part of the casing of the unit – or if there is a protective grille placed directly in line. Under these conditions about 5 dB should be added across the spectrum.

It should always be the aim to obtain from the manufacturer induction unit noise levels which appertain to the particular design to be used, and which are based upon test figures for that design.

Control of Secondary Noise

The most important factors which determine the output of secondary noise are velocity and turbulent flow. The importance of reducing both at the design stage cannot be over-emphasised.

Where noise is likely to be at all critical, the following general guidelines should be followed:

Keep velocities down to 5 m/sec in main ducts and 3 m/sec in branches.

Bends should be long radius or be fitted with turning vanes (but note conflicting advantage of high half-wavelength attenuation in mitre bends). See also Figure 3.

(Branches should be radiused or at least chamfered.

System resistance should be calculated very carefully to obviate the need for high pressure drop across balancing flow control dampers when commissioning.

It may well be found however, as with fan selection and system design, that further attenuation is required after the above guidelines have been followed. With most of the duct elements likely to produce significant secondary noise, the source is located inside the ducting system. Excess energy is then very amenable to attenuation by the simple expedient of lining the terminal branch duct, or replacing it with a small packaged splitter silencer.

The need for all significant secondary energy to be located far enough back into the system for a terminal silencer to be effective, is particularly important for flow control dampers on grilles and diffusers in the room. If it is essential to provide a damper for each grille, then it must be located on the system side of the terminal silencer, as indicated in Figure 10.

With regard to secondary noise generation at the terminal unit itself there is, almost by definition, virtually nothing that can be done to prevent the energy from being radiated into the room.

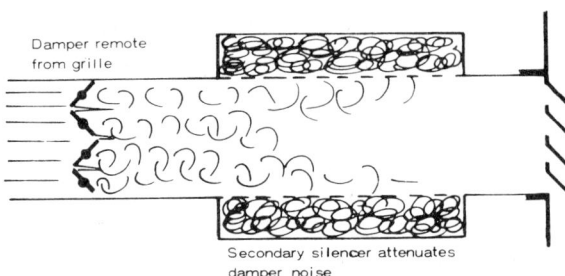

Damper remote from grille

Figure 10 Remedy for damper noise

Secondary silencer attenuates damper noise

Induction units, in particular, will almost certainly be limited in their application to noise sensitive areas. A room criteria of NR 35 is about the minimum that can be achieved with these devices.

With grilles, the only recourse is to oversize to reduce the discharge velocity to the absolute minimum consistent with the throw and distribution required. In extreme cases, air is sometimes fed to the room through a large hole covered with a wide mesh screen. Obviously though, the level of secondary noise generation there has to be balanced against a smaller low frequency reflection loss (see Figure 6).

Indirect and Flanking Transmissions

Figure 11 shows pictorially the more important indirect transmission paths which have to be examined carefully if all the work for ensuring acceptable fan noise and secondary noise levels in the ventilated space, is not to be negated. They are:

> Breakout of duct-borne fan noise energy through duct walls into rooms carrying the ducting.
>
> Radiation of fan casing and drive motor noise into the plant room and hence to adjacent area.
>
> Mechanical excitation of the duct system and the building structure.

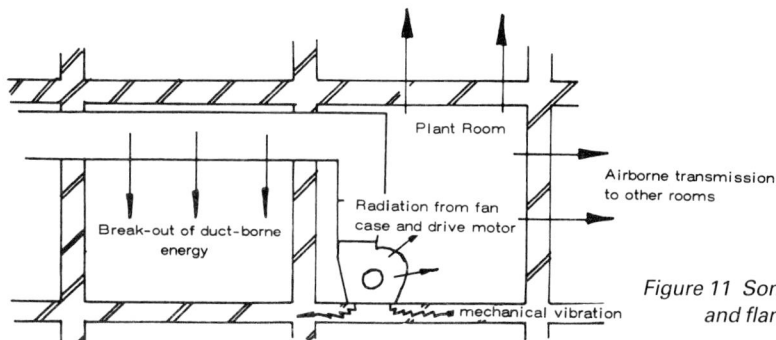

Figure 11 Some examples of indirect and flanking transmission

(i) Noise Breakout from Ducts

It was pointed out earlier in the chapter, when discussing energy losses during transmission of fan noise along plain duct runs, that the sound waves passing a fixed point in the duct vibrate the duct wall. Of the energy expended in overcoming the wall's resistance to movement, some is transformed into heat by whatever internal damping is in the material of the duct wall, and the rest is re-radiated as airborne noise outside the duct.

The amount of energy radiated through a duct wall, or, as it is commonly referred to, 'break-out' sound power level, the material of which has a Sound Reduction Index R, can be estimated from:

$$L_{WB} = L_{WD} - R + 10 \log \frac{S}{A} \quad dB$$

where L_{WD} is the sound power level inside the duct

L_{WB} is the breakout sound power level

S is the total surface area of the run of ducting concerned in m^2

A is the cross sectional area of the duct in m^2

This is known as the Allen formula, and has been used predict the sound power transmitted into a room through which the duct passes. It is not accurate for long lengths of ducting, because the sound power is not then uniform along the length. For large areas and small values of R, the value of 10 log S/A can be larger than R and so more noise can be radiated than is in the duct, which is clearly impossible. This is usually catered for by assuming that one-half the sound power is radiated and one-half remains in the duct, so that the break-out sound power is 3 dB less than the internal sound power.

(ii) Fan Casing and Drive Motor Noise

Noise radiated from the fan casing has two components. The first is from airborne excitation of the casing by the acoustic energy generated at the fan impeller, in other words a similar 'break-out' to that discussed previously for duct borne energy.

A very approximate figure for the amount of energy to be expected from this source can be obtained using the same formulas as for duct breakout, if L_{WD} is taken as the sound power level of the fan, R is the sound reduction index of the fan case, S the total area of fan casing and A the cross sectional area based upon the impeller diameter for an axial fan or the casing diameter for a centrifugal fan.

Purely airborne breakout of this kind is likely to be less than the sound power level due to the other component, mechanical excitation of the casing. Belt drives, bearings, direct drive motor support arms, all produce a degree of mechanical vibration energy at frequencies virtually across the spectrum, which transforms into airborne acoustic energy at the relatively large radiating surfaces of the fan casing. Unfortunately, the amount of mechanical vibration at the radiating surface is dependent upon the detailed mechanical design of the fan and drive motor assembly, rather than being related to overall performance, as is aerodynamic sound power. This means that resultant sound power level is almost impossible to predict, being potentially different, for example, for two identical fans with same power but different type and mounting arrangement of the drive motor, or with different residual out-of-balance tolerances.

If the fan is driven via belts by an external motor, an estimate of drive motor noise needs to be made, but it must be remembered that mechanically induced vibration of the motor support structure may prove to be the dominant source.

If a fan runs open with intake in the plant room, the impeller aerodynamic sound power level radiated from the intake will undoubtedly dominate. If, on the other hand, the intake noise has to be attenuated then it would be prudent to assume – at the design stage – that when more than 15 dB attenuation across the spectrum has been provided by the intake silencer, casing and drive motor noise will start to become significant. Any further attenuation of intake noise will then probably have to be matched by an equal amount of attenuation of the flanking radiation.

Control of Flanking and Indirect Transmission

Two options are available to achieve a reduction of the amount of sound power expected to break out through duct walls, assuming the ratio Sw/A has to remain constant. One is to reduce the sound power level inside the duct, and the other is to increase the sound reduction index of the duct wall. The first has already been discussed in the context of fan noise reduction and fan silencers.

Various methods of increasing the sound reduction index of partitions are available. Air distribution ducting is not, however, normally amenable to a straight increase in weight, even if this gave sufficient increase to provide the required decrease in outside sound power level.

It is of course possible to enclose completely a duct run across a room by, for example, boxing in with plasterboard and studding type construction. Provided sealing is adequate and internal absorption is required, it can be very effective, giving an increase of up to 20 dB average across the spectrum.

A more convenient treatment, however, is to lag the outside of the relevant duct run with 50 mm glass fibre quilt or polyurethane foam, and wrap outside this a mass skin of lead foil, Keene cement, or one of the proprietary loaded PVC sheets, of superficial density not less than 5 kg/m^2. A reduction in breakout sound power level of some 7 to 10 dB across the spectrum can be expected from this treatment.

The same treatment is eminently suitable for acoustically lagging the outside of axial flow fans, and the same order of magnitude of improvement may be expected.

For centrifugal fans, there is no alternative to at least partial, if not complete acoustic enclosure, for any significant degree of improvement in casing noise levels.

A weak point, as far as breakout is concerned, is the flexible duct connector which must be used in conjunction with vibration isolators. The traditional canvas connectors, while providing excellent vibration isolation, provide virtually zero sound reduction index. This problem can be fairly easily overcome however by manufacturing the connectors in one of the proprietary loaded PVC compounds, of superficial density up to 10 kg/m^2, which give a sound reduction index in this application about equal to that of light gauge ducting.

Cross-Talk

With a system where two separate rooms share the same air distribution duct, there is always the danger of noise in one, such as music or speech, being transmitted along the air duct to the other. This effect is commonly referred to as 'cross-talk'.

The calculation of the degree to which it is likely to occur and its control is exactly the same as for the estimation and control of fan noise described elsewhere in this chapter.

Noise Transmission to Atmosphere

It must be appreciated that fan noise can be transmitted just as readily upstream as downstream. Fan sound power levels quoted by manufacturers or estimated should be assumed to apply to either side of the fan, and the same figures should be used for calculations in both directions. In other words whether the system is for supply or extraction, it is as important to carry out the same exercise for the atmospheric side of the fan, as for the system side. The objective is of course precisely the same in that the sound power level produced at some point likely to be affected by the atmospheric intake or discharge, has to be estimated and compared with the appropriate criterion.

Two cases are considered, the first being where the atmospheric side of the system is ducted in the same way as might be the room side except that the terminal duct leads to an outside louvre; and the second being when the fan runs open in the plant room, and draws in or discharges air through atmospheric louvres in the plant room wall.

(i) Ducted System to Atmosphere

The calculation to be followed here is precisely that outlined previously for the room side. From the fan sound power level must be subtracted the attenuation provided by the various duct elements including reflection at the atmospheric louvre. The sound pressure level then at a distance r metres from the louvres and at angle θ to its axis, resulting from the sound power level L_w which actually leaves the louvres may be estimated from:

$$L_p = L_W + 10 \log\left(\frac{Q}{4\pi r^2}\right) = L_W - 20 \log r + DI - 11 \quad dB$$

DI (θ) is the Directivity Index of the radiation pattern from the louvres. For design purposes one can assume the following values for DI providing the louvre area is in excess of 1 m^2 and the louvre is located roughly centrally in a large wall.

	DI $= 10 \log_{10} Q$
0	+8
30	+6
45	+4
60	+1
75	−5
90	−15

Add 3 dB to the aforementioned figures if the louvre is near the junction of two walls or between wall and ground, or 6 dB if the louvre is near the corner formed by two walls and the ground.

(ii) Fan Running Open in Plant Room

If the fan is drawing air in directly from the plant room, the primary interest will be in the sound pressure level inside the room. From this, sound pressure levels outside or in adjacent rooms can be calculated.

The formula for plant room sound pressure level is the usual one given earlier, i.e.

$$L_p = L_W + 10 \log\left(\frac{Q}{4\pi r^2} + \frac{4}{R}\right) \quad dB$$

Here all the room parameters apply to the plant room, and L_W, the source sound power level, is that of the fan, less a reflection loss appropriate to the size of the fan inlet.

Commercial Attenuators

An attenuator can be inserted in the duct system to reduce the noise generated by the fan. Two types are illustrated in Figures 12 and 13. When supplied by manufacturers it is usual to quote tables of two parameters:

> The dynamic insertion loss is the actual noise reduction (in dB) achieved with air flowing through the device (i.e. under dynamic conditions). Attenuation is preferably measured in this way, rather than in a static condition, although in most instance there is no more than about 2 dB difference between the dynamic and static conditions unless the velocities are high (in excess of 10 m/s).

> The self-noise power is the noise generated by the air flowing through the attenuator. So the attenuator not only acts to reduce the noise level caused by the fan, but also generates its own noise by the passage of air through it. In most cases, with well designed attenuators the self noise is small, except when the air velocity is high or special considerations apply such as in a concert hall.

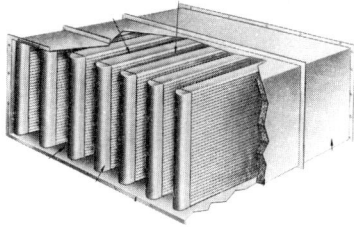

Cut-away view of a typical rectangular silencer

Cut-away view of a tubular silencer

Figure 12

Figure 13 Rectangular Attenuator manufactured of galvanised steel sheet with acoustic lining

Tables 2 and 3 show values of these parameters for a typical silencer. The values are to be taken as examples only and reference must be made to the suppliers for design data. Ratings can be given for both forward and reverse flow. To be able to take advantage of the potential insertion loss, attention must be paid to the correct mounting of the attenuator. In practice, the insertion loss is unlikely to exceed 50 dB on account of flanking transmission.

TABLE 2 – DYNAMIC INSERTION LOSS (DIL) RATINGS: FORWARD (+)/REVERSE (−) FLOW

Silencer Model No. (Lengths)	Octave Bands, Hz Silencer Face Velocity (m/s)	1 63	2 125	3 250	4 500	5 1000	6 2000	7 4000	8 8000
		Dynamic Insertion Loss in Decibels							
3S	−10	7	14	21	31	37	33	24	13
	−5	7	13	17	30	36	33	26	13
	+5	7	12	16	28	35	35	28	17
	+10	6	10	15	25	34	35	28	17
5S	−10	11	23	26	44	48	44	37	22
	−5	11	21	25	43	47	44	39	22
	+5	8	18	24	40	45	46	41	26
	+10	7	16	22	38	45	46	41	26
7S	−10	12	24	38	48	53	46	42	30
	−5	12	23	37	46	51	48	44	30
	+5	10	20	35	45	50	48	45	34
	+10	12	18	41	52	55	53	51	48
10S	−10	13	25	43	54	55	53	49	42
	−5	13	26	42	52	55	53	51	42
	+5	13	23	42	52	55	53	51	45
	+10	12	18	41	52	55	53	51	48

TABLE 3 – SELF-NOISE POWER LEVELS, DB RE 10^{-12} WATTS

IAC Silencer Model	Octave Bands Hz Face Velocity m/s	1 63	2 125	3 250	4 500	5 1K	6 2K	7 4K	8 8K
		Self-Noise Power Levels in dB							
3S	−10	68	62	61	66	61	64	67	66
	−8	62	57	57	61	59	61	60	55
	−5	54	51	50	51	54	56	52	40
5S	−4	48	46	46	46	52	53	45	29
	−3	40	40	39	36	47	48	37	20
7S	+3	36	29	35	30	31	35	22	20
	+4	47	41	42	40	40	43	32	22
&	+5	55	49	49	47	46	49	42	32
	+8	66	61	56	57	55	57	52	46
10S	+10	74	69	63	64	61	63	62	56

Chapter 4
Fan Noise

Noise generated by fans comprises broadband noise resulting from intake turbulence and vortex generation, on which is superimposed pure tone components related to fan geometry and rotational speed. The sound power generated by a fan varies between the fifth and sixth power of the fan tip speed. It is often assumed to be proportional to (tip speed)$^{5.5}$. A reduction in speed can make a considerable difference to the fan noise; thus a 50% reduction in speed results in 16 dB reduction in sound power level. Such a speed reduction would, of course, reduce the volume delivered. It is important to understand that for any given delivery volume, delivery pressure and fan type there is one speed and one diameter at which the efficiency is a maximum; if one runs the fan at any other speed, the efficiency falls and the noise increases. A larger fan running at a slower speed is not necessarily quieter.

A variety of empirical relationships have been proposed to predict the noise level, but these are very approximate and one should always rely on manufacturer's data. Noise levels should be determined by the procedure recommended in BS 848.

The overall sound pressure level can be estimated from

$$L_w = 10 \log_{10} Q + 20 \log_{10} P + 60 + K \quad \text{dB re } 10^{-12} \text{ W}$$

Q is the flow rate in m^3/s and P is the static pressure in Pa and K is a correction to be applied for specific kinds of fans to give the sound power in the frequency bands 63 Hz to 8 kHz. Values of K are given in Table 1.

Another formula which is given in BS 848 is:

$$L_w = L_g + \{10 (6 + a) \log_{10} n] + \{10 (8 + 2a + b) \log_{10} D_R\}$$
$$+ F(p,t) \text{ dB}$$

where L_g is a correction for the frequency spectrum, a and b are experimental indices, n is the rotational speed in r/s, D_R is the diameter in m and F(p,t) is a correction for pressure and temperature. In this form it is presented as a more accurate means of predicting fan noise when sound measurements of similar fans are available. L_g, a, b and F(p,t) are obtained from experimental results of fans having geometric similarity. For a full explanation of the formula refer to BS 848: Part 2: 1985.

The noise level of a fan can be better predicted, if the level for another fan of similar geometry and operating conditions is known. The assumption to be made is that each fan is operating at the same efficiency. A relationship derived from dimensional analysis is:

$$L_{w1} - L_{w2} = 50 \log_{10} \frac{n_1}{n_2} + 70 \log_{10} \frac{D_{R1}}{D_{R2}}$$

where suffix 1 relates to the unknown fan and suffix 2 to the known fan.

In addition to the broad-band spectrum produced by the use of these equations, there is a further source of noise whose value has to be added to the particular octave band in which it occurs. The frequency at which this occurs is given by:

$$f + \frac{n N_1}{60}$$

where n is the rotational speed in r/min and N_1 is the number of blades. The magnitude of the extra noise is given in Table 1 as the blade frequency increment BFI.

The noise that is generated is radiated partly downstream and partly upstream. On the assumption that these two components are equal, then 3 dB should be deducted from the overall noise to determine the level in one direction.

If the fan operates other than at its peak efficiency, a further addition has to be made to the sound power level as in Table 2, however manufacturers should have available curves of efficiency and the corresponding noise levels.

Vortex Noise

Considering, first, the mechanics of fan noise generation, it is a characteristic of translational airflow over any solid surface that a shear gradient will be established in which the velocity varies from zero infinitesimally close to the surface, to the free stream velocity of the bulk of the air at some distance away from it. The region in which such a shear gradient exists is known as the boundary layer. At some distance along the surface, the velocity shear gradient becomes such that a laminar flow can no longer be maintained and the flow breaks down to form a turbulent boundary layer. If the surface is of finite length, the turbulent boundary layer leaves the top and bottom surfaces to form the wake. Figure 1 illustrates these flow regimes diagrammatically for an aerofoil shape typical of the blade section of a wide variety of fans.

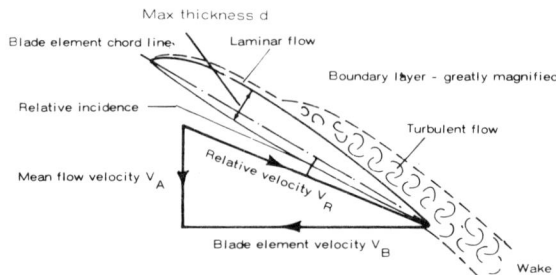

Figure 1 Flow around aerofoil blade section
– some definitions

The passage of this turbulent flow over the after end of the aerofoil and into the wake sets up fluctuating forces on the aerofoil surfaces which in turn act as sources of acoustic energy. Because the turbulent flow is random in the nature of its eddy size and velocity variations, the pressure fluctuations are similarly random, as is the generated sound power. The result is a frequency spectrum which covers a considerable range of the audio spectrum. It is not flat as would be the case if the process were truly random but has a maximum value extending over perhaps as much as two full octaves, and sloping off at about 5 dB per octave on either side. The frequency of maximum acoustic energy generation is governed by the velocity of air relative to the section and the thickness of the section projected in the direction of airflow. This maximum frequency is given by:

$$f_{max} = 200 \ \frac{V_r}{d}$$

with V_r in m/sec and d in mm as defined in Figure 1.

Figure 2 Impeller blade section traversing turbulent flow

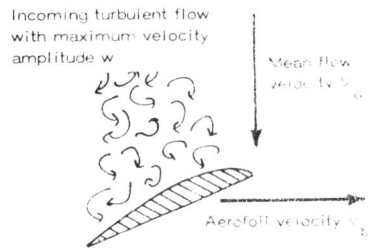

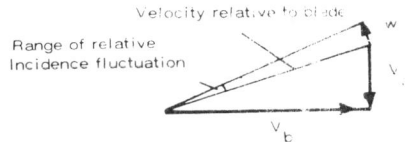

Intake Turbulence

The basic mechanism described previously by which any surface in a flow generates noise will occur to some degree or other in a fan, even if the entry conditions to the impeller are perfect. It is frequently the case however that in industrial installations, the intake flow to the impeller is itself turbulent. If we assume the turbulent region to be one in which a particle of air has a mean transport velocity Va, and an instantaneous perturbation velocity superimposed on it, which at any instant can be of magnitude up to a maximum of say w in any direction, the effect upon the velocity vector diagram is as shown in Figure 2. The addition of the perturbation vector alters the direction rather than the magnitude of the relative velocity vector V. Remembering that W is continuously varying in magnitude and direction, it is clear that the blade section relative incidence is also varying, as then must be the total lift on the section. Because the initial perturbation in the turbulent flow is random in nature, the sound power resulting from the lift fluctuation must also be random. Hence the frequency spectrum is broad band.

Again the spectrum is not flat but tends to a maximum at a frequency similar to that of vortex noise, as estimated from the previous expression, with V_r equal to the relative velocity of air past the blade, and d the characteristic size of eddy in the turbulent field. Unfortunately d is not easy to determine. In a very simple case, such as turbulence induced by an up-stream duct support or instrumentation probe, it is not too much of an error to take d as being equal to the cross-stream dimension of the obstacle. In many ventilating fan installations, however, of which Figure 3 shows quite typical examples for an axial fan, the characteristic size of eddy passing over the blade tip is much more difficult to estimate.

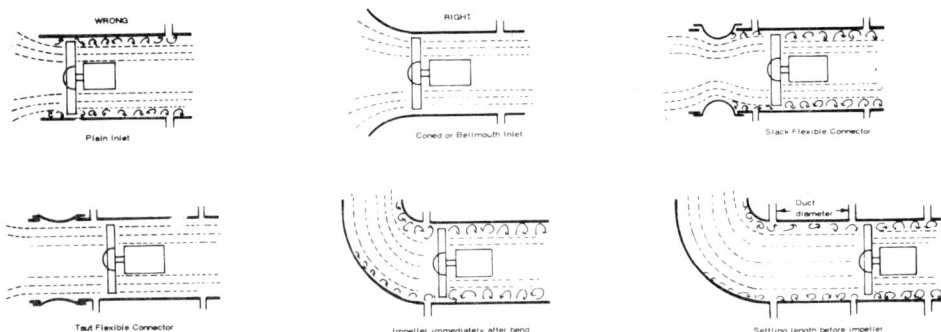

Figure 3 Examples of fan installations producing intake turbulence

As a practical guide, it is a matter of observation that the additional noise due to this mechanism nearly always occurs at frequencies up to and including the 500 Hz band. Estimates given later for excess in sound power level due to intake turbulence apply to these bands unless otherwise stated.

One reason for the preponderance of low frequency noise is that the lift fluctuation of a blade section is strongest for eddy sizes which are comparable with, or greater than, the blade chord. To take a very simple model, if eddy size is much smaller, then at any instant a number of eddies will be spread across the blade chord. Since the individual velocity and pressure perturbations exerted locally on the aerofoil are varying randomly both in amplitude and in phase relative to each other, the net or effective force fluctuation on the surface which produce the acoustic energy will statistically average out to zero, or at least to a very small residual. This effect can be easily demonstrated by observing the difference in additional noise output when turbulence is introduced first by, say, a very fine gauge mesh, then by a flat bar of cross stream dimension equal to the blade chord.

Interaction Noise

The physical mechanism of generating this component of the spectrum is virtually the same as that for intake turbulence. If the flow onto the impeller blading contains non-uniformities which are random in nature such as turbulent flow then, as stated earlier, the resulting additional noise it generates has a frequency spectrum which is broad band. If on the other hand the flow disturbance is periodic, the resulting noise generated will appear in the form of pure tones at the fundamental blade passage frequency and its harmonics where,

$$\text{Blade passage frequency} = \frac{\text{rev/min}}{60} \times \text{number of impeller blades}$$

One way of generating such an intake flow is illustrated in Figure 4 for the specific case of a rotor blade passing behind a single fixed obstacle such as an upstream bearing support, or a silencer pod support. Immediately behind the obstacle there is a decrement in the mean axial velocity of oncoming air in the region of the wake. The effect upon the impeller blade velocity triangle of a change in axial component, can be seen to be very similar to that of the addition of a turbulent velocity as in Figure 2. Again the effect is to vary the direction of the relative velocity onto the moving blade with a consequent change of relative incidence and lift on the section, and hence acoustic energy is radiated.

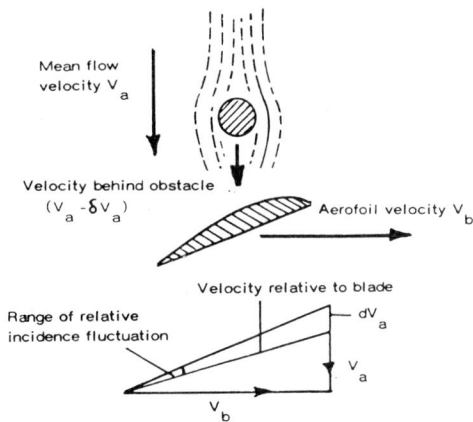

Figure 4 Impeller blade section traversing wake of fixed obstacle

As mentioned, the strongest component is nearly always produced at the fundamental blade passage frequency, with harmonic content depending upon the pressure wave form of the variation of axial velocity across the region of the disturbance. With industrial fans this mechanism of discrete frequency tone generation is nearly always due to the addition of inlet or outlet guide vanes. It is also prevalent in contra-rotating, two-stage axial fans. Occasionally it can be introduced by setting such upstream duct components as silencers too close to the impeller.

One other interactive effect producing discrete frequency noise, but at levels probably 10–15 dB higher than the wake effect described earlier, is the interaction of potential flow fields. If moving and fixed surfaces are very close together, say within one third of their chord, each will receive an impulsive loading from the static pressure field of the other as they pass. The magnitude then of the force change which produces the acoustic energy output, is usually very much larger than that induced by the relative incidence change described previously.

The effect is marked in multi-stage axial compressors of the gas turbine type, high pressure developing centrifugal blowers, and 'non-streamlined' bladed machines such as radial paddle-bladed fans primarily for dust and other particulate removal. In the last two, the reaction is between the blade tips and the scroll cut-off.

Corrections for Installation Design

(i) Intake Turbulence

Intake designs which characteristically produce turbulence onto the fan impeller have already been shown for the specific case of an axial fan in Figure 3. If any one of these effects, or the installation of a fan silencer immediately upstream of the fan without any settling length as described in the chapter on *Air Distribution Systems*, are present, 6 dB should be added to each octave band spectrum.

(ii) Form of Running

The form of running for an axial fan is conventionally designated A if the flow passes over the motor before the impeller, and B if it passes over the impeller first.

Clearly in the light of the earlier descriptions of noise generating mechanisms, Form A might be expected to be the noisier because of the turbulent flow induced by the motor and its supports. The extra noise will probably be broad band but in designs where motor support struts are within one strut diameter of the rotor, some discrete frequency interaction noise may also be present.

In the absence of specific data to the contrary from the fan manufacturer, it is prudent to assume their published figures are for Form B running, and if Form A is to be installed, add 5 dB for each octave band up to and including 500 Hz.

(iii) Inlet Guide or Damper Vanes

One way of upgrading the performance of a fan is to fit a row of guide vanes to the impeller inlet; these have the effect of pre-swirling the air onto the impeller. For maximum efficiency, the blades need to be as close as possible to the moving blades, a condition which unfortunately maximises the interaction noise they produce, which appears at the rotor/blade passage frequency and its harmonics.

A similar effect can occur if flow control dampers are placed close to the impeller. Where the added guide or damper vanes are located within a distance from the impeller equal to the length of one stationary vane chord, add 6 dB in the octave bands up to and including 500 Hz.

(iv) Multi-staging or Standby Axial Fan Configuration

Another way of upgrading a system is to add a second fan stage in series with the first. In centrifugal fan systems, there should be no noise penalty, but in axial fan systems where the second stage will usually be contra-rotating, there almost certainly will be added noise due to the downstream impeller operating in the wake of the upsteam one.

A similar situation arises in standby systems. Even if the upstream impeller is not powered, its windmilling will cause sufficient disturbance of the airflow onto the downstream one to generate the additional noise components.

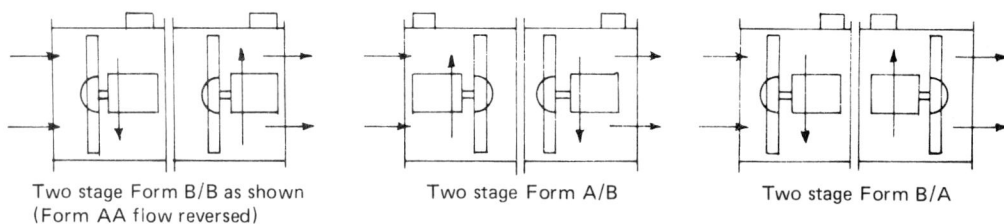

Two stage Form B/B as shown   Two stage Form A/B   Two stage Form B/A
(Form AA flow reversed)

Figure 5 Forms of running for multi-stage fans

The amount to be added to the sound power level of the single fan depends upon the forms of running of the industrial fans, with Form A/B being the worst. Possible configurations are shown in Figure 5. Again, in the absence of specific data from the manufacturer it would be safe to assume the following additions to apply over all octave band levels:

> Form A/B + 10 dB
> Form B/B + 6 dB
> Form A/A + 6 dB
> Form B/A + 4 dB

It must be recognised that Form B/A may not be acceptable to give full multi-stage performance, the impellers being too far apart. On the other hand, it should be a perfectly acceptable arrangement for standby configurations where the fans will in any case be operating as individual units.

Designing for Minimum Noise

In most commercial fan systems, it is true to say that the most common method of achieving a required reduction in fan noise is by the addition to the system of a fan silencer or silencers. The system designer can go some way, however, towards minimising the size, and hence cost of the silencer that may be required, if not obviating the need for one altogether, by careful design and selection of both fan and system.

It is also true to say that designing the fan to produce minimum noise is more properly the preserve of the machine manufacturer. Some of the following points however may prove useful to the fan designer, and in any case the system designer will have to be aware of possibilities, not so much to design a quiet fan, but at least to ensure that at the end of the day, the required duty is being performed by the system with the least possible noise as a side effect.

Vortex Noise

It has already been pointed out that every fan will produce a certain minimum level of aerodynamic noise, the amount of which is determined primarily by its duty, and by implication by its size and speed. This noise will be at a minimum when:

(a)   the only mechanism operating is that of vortex noise.

(b)   the impeller blading is operating at its maximum aerodynamic efficiency.

Condition (a) requires of course that there is no induced turbulent flow onto the impeller and that the fan has been designed to avoid any discrete frequency noise. Avoidance of these is discussed later.

Condition (b) requires that the combination of volume flow through the impeller and blade speed is such that each blade element is at its optimum relative angle of incidence. Figure 6(a) shows diagrammatically how the flow will pass over the blade element under such conditions. Here the main flow follows the top and bottom surfaces well, growth of the turbulent boundary layer is small with turbulent intensities being correspondingly small. The consequent lift fluctuations are also lower in magnitude and an appropriately lower level of sound power is generated.

Comparing this with Figure 1 and noting the velocity vector diagram, it can be seen that if for any reason, volume flow and hence the velocity vector V_a is reduced, the relative velocity V_r which the blade 'sees' is altered not so much in magnitude, but in direction. The relative angle of incidence is in fact increased for a decreased volume flow. In Figure 6 (b) the effect of high incidence upon boundary layer flow is indicated, again diagrammatically. It can be seen that the boundary layer has grown considerably thicker. Turbulent velocity and pressure intensities behind the transition point are correspondingly higher, as is the magnitude of lift fluctuation and hence sound power generated.

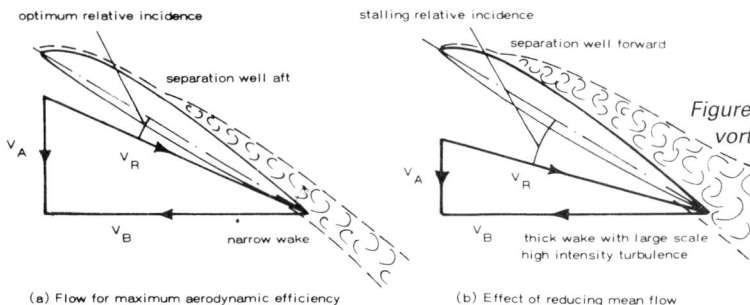

Figure 6   Effect of mean flow on vortex formation and noise

(a) Flow for maximum aerodynamic efficiency   (b) Effect of reducing mean flow

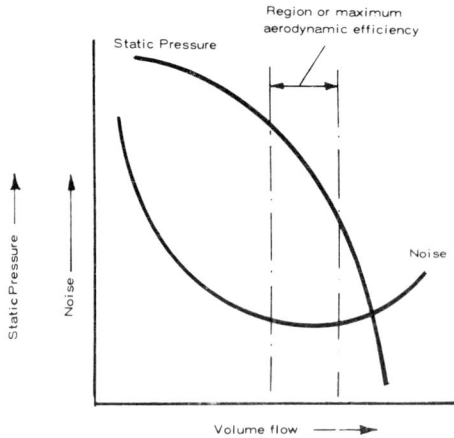

Figure 7 Effect of fan duty

The effect is summarized diagrammatically in Figure 7 which shows both pressure develop-ment and overall sound power produced by a given fan at a given speed, as a function of flow through the fan. The curve marked pressure is the normal pressure – volume characteristic which is produced by all fans. The noise curve is also typical, showing that as volume is reduced, a slight fall-off is evident from free air conditions to a well defined minimum at maximum aerodynamic efficiency; followed by a sharp rise as volume flow decreases further.

The relative importance of the individual sections of the noise curve varies from one type of fan to another but the general shape is common to all fans.

It is also to be noted that the thick boundary layers producing the noise increase generally mean a larger characteristic eddy size and hence lower frequency of noise generation. Such fans as large chord backward curved centrifugal, sheet metal bladed propeller, and mixed flow all pro-duce characteristically low frequency noise increases when called upon to work against pressure.

The lessons to be learned here are:

(i)   Always choose the fan type, size and speed which produces the system duty point closest to the fan's maximum aerodynamic efficiency.

(ii)   If off-design operation is at all likely, ensure the manufacturer gives the range of sound power produced at each end of the duty range likely to be encountered.

Intake Turbulence Noise

Noise generated by the action of a turbulent intake flow has already been stated to be entirely additional to whatever vortex noise the blade was producing for the same mean velocity smooth flow. The object must therefore be to avoid completely any turbulence at the intake. A check list for the installation designer might be as follows:

(i)   Fit coned or bell mouth intakes to all open running fans, or fans drawing direct from plenums.

(ii)   Check that all flexible connections are taut, and are unable to form 'bellow', especially on the suction side.

(iii)   Axial fans should always be run Form 'B' wherever possible.

(iv)   Avoid locating fans – particularly axial flow fans – immediately behind heater batteries, cooling coils, transformation sections or any in-flow components producing turbulent wakes or separated flow. Always aim for at least one fan diameter between turbulence generator and impeller.

(v) Upstream silencers should always incorporate settling lengths.

For the two-stage or standby axial fan installation, the aim should be Form B/A if possible (see Figure 5), or at most Form A/A.

Interaction Noise

Although much has been said about the mechanism and amount to be expected of discrete frequency interaction noise, it is not so much of an installation problem as are the broad band turbulence effects, at least with industrial fans. Its avoidance is a fairly straightforward procedure of ensuring maximum possible spacing in the flow direction between fixed and moving surfaces. In particular, to avoid the acoustically disastrous effects of potential flow interaction between rotor and stationary guide or control vanes in axial fans, and between impeller blades and scroll cut-off in centrifugal blades, separations should never be less than half the blade chord.

Beyond this, the wake effect previously described will predominate, and to ensure its contribution to overall sound power is as low as possible a separation of more than two chords of the wake producer should be the aim.

TABLE 1 – SPECTRUM CORRECTION FACTORS TO ESTIMATE FAN SOUND POWER LEVELS (dB)

Fan type	Diameter (m)	Octave band centre frequencies								
		63	125	250	500	1k	2k	4k	8k	BFI
Aerofoil										
Backward curved	>0.75	25	25	24	19	15	8	4	2	3
	<0.75	30	30	28	24	19	13	9	4	3
Radial fan										
Low pressure	>1	41	32	28	24	22	17	14	11	7
	<1	52	44	38	28	27	24	19	16	7
Medium pressure	>1	43	39	30	27	23	18	14	11	8
	<1	53	48	36	33	31	26	22	19	8
High pressure	>1	46	43	38	33	31	29	26	23	8
	<1	56	52	44	33	31	29	26	23	8
Forward curved		38	38	28	21	21	16	11	6	2
Vane axial										
hub ratio 0.3 – 0.4		34	28	28	33	22	30	23	19	6
hub ratio 0.4 – 0.6		34	28	31	28	26	21	15	13	6
hub ratio 0.6 – 0.8		38	37	36	36	34	32	28	25	6
Tube axial	>1	36	31	32	34	32	31	24	22	7
	<1	33	32	34	38	37	36	28	25	7
Propeller		33	36	43	41	40	37	31	27	5
Predicted accuracy ± dB		8	4	2	2	2	2	2	4	

Notes: Low pressure = 1 – 2.5 kPa
Medium pressure = 2.5 – 5 kPa
High pressure = 5 – 15 kPa
BFI = blade frequency increment

TABLE 2 – ADDITIONS TO BE MADE TO L_w TO ALLOW FOR OPERATING OFF PEAK EFFICIENCY

Increase dB	Efficiency (%)		
	Aerofoil centrifugal and vane axial	Backward curved centrifugal	Forward curved centrifugal
3	68 – 71	64 – 66	55 – 57
6	60 – 67	56 – 63	49 – 54
9	52 – 59	49 – 55	42 – 48
12	44 – 51	41 – 48	36 – 41

Chapter 5
Acoustic Glazing

Windows are a particular architectural feature that require separate evaluation. Two situations in which the acoustic properties of windows need to be examined are first where the sound is external to the building and the occupants wish to be protected from it and secondly where an enclosure is constructed with a noisy machine or process inside into which inspection panels have to be inserted. In the first case, the external sound may come from traffic, aircraft or a factory, and there is little the occupants are able to do to reduce the level of the incident sound. In the second case, the application of an absorbent treatment to the interior of the enclosure is likely to be of value in reducing the level of the incident sound; this is discussed in the chapter on *Sound Insulation*.

Double or triple glazing may be advantage in improving the sound insulation of a window, but it should be noted that most double glazing in domestic use is more likely to be installed for heat insulation that for sound insulation. There may be a design conflict between the two requirements.

The main significance of windows and glazed areas in buildings is their permeability to outside sounds. External sound will normally be random incident on windows, so there can be marked spread of sound through a window area, further diffused by reflections from interior walls. Thus the ratio of window area to facade area is significant. The most marked loss occurs within the window/facade area up to 20%, after which increasing the window area has a much smaller effect. The 'turning point', area 20%, is significant in that the insulation of the whole facade is not very much better than that of the window area itself – see Figure 1.

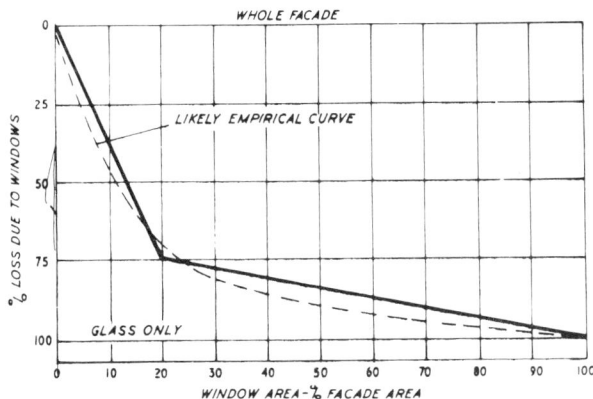

Figure 1

The following rules apply:

(a)   Small windows are unlikely to be effective in maintaining the sound insulation of the whole facade.

(b)   Altering the shape of the windows will have little appreciable effect on sound insulation as it is window area which is significant.

(c)   Large window areas do not automatically mean additional 'transparency' to sound. The damage has largely been done in this respect once the window area exceeds 20% of the total area.

(d)   The insulation of the whole facade can only be maintained by improving the sound insulation offered by the windows. The size and shape of the windows can be chosen from considerations of the daylight, aesthetic, view or heating requirements.

(e)   An alternative, or additional possibility exists, namely reduction of the noise incident upon a window. This may be achieved by siting (i.e. increasing the distance from the noise source), the use of barriers and buildings as noise screens, locating noise tolerant rooms on the noisy facades, and rooms requiring silence on the quieter sides of the building. This can be effective in reducing the sound insulation required of the windows.

The sound insulation values of window glass can be predicted with reasonable accuracy from the Mass Law curve (Figure 2). In other words with single glazing glass thickness (or more accurately glass weight) is the significant parameter. Sound insulation is, in fact, increased by about 5 dB for each doubling of the mass per unit area of the glazing. A typical figure for average constructions is a sound insulation figure of about 25 dB or rather less, depending on how well the window seals.

As a general rule glass has rising characteristics of sound insulation with increasing frequency, although this is modified by resonance effects. As a result, increasing the thickness of glass does not automatically ensure that the overall effect will be that much better. The thicker the glass the lower the coincident frequency. Thus 12 mm float glass with its initial frequency of 1 kHz has sufficient mass to be a good barrier to traffic noise. Doubling the glass thickness would give

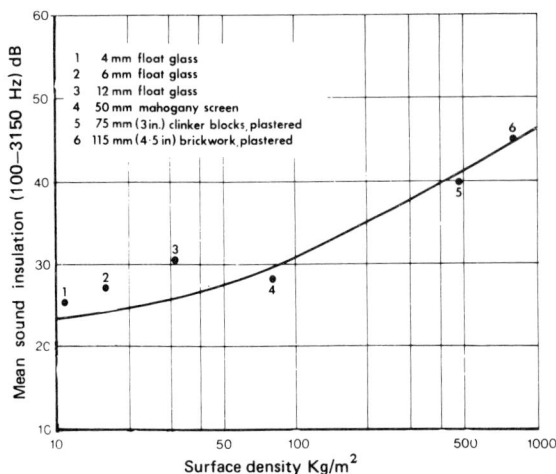

Figure 2 Sound insulation values of window glass and other materials related to Mass Law

an overall improvement in sound reduction of 5 dB but reduce the critical frequency to 500 Hz, at which frequency typical traffic noise has strong components.

The critical frequency of glass can be calculated as:

$$\text{Critical frequency } f_o = \frac{15200}{t}$$

where t is the glass thickness in millimetres

The real acoustic performance of single, monolithic glasses are displayed in Figure 3, which shows the results of a statistical analysis of data originating from laboratories all over the world. While the frequencies at which wave coincidence occurs can be predicted with accuracy, the actual fall in sound insulation and bandwidth over which it operates depend on the acoustic damping of the glass.

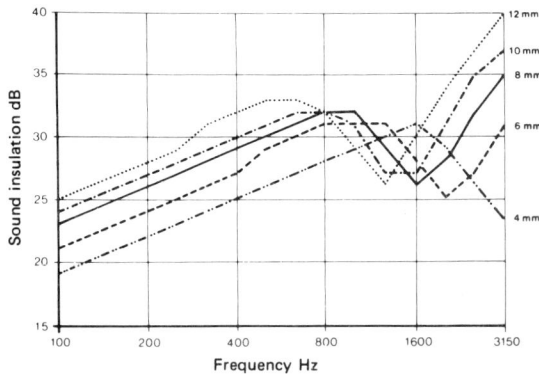

Figure 3 Acoustic performance of single monlithic glasses

Double Glazing

Double windows, employing two glasses with an intervening air space (known as double glazing) offer a direct solution to increasing the mass per unit of a window without necessarily going to excessive glass thickness, combined with further insulation offered by separating the glass into two leaves with an airspace between. To be effective as a *sound* insulator, however, the air space must be generous. A minimum airspace of 100 mm is normally necessary for any realistic improvement in sound insulation, with an optimum of about 300 mm. With greater separation there is very little improvement in sound insulation. For practical (and aesthetic) reasons, the normal maximum adopted is 200 mm. Airspaces of less than 100 mm may also be used; but less than 50 mm airspace gives sound insulation directly comparable to the same (total) weight of single glazing.

The sound insulation performance of double glazing can be improved by incorporating absorbent material in the sides. This is widely used to offset the less than optimum performance realized with smaller air gaps. Ultimately, the overall performance will depend very largely on the integrity of the panel and frame seals. Typical performances of good double glazing using differing thickness of glass are shown in Figure 4.

Sealed double glazing units do not normally offer really high sound insulation because: (a) the relatively narrow airspace is insufficient to decouple the movements of the component glasses and (b) sound energy is transferred via the peripheral sealing strip. Generally, they offer an improvement of up to 3 dB in sound insulation over that provided by the thicker of the component

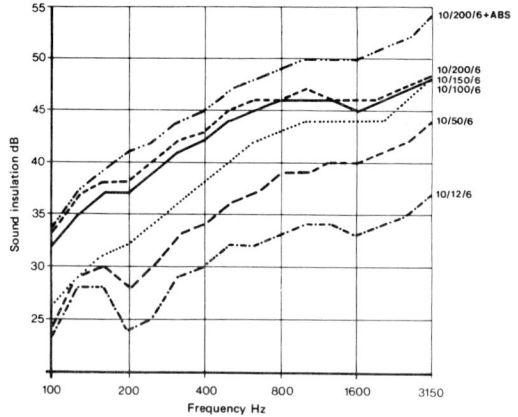

Figure 4 Sound insulation of double glazing.
(Pilkington Environmental Advisory Service)

glasses alone. Nevertheless, by using a thick glass in such a unit, an acceptable compromise is often achieved between reasonably high sound insulation, coupled with the benefits of thermal insulation, and compactness.

It is usual to specify glasses of differing thicknesses so that sympathetic resonances are avoided. The coincidence frequencies of the two glasses should be displaced so that while one is resonating (and consequently providing lower sound insulation than the norm), the other is unaffected, and vice versa. By this means, smoother overall sound insulation is obtained; a thickness difference of about 30% is usually sufficient to secure this. Thus, when a total glass thickness of 16 mm is to be installed, it is better to avoid using two leaves of 8 mm glass; 10 mm and 6 mm glasses (or 12 mm and 4 mm glass) are to be preferred. Even greater control is possible using laminated glass or triple glazing.

Double glazing in common with other twin panel constructions exhibits two other resonances, Figure 5. Mass–air–mass resonance is the lowest with a resonant frequency

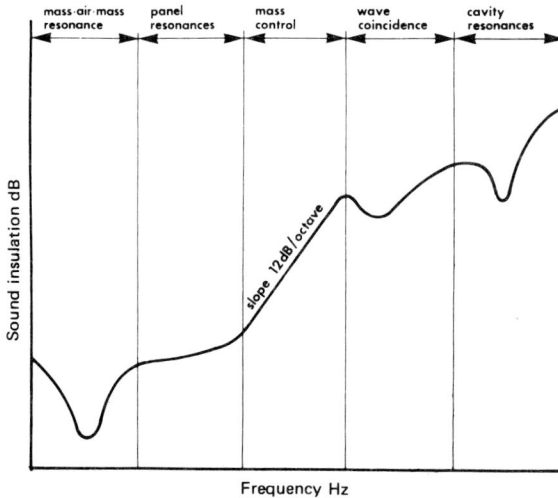

Figure 5 Typical double-glazed window
characteristics

$$f_m = 80 \sqrt{\left(\frac{m_1 + m_2}{d\, m_1 m_2} \right)}$$

m_1 is the surface density of the first layer, m_2 is the surface density for the second layer and d the gap width in m.

Cavity resonances occur because of standing waves in the air space between the two panes. This limiting frequency is:

$$f_1 = \frac{c}{2\pi d} = \frac{55}{d} \quad Hz$$

c is the speed of sound in m/s and d the cavity thickness in m.

With double glazed windows, where the two panes are independently supported, so there is no mechanical or acoustic coupling, it is possible fairly accurately to predict the transmission loss throughout the frequency range, as follows

Calculate the sound reduction indices R_1 and R_2 for the two panes separately with the relationship:

$$R_1 = 20 \log_{10}(m_1 f) - 48 \text{ dB}$$
$$R_2 = 20 \log_{10}(m_2 f) - 48 \text{ dB}$$

Calculate R_c as if the two panes acted together, i.e.

$$R_c = 20 \log_{10}[(m_1 + m_2)f] - 48 \text{ dB}$$

Then R as a function of f is

$$R = R_c \quad \text{dB} \qquad\qquad\qquad\qquad \text{for } f < f_m$$
$$R = R_1 + R_2 + 20 \log_{10}(fd) - 29 \quad \text{dB} \qquad \text{for } f_m \leqslant f < f_l$$
$$R = R_1 + R_2 + 6 \quad \text{dB} \qquad\qquad\qquad \text{for } f \geqslant f_l$$

In order for the full insulation effects of double glazing to be achieved, not only must the panes be separately supported but the sides of the air space should be lined with sound absorbent material, Figures 6, 7 and 8.

Tables 1 and 2 gives some typical values of sound reduction for various units.

Laminated Glasses

Laminated glasses, two or more glasses bonded together by transparent plastic interlayers (usually of polyvinyl butyral), offer significant advantages in sound insulation. Figure 9 shows that a single, solid glass has a dip in the sound insulation curve at the coincidence frequency, which, for common glass thickness, occurs in the 1000 Hz to 2000 Hz region. This dip is smoothed out by using a laminated glass of the same overall thickness as the single glass. There is thus an improvement at the coincidence frequency, but at other frequencies the very small improvement achieved must be weighed against the cost.

Sealed double glazing units, incorporating laminated glass, are sometimes used where space limitations do not permit the use of dissimilar glass thicknesses to suppress coincidence effects. Here two panes of the thinner substance may be employed, one being laminated. Little further improvement is gained from laminating the second glass also. Laminated glass offers no

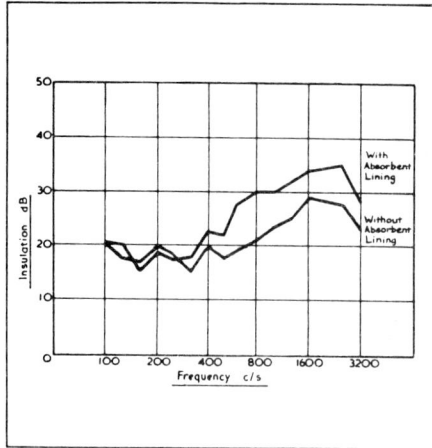

Figure 6 The effect of absorbent
lining on sound insulation

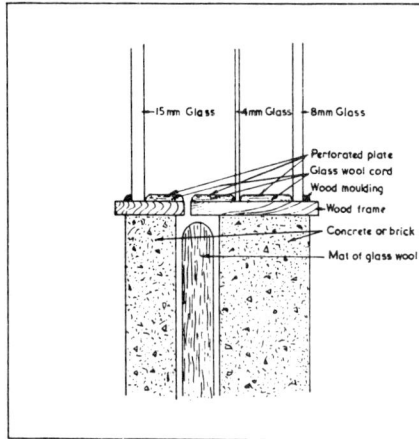

Figure 7 Panes of different thicknesses

acoustic advantage in wide airspaced double windows, because in these arrangements it is mainly the decoupling effect of the airspace itself which secures the resulting high sound insulation.

Special Window Designs

Virtually any required sound reduction can be achieved by suitable design of glass windows and/or glass screens, following the general principles outlined above, buy due consideration must also be given to other factors such as space restrictions, load bearing capacity of structures and economics, etc. Thus there is unlikely to be any 'ideal' design for a particular application. It has to be an acceptable compromise. The value of the special window designs also depends on

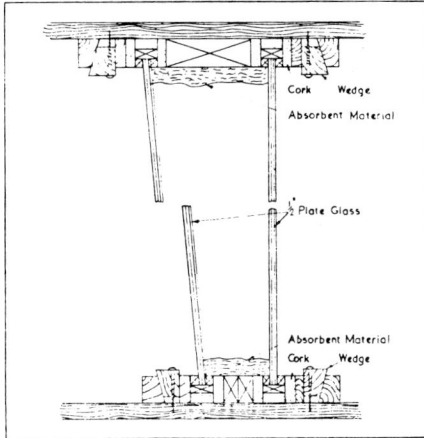

Figure 8 Non-parallel panes

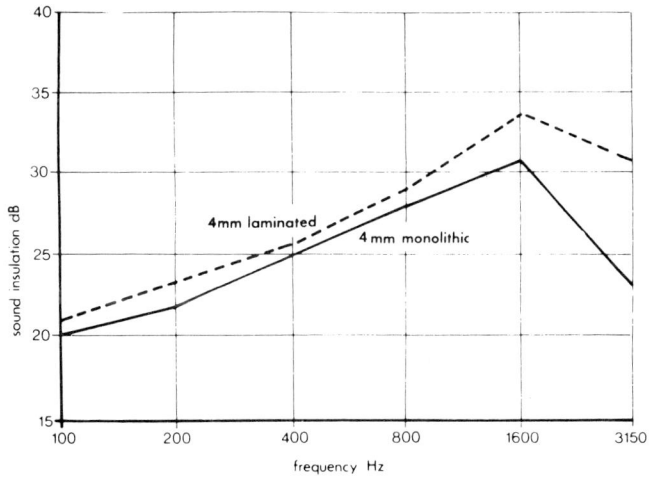

Figure 9 Comparison of monolithic and laminated glasses

the magnitude of flanking transmissions – e.g. the performance of the window may be greatly improved, but actual improvement in sound insulation may only be marginal because sound is entering the insulated space by other paths.

Glasses of different thicknesses are not difficult to fix and may be considered for building applications, but setting one glass at an angle to the other and the use of flexible mountings are devices that are usually reserved for the very special observation windows of broadcasting studios, audiology laboratories and the like.

The sound insulation required of a facade is best found by subtracting, at each frequency, the acceptable indoor noise from the existing outdoor noise. This gives a criterion curve against which the properties of various constructions can be matched and a selection made. A simplification can be used for selecting the windows when traffic noise is the dominant problem.

TABLE 1 – TYPICAL VALUES OF SOUND REDUCTION INDEX (SOUND TRANSMISSION LOSS) IN dB FOR TYPES OF GLAZING

Thickness (mm)	Frequency (Hz)					
	125	250	500	1 k	2 k	4 k
Single Glazing						
6	11	24	28	32	27	35
8	18	25	31	32	28	36
9	22	26	31	30	32	39
16	25	28	33	30	38	45
25	27	31	30	33	43	48
13 laminated	23	31	38	40	47	52
Double glazing A = absorbent reveals S = sealed unit						
2.4 + 7 + 2.4 S	22	16	20	29	31	27
9 + 50 + 9	25	29	34	41	45	53
6 + 100 + 6	28	30	38	45	45	53
6 + 188 + 6	30	35	41	48	50	56
6 + 188 + 6 A	33	39	42	48	50	57
6 + 200 + 9 A	36	45	58	59	55	66
3 + 55 + 3	13	25	35	44	49	43
6 + 55 + 6	27	32	36	43	38	51
6 + 100 + 5	27	37	53	55	50	55
6 + 100 + 8	35	47	53	55	50	55

TABLE 2 – SOUND TRANSMISSION CLASS FOR NON-OPENABLE DOUBLE GLAZED UNITS WITH VARIOUS AIR SPACING THICKNESSES, d mm

STC	Single glazed thickness mm L = laminated	Double glazing 3 mm glass d mm	6 mm glass d mm	6 mm glass d mm
48				100
46			120	60
44		150	80	40
42		100	50	25
40	20 L	70	30	16
38	12 L	50	20	10
36	12	30	13	
34	6 L	20	8	
32	6	10		
30	3 and 4	6		

Chapter 6
Commercial Buildings

Noise in Commercial Buildings

Sources of noise in office blocks and commercial buildings are:

Ambient internal background level, or the irreducible noise level in an unoccupied office normally associated with the air conditioning system. This is usually constant in level and frequency content.

Occupation noise, being the sound generated in an office during the normal working day. Ideally it should be a steady 'hubbub' arising from conversations, office machinery, personnel movements, *etc.*, but it could also include undesirable components such as bangs and thumps from accidental impacts, loud voices, or the monotonous whine of a fan or pump. Occupational noise can be highly variable in level, character and duration, and include talk, typewriters, telex and office machines, footsteps, slamming of doors, lift gates and lift working, *etc.*

Noise from *external* sources, such as traffic noise outside the building, and occupation noise in adjacent rooms. The proportion which enters a building structure is moderately variable in level and spectral content.

The consequence of sound on the noise environment in individual offices is exaggerated by the fact that modern buildings of this category are designed to provide the maximum accommodation and occupancy on a minimum area of land, favouring the use of lightweight structures and open-plan offices. This can be offset by sub-dividing the floor area into individual offices, but there may be limits imposed by the structure on the weight of inter-office walls or partitions which can be used.

Basic Requirements

Basic requirements for satisfactory office working are:

Good communication conditions must be provided in the vicinity of the speaker.

There must be privacy outside the group work area.

Background noise levels which are comfortable and non-intrusive should be maintained throughout the office.

Occupational noise levels should be free from potential irritations such as impulsive noises, *e.g.* thumps and bangs from slammed doors, loud and distinctive voices outside the group area, or periods of high noise activity from machinery elsewhere in the office.

Ideally there should be a steady background noise in the office; it should be pleasant and easy on the ear; it should be capable of 'masking' distant conversations so that they are relatively unintelligible, and the long distance propagation of speech and noise should be controlled by the room acoustics or the room furnishings.

Acceptable Office Noise Levels

The acceptability of background noise level is essentially a subjective response, but is closely related to the working environment. Thus, in a busy office, including typewriters, a relatively high noise level would be regarded not only as acceptable, but a normal working environment. It would be regarded as intrusive if it interfered with a particular working requirement – *e.g.* made speech or the use of a telephone difficult. In fact, the overall background noise level may be higher than desirable, but one to which the individual occupants become conditioned. It is desirable to set maximum recommended levels, as conditioning to unduly high levels can be accompanied by loss of working efficiency. Such recommended levels are, therefore, set with regard to maintaining good working efficiency, but at the same time, must be economically realistic .

TABLE 1 – TYPICAL AVERAGE NOISE LEVELS FOR OFFICES

Type of Office	Typical Background Noise Level dB(A)	Recommended Maximum Acceptable Noise Level dB(A)
Board room	30–35	30–35
Conference rooms – small	35–45	35
Conference rooms – large	35–50	30
Managing Director	35	
Managing Director's secretary	40–45	40
Other Director's offices	35–45	40
Personnel Manager	45–50	
Plant Manager	50–75	50
Officer Manager	50–70	50
General Office		
over 100 people	60–70	50
3–10 people	52–64	50
less than 3 people	50–60	45
Clerical Office	55–65	60
Drawing Office	40–55	50
Large busy office up to	70	70
Very large, very busy office up to	80	60
Office machine room	75	

For commercial buildings an index known as a Noise Criterion (NC) level is often used. It has some relation to the well known A-weighting in that it takes a different account of the frequencies throughout the spectrum. An NC readout is not normally available on sound level meters, but it can be related to dB(A) by use of the approximate equation

$$NC = dB(A) - 5$$

NC curves are widely used for the selection of suitable criteria for the heating and ventilating industries. For further details of the NC method, refer to the chapter on *Noise Scales and Ratings*.

Table 2 gives NC ratings for a range of commercial buildings, specifically intended for the planning of ventilation systems.

TABLE 2

Background Noise Criteria for Ventilation Noise Control	
Section 1	*NC*
Studios and Auditoria:	
Sound Broadcasting (drama)	15
Sound Broadcasting (general), T.V. (general), Recording Studio	20
T.V. (audience studio)	25
Concert Hall, Theatre	20–25
Lecture Theatre, Cinema	25–30
Section 2	
Hospitals:	
Audiometric Room	20–25
Operating Theatre, Single Bed Ward	30–35
Multi-bed Ward, Waiting Room	35
Corridor, Laboratory	35–40
Wash Room, Toilet, Kitchen	35–45
Staff Room, Recreation Room	30–40
Section 3	
Hotels:	
Individual Room, Suite	20–30
Ballroom, Banquet Room	30–35
Corridor, Lobby	35–40
Kitchen, Laundry	40–45
Section 4	
Restaurants, Shops and Stores:	
Restaurant, Department Store (upper floor)	35–40
Night Club, Public House, Cafeteria, Canteen, Retail Store (main floor)	40–45
Section 5	
Offices:	
Boardroom, Large Conference Room	25–30
Small Conference Room, Executive Office, Reception Room	30–35
Open Plan Office	35
Drawing Office, Computer Suite	35–45
Section 6	
Public Buildings:	
Court Room	25–30
Assembly Hall	25–35
Library, Bank, Museum	30–35
Washroom, Toilet	35–45
Swimming Pool, Sports Arena	40–50
Garage, Car Park	55
Section 7	
Churches and Academic Buildings:	
Church	25–30
Classroom, Lecture Theatre	25–35
Laboratory, Workshop	35–40
Corridor, Gymnasium	35–45
Section 8	
Industrial:	
Warehouse, Garage	45–50
Workshop (light engineering)	45–55
Workshop (heavy engineering)	50–65
Section 9	
Private Dwelling (Urban):	
Bedroom	25
Living Room	30

Speech Intelligibility

A measure of speech intelligibility, widely used in the USA, is the Articulation Index, which is related to the level of speech above a background level. The calculation of AI is complex, and requires the speech noise ratio to be determined in 20 frequency bands, although a simpler but less accurate method based on one-third or octave bands can be used. Basically the method requires the peak speech noise level and the rms noise level to be measured in each band of interest, and the S/N ratio (in dB) calculated. A weighting factor is then applied to the separate S/N ratios and the summation found. The weighting factors and a full description of the method is to be found in ANSI S3.5. The Articulation Index is expressed as a number between 0 and 1.0 as shown in Table 3.

TABLE 3

SPEECH PRIVACY	AI	COMMUNICATION
	1.0	
	0.9	Excellent
Nil	0.8	
	0.7	
	0.6	Good
	0.5	
Very Poor	0.4	Fair
Poor	0.3	Poor
	0.2	
Acceptable	0.1	Very Poor
Excellent	0	Nil

Speech Interference Level (SIL) is another index designed for evaluating the interfering effects of noise on speech. The method is described in ANSI S3.14 and is much simpler to use than AI.

An objective method for assessing speech intelligibility is the Rapid Speech Transmission Index (RASTI) which is explained in BS 6840: Part 16 (IEC 268 – 16). This requires the measurement of a transfer function (effectively the signal/noise ratio) between a noise transmitter situated at the speaker's position and a receiver situated at the listener's position. Only two octave bands are employed to determine the RASTI (500 Hz and 2 kHz). A Speech Transmission Meter (Figure 1) generates an intensity modulated noise signal simulating the human voice. The receiver analyses the signal and calculates the RASTI value from a measure of the modulation reduction factor. The RASTI is similar to the Articulation Index of Table 3.

Provides a Speech Intelligibility Index by the RASTI Method in less than 10 seconds

Types of Commercial Buildings

Each type of building has its own requirements and an appropriate criterion for noise level has to be determined. For internally generated noise as in offices, the criterion could easily be that of

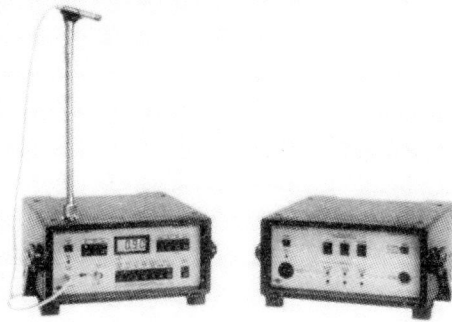

Figure 1 Speech Transmission Meter

intelligibility, but for buildings such as educational establishments, the problem should be tackled at the design stage. For these buildings, sound insulation is required to ensure that the sound producing areas are sufficiently well isolated from the quiet rooms. Table 4 gives the recommended insulation properties of the various kinds of rooms. BS 8233 gives further guidance.

TABLE 4 – RECOMMENDED MINIMUM SOUND INSULATION FOR ROOMS IN EDUCATIONAL ESTABLISHMENTS

Class	Type of room	Minimum insulation (dB)	
		D	D_w
A	Noise producing: workshops, kitchens, dining rooms, gymnasia, indoor swimming pools and boiler rooms	25	28
B	Noise producing but needing quiet at times: assembly halls, lecture halls, music rooms, commerce and typing rooms	45	48
C	Average: general classrooms, practical rooms, laboratories and offices	35	38
D	Rooms needing quiet: libraries, division rooms, studies, medical rooms and staff rooms	35	38

Notes: The minimum sound insulation is the average over the sixteen one-third octave bands from 100 Hz to 3150 Hz.
D = sound level difference,
D_w = weighted sound level difference (see chapter on Sound Insulation)
When a room has more than one use, the higher insulation value is to be used.
The minimum sound reduction index between rooms in different classes should be 45 dB.

Sound Fields in Large Offices

In a room having reflecting surfaces on the floor, ceiling and walls, a 'reverberant' sound field is created which has a constant sound level at all points, irrespective of the distance from the source. This is the reason why offices with plastered walls and ceilings and uncarpeted floors sound noisy and acoustically unsatisfactory.

If the acoustically 'hard' surfaces can be replaced by absorbents, a 'semi-reverberant' sound field will be created which will give an improved rate of fall-off in level with distance. In open-plan offices it is obviously desirable that the theoretical maximum rate of fall-off with distance, (6 dB per distance doubling), should be approached, and a value of at least 5 dB per doubling is required. This can only be achieved if all the available surfaces are treated with highly efficient sound absorbents.

The sound absorbing properties of the floor covering should approach the maximum feasible performance. A good quality carpet laid over a felt or foam underlay should ensure that a target absorption value of 0.5 is obtained over the frequency range 500 to 2000 Hz. The use of a carpet also helps to control surface noise (the noise of footfalls and furniture movements), and impact sound transmission to rooms below.

The ceiling which ordinarily constitutes the principal sound reflection path should be surfaced with a highly efficient sound absorbing material, especially if it is plain unbroken surface. Further improvements in ceiling performance can be achieved if the surface is broken up by coffering or baffles.

The ceiling structure should have an average sound absorption coefficient of at least 0.80 over the range 500 to 2000 Hz.

In most purpose-designed large open-plan offices the wall area is small in comparison with the plan area and in consequence does not have a measurable effect on the overall sound field in the room.

Nonetheless, where possible, the walls should be treated with acoustically absorbing materials, particularly if a noise source such as piece of office machinery is located close to it, otherwise there would be a tendency for the acoustically 'hard' surface to reinforce the reflected sound field.

Sound reflections off windows and glazed walls can be controlled by the use of drapes and blinds.

In small open offices where the walls are less than 15 m apart and consequently comprise a large proportion of the total room area, it is essential that a sound absorbent surface treatment is applied.

This may take the form of sprayed acoustic plasters, acoustic tile on battens, or slabs of rockwool or glassfibre faces by slotted wood or plastic covers.

Under no circumstances should two acoustically reflective surfaces faces each other, otherwise an objectionable 'flutter' echo will occur between them.

Problems with Common Suspended Ceilings

Modern building technique which conceals mechanical services above a lightweight suspended ceiling can lead to problems of flanking sound transmission through the ceiling void. This can nullify the effects of good sound insulating positioning. In such cases, particularly where demountable partitions are used for flexibility of layout, special attention may have to be given to the ceiling, even to the extent of installation of a plenum barrier above the partitions.

Screens

Screens are a significant feature of most open-plan offices. Their primary function is to break up sightlines and thereby provide privacy. This can sometimes be provided by furnishing such as files and bookshelves or even by large indoor plants. More commonly, however, purpose-built free standing screens are provided which, like other rigid objects in the furnishing scheme, can provide acoustic screening.

All too often in the past, much dependence has been placed on screens as remedial correction for a lack of acoustic privacy, often with disappointing results. To be effective, a screen should be located as close as possible to the source or to the recipient of sound. A screen placed within 50 cm of a noisy typewriter could give up to 15 dB sound reduction for an observer situated on the other side of the screen.

Screens erected in open offices to provide visual privacy seldom ensure acoustic privacy. They are normally used as the demarcation of territorial boundaries within the open plan and, as such, tend to be placed mid-way between the observer and an identified noise source. If the typewriter and observer referred to in the example above were separated by 8 m and a single free standing screen placed on the sightline, the reduction in sound level would amount to approximately 2 dB – a change that is barely detectable to the human ear.

Use of Enclosures

The usual formula for determining the sound level inside the enclosure applies:

$$L_p = L_w + 10 \log_{10} \left(\frac{Q}{4\pi r^2} + \frac{4}{R_c} \right) \quad \text{dB}$$

A typical circumstance is when the source is mounted on a reflective floor, when $Q = 2$. R_c is the room constant inside the enclosure, whose value is given by:

$$R_c = \frac{S\alpha}{1 - \alpha}$$

S is the surface area of the the inside of the room. When examining the effect of the enclosure on the sound level outside (i.e. in the room in which the enclosure is built), two extremes can be considered: when the outer room is effectively equal to a free field, and when it is wholly reflective.

When the enclosure radiates into a large room or one with very absorbent surfaces or into the open air, the sound pressure level outside is

$$L_{po} = L_p - R - 6 \quad \text{dB}$$

L_{po} is the sound level immediately outside the enclosure and R is the sound reduction index. The total sound power radiated from the enclosure is

$$L_w = L_{po} + 10 \log_{10} S_o \quad \text{dB}$$

S_o is the surface area of the enclosure. At a distance r metres away from the enclosure the sound pressure level is

$$L_{pr} = L_w - 20 \log_{10} r - 8 \quad \text{dB}$$

When the enclosure radiates into a reverberant room (the other extreme),

$$L_p - L_{po} = R - 10 \log_{10} \frac{S}{S_o \alpha}$$

In order for an enclosure to be effective at reducing the sound pressure outside, it is important that there should be sufficient absorption inside it, otherwise the sound level inside would increase through reverberation and vitiate the effect of the sound reduction of the walls. If the inside of the enclosure were fully reflective, it would offer no sound reduction at all.

Partial Enclosures

There are many instances where a partial enclosure is effective at reducing the sound pressure level. These can vary from a simple partition between workstations to one in which an item of machinery can be almost totally enclosed with just a small opening directed away from the most sensitive noise path. Provided that sufficient sound absorption is incorporated in the screening and in the walls and ceiling of the main room, a significant reduction is possible, depending on the percentage open area. For a 50% open area, 3 dB is the maximum, and for a 1% open area 20 dB is possible. In practice a partial enclosure is unlikely to give better than about 15 dB in the best direction. The sound transmission from the source to the listener can be direct or indirect. For direct transmission, the sound reduction depends on the sound reduction index of the material of the screen or enclosure. For indirect transmission, the sound received by the listener depends primarily on the reflective properties of the wall surfaces in the main room.

For open plan offices as illustrated in Figure 2, a lightweight screen is perfectly adequate in reducing the direct sound. There is no point in building a heavy screen with a high sound reduction index, when most of the sound is indirect. For this kind of installation, an index of about 20 dB is adequate, which could be achieved by a panel with a superficial weight of about 5 kg/m^2. A thin sheet of plywood covered by a sound absorbent material and mounted in a suitable frame would

Figure 2 Privacy screens in a drawing office

be sufficient, but in practice, aesthetic considerations may require an attractive proprietary screen with sufficient weight for its own stability. To be effective the use of screens needs to be combined with acoustical treatment to the walls, floors and ceilings.

The following general recommendations apply to the use of screens in open plan offices.

> Height and width at least 2 m
>
> Distance of bottom edge from floor less than 100 mm
>
> Sound reduction index (or STC) better than 20 dB
>
> Absorption coefficient better than 0.8
>
> Acoustic treatment to be applied to ceilings by acoustic tiles and if necessary by installing a false ceiling
>
> Treatment of floors by use of carpets, which not only reduces the reflections but suppresses noise of movement
>
> Use of curtains at the windows – the heavier the better; Venetian blinds have little effect
>
> Use of soft rather than hard furnishings
>
> Consider use of sound conditioning systems

Individual Offices

Sound reduction treatment for individual offices involves specific attention to sound insulation of the boundary surfaces, the elimination of flanking transmissions and structural-borne sound, and attenuation of the ventilating system.

The typical average noise level in an office will tend to increase with the number of people working in the office. It may be desirable, or even necessary, to apply noise control at source, such as by the erection of partitions or partial barriers, or similar sound absorption treatment.

Some possibilities are:-

> *Acoustic treatment of ceiling* – can be quite effective for simple treatment in large offices.
>
> *Partial barriers* – locate cabinets, *etc.* to mask local sound sources (most effective with acoustically treated ceilings and walls).
>
> *Room dividers* – effectiveness depends on sound absorption realised. Simple glass partitions are relatively ineffective.
>
> *Part-height partitions* – limited effectiveness.
>
> *Full partitions* – more effective but may have local weaknesses.
>
> *Ceiling hung blankets* – effective, but not usually desirable or practical.
>
> *Enclosure of quiet area* – most effective with satisfactory acoustic treatment but may have local weaknesses (e.g. at window or door).
>
> *Doors* – gasket seals around door area most effective. Avoid use of louvred doors.
>
> *Acoustic traps* – may be applied to louvred doors or wall openings.
>
> *Wall panels* – see Figure 3, 4, and 5.

Figure 3 Wall panels used to improve the acoustic environment

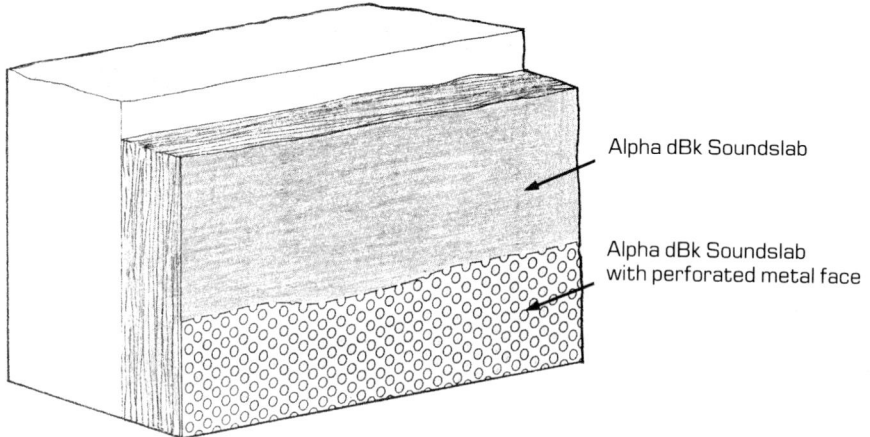

Alpha dBk Soundslab

Alpha dBk Soundslab with perforated metal face

Figure 4 Wall treatment using mineral fibre slabs on a rigid backing

Intrusion of External Noise

However good the sound treatment of internal strucures, performance is frequently modified by the presence of opening windows, where, on noisy sites, there is often an unfortunate choice between noise exclusion and ventilation. Here planning can provide the answer by foreseeing the problem and providing sealed noise-excluding windows and air conditioning in offices on the sides of the building exposed to high ambient noise.

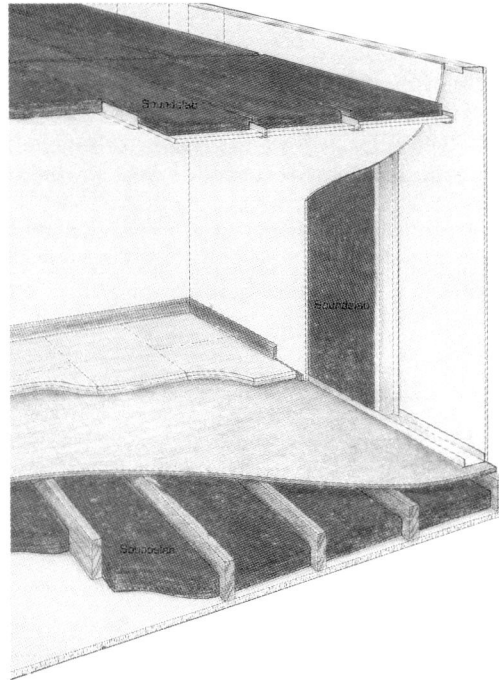

Figure 5 Mineral fibre slabs in a variety of applications

Traffic noise, in particular, can be intrusive in city and urban areas, where even a fully sealed facade may not provide a complete answer. This is because although traffic noise is excluded (or at least reduced to insignificant levels) by the noise insulation partition of the facade, traffic vibrations are present in the building.

Sound Reduction Between Rooms

Modern offices make much use of demountable partitions to give maximum flexibility of space usage. The assessment of the effectiveness of a partition treats the room on one side of a partition as a source and the room on the other as a receiving room.

In calculating the sound insulation required, the assumption can be made that all the sound passes through the partition. If the level in the source room is L_s and in the receiving room, L_r

$$L_s - L_r = R - 10 \log_{10} \frac{S}{A_r}$$

R is the sound reduction index for the partition, S is the common area of the partition and A_r is the total sound absorption of the room. $A_r = S_r \alpha$, where S_r is the area of the receiving room and is its average sound absortion coefficient. To be effective the same considerations apply as to any sound barrier: there must be no flanking transmission, all gaps must be sealed and doors must be well fitting. A wide variety of partitions are available. A sound reduction of the order of 40 dB is readily obtainable in lightweight partitions, where a solid wall would be capable of about 50 dB.

Sound Conditioning

The basic principle is very simple, with sound being generated electronically and introduced into the open–plan area through a number of loudspeakers. The low level background sound which is produced is a combination of frequencies carefully selected to mask speech and other noises.

The sound-generating electronics in the central control unit could be installed in any convenient service area. The loudspeaker system is usually installed within a false ceiling void, with the speaker cones facing upwards, in order to produce a diffuse field in the office area below. Grouping the loudspeakers together allows the background sound to be adjusted to suit the individual requirements. Speaker density should be one unit for every 10 m² of area covered; with a lower density, the directional effect becomes apparent, and the system will not be as effective in providing speech privacy. Figure 6 shows the system.

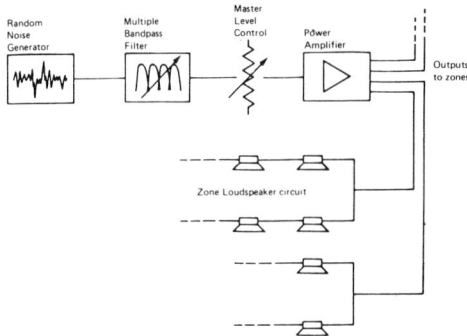

Figure 6 Diagrammatic representation of a typical sound conditioning system.

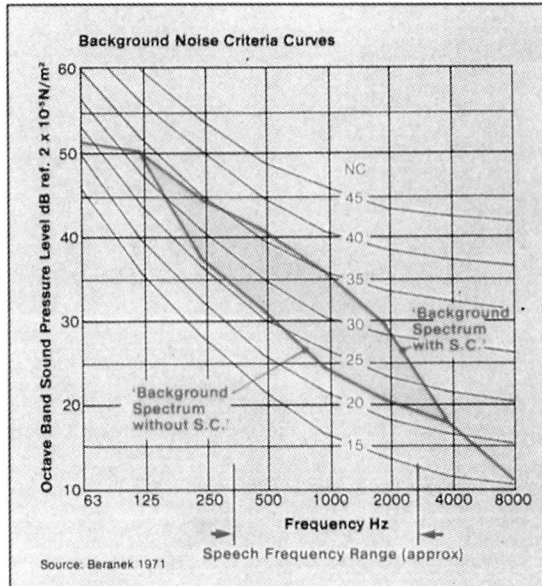

Figure 7 Typical background noise spectrum for an office area equating to NC34, and rich in low frequencies

Some of the considerations involved in the installation are:

A sound survey should be carried out to establish existing noise criteria.

Other features which can be incorporated, such as a public address system, should be established at the outset.

Final background should be as shown in Figure 7.

Not more than 10 dB should be added to any individual frequency, and sound levels should be kept to a maximum of NC 45.

Install the system before the false ceiling is installed.

Complete floors should be treated, since this leads to greater flexibility.

Natural acoustic thresholds such as doors should be used to define the limits of the sound-conditioned area.

Zone size should be limited to 15 speakers to optimise flexibility and control.

Central control should be in a secure place to prevent unauthorised adjustments.

The system should be left on 24 hours a day. Power consumption is low and continuous running avoids 'acoustic shock'.

Chapter 7
Noise in Domestic Buildings

The same principles of noise control apply to domestic buildings as to any other building. The background noise in a house is determined either by the outside noise from traffic, aircraft, nearby factories etc., or from other units in the same structure. Maisonettes, flats, semi-detached or terrace houses present their own problems, and there may be a requirement to isolate one or more rooms in the same dwelling so that the activities inside do not offend other occupants. Noise can be either airborne or can originate from vibration.

Modern buildings in UK are required to meet certain requirements in respect of acoustic treatment, and much useful information can be obtained from a study of the appropriate Building Regulations. Table 1 lists the basic requirements for sound insulation for floors and walls. The regulations are primarily concerned with ensuring sufficient acoustic isolation in buildings under multiple occupancy. Treatment may be desirable between rooms in the same occupancy e.g. for bathrooms and music rooms, but this is not required by the regulations.

TABLE 1 – UK BUILDING REGULATIONS: SOUND INSULATION

	Individual values dB	Mean values dB	
		Tests involving 4 pairs of rooms	Tests involving 8 pairs of rooms
Airborne sound (minimum)	49 dB walls 48 dB floors	53 dB walls 52 dB floors	52 dB walls 51 dB walls
Impact sound (minimum)	65 dB	61 dB	62 dB

Notes: Airborne sound is the single valued weighted standardised level difference $D_{nT,w}$. Impact sound is the weighted standardised sound pressure level $L'_{nT,w}$.

The parameters of Table 1 are defined in BS 5821 and are derived experimentally by the method described in BS 2750 and reviewed in the chapter on *Sound Insulation*. 1 dB is usually within experimental error, so one wonders why such a small difference is thought to be significant in the table.

In practice, the building regulations offer a variety of standard constructions which are deemed to satisfy the requirements, although a novel construction may still require tests to be performed.

In the case of multiple-occupancy buildings, concrete floors provide the best performance as regards resistance to the transmission of airborne and impact sound, although variations in design are numerous. Those may range from a basic floor slab through a resiliently supported slab with an air space to a fully floating floor with spring isolators and a resiliently suspended ceiling. Figure 1 shows the relative performance of a basic structural floor slab, a floor slab with an extra 100 mm of concrete cast on top and the same structural floor with a 100 mm slab resiliently supported to provide a 100 mm air space between the two air masses. The enormous increase in performance possible when separating the two masses is clearly illustrated.

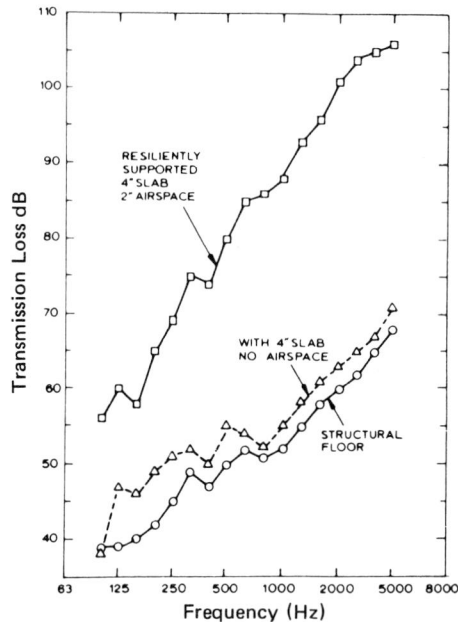

Figure 1

By comparison Figure 2 shows the attenuation provided by various building board constructions for suspended ceilings and partitions. Further data on the performance of wall and partition constructions are summarized in Table 2.

Insulation Between Houses

The sound insulation required between two attached dwellings depends on three basic criteria. These are the general background noise level in the area where the dwellings are situated, the noise levels and nature of noise created by the neighbours, and the sensitivity of the occupants to the noise from their neighbours.

The UK Building Regulations attempt to simplify this situation by setting a minimum standard of sound insulation on the basis that if this standard is achieved it will satisfy a good percentage of occupants. Thus, even if this standard was always met there would be a number of dissatisfied people demanding improved sound insulation. The regulations then attempt to simplify the position further by giving a 'Deemed to Satisfy' list of specific constructions which may be used

Partition system	Sound reduction index of partition to BS 2750 100-3150 Hz	Ceiling system Sound transmission loss Avg 100-3150 Hz		
		1 Gyptone tile	2 Plasterboard single	3 Plasterboard double
Metal stud Single layer	35.6 dB	30 dB	32.5 dB	
Double layer	42.8 dB	32 dB	39 dB	
Timber stud Single layer	33.7 dB	32 dB	32.5 dB	
Double layer	39.3 dB		37.5 dB	
Dry partition	26 dB	26 dB	24 dB	
	44.3 dB			41.3 dB

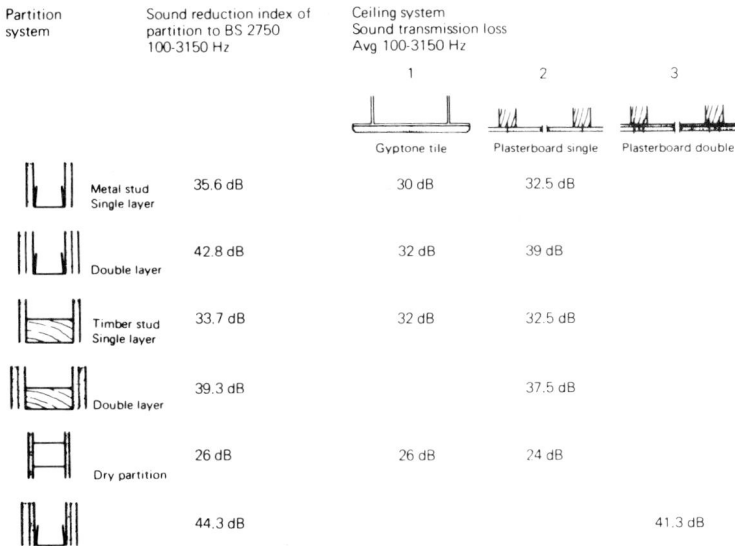

Figure 2 Sound transmission through various combinations of suspended ceiling/partition constructions

without the need to test whether or not the performance standard is achieved. Figure 3 gives a graphical presentation of the 'Deemed to Satisfy' list. The fact is that apart from the dense concrete block walls, all the other constructions cited will have varying degrees of risk of failing to meet the performance standard.

If the required acoustic insulation is not attained, the cause may be due to an inadequate basic construction, to poor workmanship or both. Even if the construction of the wall is correct, there may be unexpected paths for the sound to pass through. These may be leakage through windows, gaps around the skirting, insufficient care in installation, plumbing or electrical work and transmission through the ceiling void.

An examination of the sound insulation curve may give some idea of the source of the problem. Most standard constructions have their own characteristic shape and a deviation from it may give some clue as to the problem.

One possible diagnostic approach is to measure vibration of the walls in an attempt to estimate the sound radiated by them. A formula which can be used for this is:

$$L_p = 20 \log_{10} (a/f) + 10 \log_{10} \left(\frac{ST\alpha}{V} \right) + 144 \, dB$$

where a is the acceleration at the frequency f, S is the surface area of the wall, T is the reverberation time, α is the radiation coefficient of the surface and V the volume of the room.

There are a number of problems inherent in the use of this formula, one of the main ones being that α is frequency dependant. At best it can only be used as rough guide. A much better technique, if the equipment is available, is to take sound intensity measurements over the wall surface

TABLE 2 – PERFORMANCE OF WALLS AND PARTITIONS

Construction	Approximate Weight kg/m²	Average Sound Reduction (100–3200 c/s) dB
Single-Leaf Walls or Partitions		
2 inch compressed straw slab	56	28
2 inch compressed straw slab with skim-coat plaster both sides	88	30
2¼ inch hollow slab consisting of two sheets of ⅛ inch plasterboard joined by cardboard egg crate core	64	26
2½ inch hollow slab consisting of two sheets of ½ inch plasterboard joined by cardboard egg crate core	96	29
4 inch hollow slab formed with two layers of ½ inch plaster joined by plaster honeycomb core	192	27
2 inch wood-wool slab unplastered	96	8
2 inch wood-wool slab plastered ½ inch both sides	224	35
¾ inch plasterboard with 1/8 inch thick plaster both sides (total thickness 2 inches)	208	34
2 inch thick solid gypsum plaster (reinforced)	320	35
2 inch hollow clay block (unplastered)	144	28
2 inch hollow clay block with ½ inch plaster both sides	272	35
3 inch hollow clay block with ½ inch plaster both sides	352	36
4 inch hollow clay block with ½ inch plaster both sides	400	37
2 inch clinker block with ½ inch plaster both sides	352	37–38
3 inch clinker block (unplastered)	320	23
3 inch clinker block with ½ inch plaster one side	400	39
3 inch clinker block with ½ inch plaster both sides	480	41
4 inch clinker block with ½ inch plaster both sides	600	43
8 inch hollow clinker block with ½ inch plaster both sides	560	42
8 inch hollow dense concrete block with ½ inch plaster both sides	800	45
4½ inch brick (unplastered)	720	35–40
4½ inch brick with ½ inch plaster both sides	880	45
9 inch brick with ½ inch plaster both sides	1600	50
13½ inch brick with ½ inch plaster both sides	2320	53
18 inch brick with ½ inch plaster both sides	3000	55
7 inch dense concrete with ½ inch plaster both sides	1520.	50
10 inch dense concrete with ½ inch plaster both sides	2080	52
15 inch dense concrete with ½ inch plaster both sides	3000	55
Stud Framed Partitions		
½ inch fibre insulation board both sides	24	20–22
½ inch hardboard both sides	21	23
¼ inch plywood both sides	24	24
¼ inch asbestos wallboard both sides	48	28–30
¾ inch blockboard both sides	72	30
½ inch plasterboard both sides	64	30
½ inch plasterboard and plaster skim-coat both sides	96	32
½ inch plasterboard and ½ inch plaster both sides	224	35
3-coat plaster on wood or metal lath both sides	256	35–37
½ inch plaster on 1 inch wood-wool slab both sides	256	37

TABLE 2 – PERFORMANCE OF WALLS AND PARTITIONS
(contd.)

Construction	Approximate Weight kg/m^3	Average Sound Reduction (100–3200 c/s) dB
2 inch compressed straw slab both sides	112	33
2 inch compressed straw slab both sides and skim coat	144–160	35
Double-Leaf Walls or Partitions		
Double 2 inch wood-wool slab with 2 inch cavity Thin wire ties. $^1/_2$ inch plaster both sides	320	42
Double 2 inch clinker block with 2 inch cavity Thin wire ties. $^1/_2$ inch plaster both sides	600	47
Double 2 inch clinker block with 6 inch cavity No ties. $^1/_2$ inch plaster both sides	600	50
Double 3 inch clinker block with 2 inch cavity Thin wires. $^1/_2$ inch plaster both sides	800	50
Double 3 inch clinker block with 3 inch cavity Thin wires ties. $^1/_2$ inch plaster both sides	800	50
Double 4 inch clinker block with 2 inch cavity Thin wires ties. $^1/_2$ inch plaster both sides	1000	50
Double 4$^1/_2$ inch brick with 2 inch cavity Thin wire ties. $^1/_2$ inch plaster both sides	1600	50–53
Double wall with 9 inch brick leaf and 4$^1/_2$ inch brick leaf. 2 inch cavity. Thin wire ties. $^1/_2$ inch plaster both sides	2300	53
Double 9 inch brick wall with 2 inch cavity. No ties. $^1/_2$ inch plaster both sides	3000	55

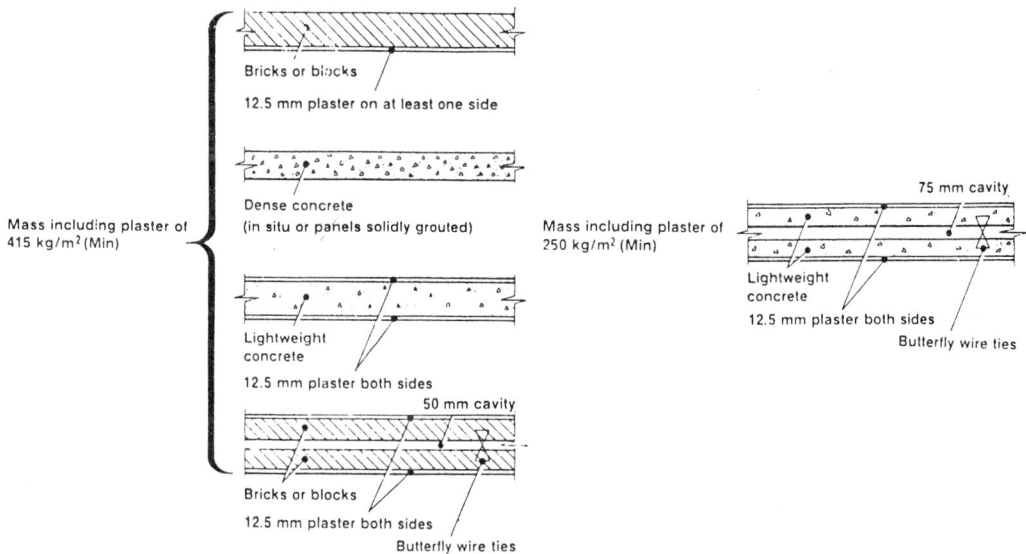

Figure 3 'Deemed to Satisfy' wall constructions. (Building Regulations)

and so determine the sound power radiated by the wall. Any difference between the sound level from this source and the total sound measured in the room can be put down to leakage.

If one particular surface is shown to be governing the sound pressure in the room, that is the one to treat first with the appropriate corrective measures. If two surfaces are of equal significance then both must be treated.

Constructional Details of Masonry Walls

The following details should ensure maximum sound insulation in solid walls:

Avoid forming recesses in the wall. If electrical sockets are to be inserted, they should not be placed back-to-back.

Ensure complete filling of the mortar joints, particularly when the wall is left unplastered. Brick walls should always be laid frog up, so that the frogs are filled.

Wall construction should be carried up into the loft space and rendered there. All gaps should be sealed as in the main room.

Avoid penetration of the walls by joists. Use joist hangers where possible.

In walls made from coarse poured materials, the surface should be sealed, including the surface beneath suspended floors.

The minimum number of interleaf connections should be used, consistent with structural stability. Lightweight 'butterfly' wire ties are preferred to heavy metal strips, and flexible plastic ties to wire.

The cavity should be filled with an acoustic foam, even if a filling is not required for heat insulation

Care must be taken that the cavity is not bridged by mortar or rubble.

Air bricks in adjacent buildings should be separated by as much distance as possible to avoid external paths for sound.

The reveals of windows should be well sealed to reduce transmission across the cavity.

Masonry separating walls should be bonded or tied to the inner leaf and not joined by a untied butt joint.

Figure 4 shows the use of rockwool in various kinds of construction. Rockwool, a mineral fibre can be used as a blanket or made into self-supporting rigid slabs in a variety of thicknesses. As discussed in the chapter on *Sound Absorption* materials, it can be used either as a sound absorbing surface or inserted into partitions to suppress the air movement in the space between the leaves.

Improvements to Existing Construction

Permanently open window vents, louvred windows or air vents in walls may be the cause of a problem. If these are present they should be temporarily blocked off to see whether any improvement is perceived. If this is the case then action should be taken to reposition or acoustically attenuate the openings if these are essential for ventilation purposes.

Any cracks at ceiling, wall angles and window frame/masonry junctions should be sealed using a flexible acrylic sealant or plaster with a paper reinforcement.

Separating and external wall skirtings should be removed and gaps below plaster or linings filled with plaster or a flexible acrylic sealant.

2 layers plasterboard
minimum 30 mm
overall

60 mm faced
and flanged
Rockwool Party
Wall Quilt

Light metal ties

80-90 mm unfaced
Timber Batt acts
also as continuous
fire barrier

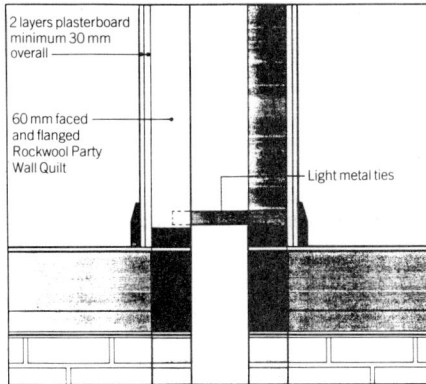

Rockwool Party Wall Quilt as acoustic insulation to timber framed party wall

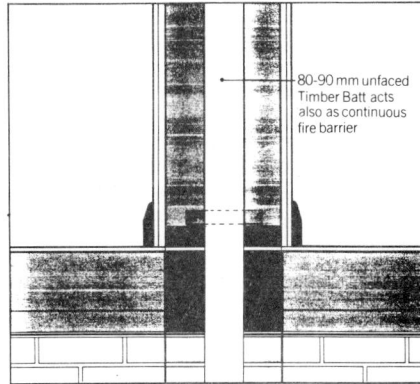

Alternative method with Rockwool Timber Batts between studs

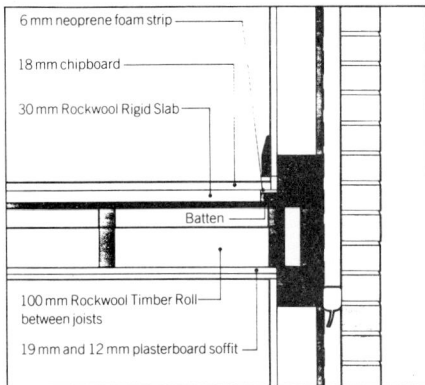

6 mm neoprene foam strip

18 mm chipboard

30 mm Rockwool Rigid Slab

Batten

100 mm Rockwool Timber Roll
between joists

19 mm and 12 mm plasterboard soffit

Rockwool Rigid Slab and Timber Roll for airborne and impact
noise control in a timber joisted floor/timber frame construction

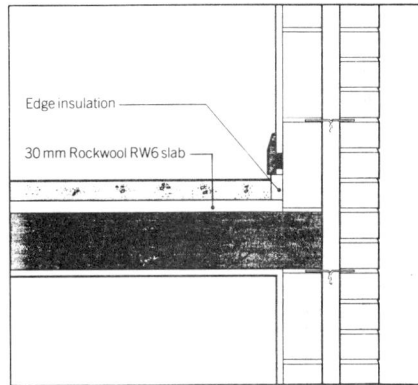

Edge insulation

30 mm Rockwool RW6 slab

Rockwool RW6 Slab as impact noise insulation to screeded concrete
floor slab

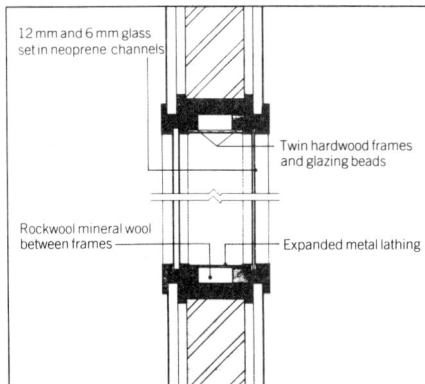

12 mm and 6 mm glass
set in neoprene channels

Twin hardwood frames
and glazing beads

Rockwool mineral wool
between frames

Expanded metal lathing

Mineral wool lining to acoustic window — block partition

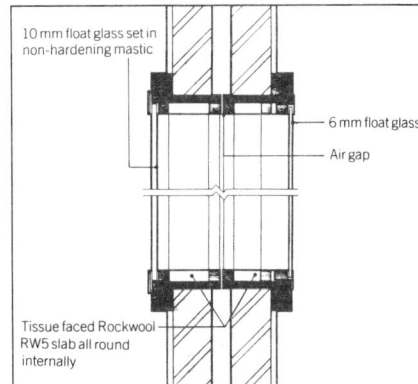

10 mm float glass set in
non-hardening mastic

6 mm float glass

Air gap

Tissue faced Rockwool
RW5 slab all round
internally

Acoustic window treatment — double skin wall

Figure 4 Recommended construction details using mineral wool slabs

Where floor joists run into the separating wall or external wall, the section of floor boarding adjacent to the wall should be removed and an inspection made of the joint/wall junction. If gaps are present around the joists then these should be filled with a sand/cement mortar. In order to avoid any possibility of the wall being porous then the whole area should be rendered with a sand/cement mortar or plaster.

Where floor joists run parallel to the separating wall or external wall the floor boarding/wall junction should be sealed with a flexible acrylic sealant.

If ceilings are found to be lightweight boarding, for example, foamed plastic cored board, they should be replaced or underlined with plasterboard or a material of similar surface mass (about 10 kg/m^2). An examination of the separating wall in the roof space will reveal any obvious holes or cracks which should be filled with sand/cement mortar and if necessary bricks or blocks. An amount of glassfibre mat laid on the back of the ceiling for thermal insulation purposes will help to dissipate the sound energy within the roof space.

If on checking and, where appropriate, rectifying the aforementioned details, the situation has not been improved it is highly likely that the separating wall itself, the external wall or both in combination are not capable of providing the level of sound insulation required. If an objective assessment is available then it may be apparent which surface is the problem, but if not, then on the assumption that the houses are of traditional construction, it is generally found that improvement of the separating wall element is beneficial.

There are many ways of increasing the sound insulation of a masonry wall; the most efficient and cost effective employs the principle of erecting a lightweight independent lining adjacent to one face of the wall.

The requirements for the lining are:

It should be capable of providing a reasonable amount of sound insulation in its own right and should therefore have a mass in the region of 20–40 kg/m^2.

It should be structurally stable when free standing, i.e. fixed only at top and bottom such that there is no mechanical connection between it and the wall surface.

It should be well sealed at the perimeter and at joints between panels.

It should be set far enough away from the wall to minimise the effects of mass-airspring-mass resonances. In theory the spacing should be such that the resonance occurs well below 100 Hz but in many cases this is not possible due to practical constraints.

The use of a glassfibre mat hung in the cavity between the lining and existing wall has the effect of damping the airspring at resonance, thus minimising its influence on the final result. The damping of the airspring also occurs above the resonant frequency resulting in improvements across the whole range.

A typical lining system is shown in Figure 5. This involves erecting a timber study frame adjacent to, but not touching, one side of the separating wall. The frame is fixed to the floor and ceiling only and is set at a distance from the wall that will give the desired cavity width. A glassfibre mat is hung in the stud frame cavities and the frame is then lined with two layers of plasterboard with joints staggered between layers. All external joints and perimeters are sealed by a method which is unlikely to result in subsequent cracking occurring. Figure 6 shows the sound insulation before and after treatment in a case where the checks outlined earlier were carried out prior to the remedial work.

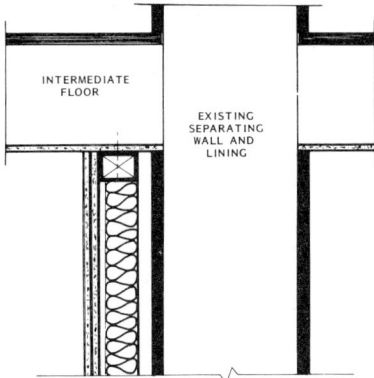

Figure 5 Independent timber frame method of remedial treatment

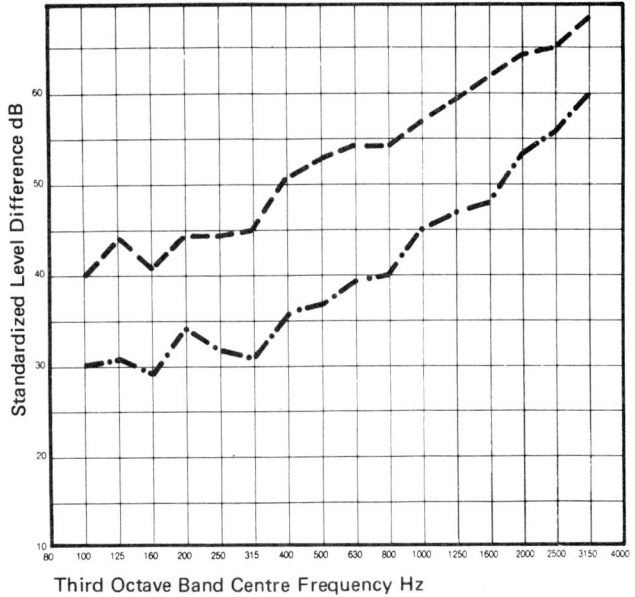

Third Octave Band Centre Frequency Hz

Figure 6 Improvements in sound insulation due to the independent timber frame method of remedial treatment

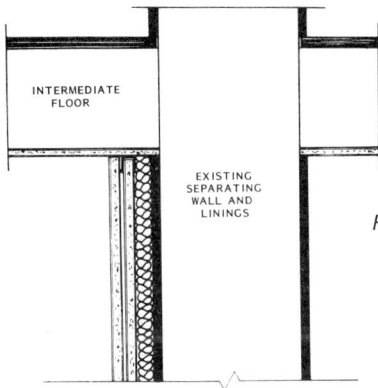

Figure 7 Laminated plasterboard method of remedial treatment

Figure 7 shows another method of remedial treatment. This comprises a double layer of 19 mm plasterboard, the first layer being fixed at the top, bottom and sides in metal channels and the second layer laminated to the first with a plaster adhesive. In an exercise to evaluation this lining, a number of walls of the same basic construction were treated the cavity between lining and wall varying rom 25–100 mm. In each case a 25 mm glass fibre mat was hung in the cavity. No other examinations or remedial treatment took place prior to erecting the linings. In this instance improvements of similar magnitude were made in all cases, there being no evidence to suggest that the larger cavity was beneficial. Figure 8 shows the typical performance before and after treatment for a number of walls where the lining was set with a cavity width of 25 mm.

If it is found necessary to line the external wall then linings similar to those recommended for the separating wall may be used.

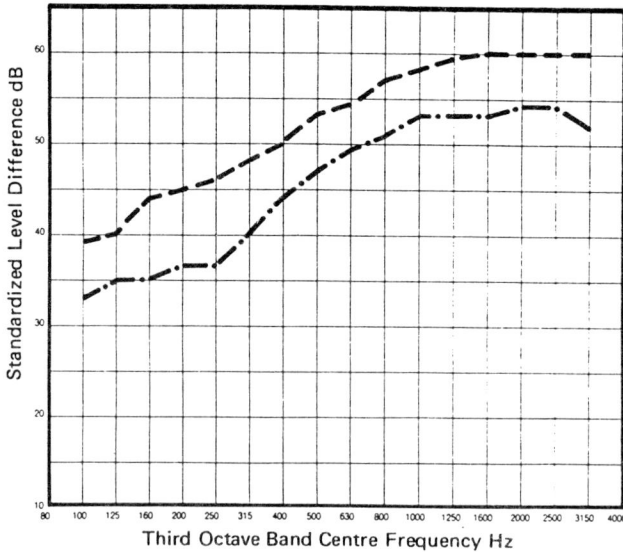

Figure 8 Improvements in sound insulation due to the laminated plasterboard method of remedial treatment.

Internal Treatment

The sound level in any room may be appreciably modified by the presence of furniture, curtains and floor coverings. Carpets and other coverings reduce the vibration caused by footsteps. They act like thin porous absorbers and are effective at high frequencies only. An underfelt, by adding extra thickness improves the absorption at lower frequencies.

Table 3 gives some typical absorption coefficients that might be expected, but tests are always to be preferred.

TABLE 3 – ABSORPTION COEFFICIENTS OF FLOOR COVERINGS

Material	Absorption coefficient at frequency (Hz)				
	250	500	1k	2k	4k
Carpets laid on floor					
Axminster, thin	0.5	0.1	0.2	0.45	0.65
Axminster, medium	0.5	0.15	0.3	0.45	0.55
Turkey	0.1	0.25	0.5	0.65	0.7
Underlay					
Needleloom	0.05	0.15	0.3	0.5	0.6
Sponge rubber	0.05	0.05	0.2	0.2	0.15
Carpets on underlay					
Axminster medium on needleloom	0.15	0.4	0.6	0.75	0.75
Axminster medium on sponge rubber	0.05	0.2	0.4	0.6	0.65
Turkey on needleloom	0.3	0.55	0.65	0.65	0.65
Turkey on Axminster on needleloom	0.45	0.65	0.6	0.65	0.65
Cork carpet (12 mm), laid	0.05	0.05	0.3	0.1	–
Cork tiles (14 mm), stuck on floor	0.05	0.15	0.25	0.25	–
Rubber sheet stuck to floor					
Hard (6 mm)	0.05	0.05	0.1	0.05	–
Soft (6 mm)	0.05	0.1	0.05	0.05	–
Sorbo (5 mm) hardwearing surface	0.05	0.1	0.1	0.05	–
Sorbo (8 mm) hardwearing surface	0.05	0.15	0.1	0.05	–
Woodblock	0.05	0.05	0.5	0.05	–

Chapter 8
Auditoria

Auditoria and Concert Halls

Sound treatment applied to a concert hall, auditorium, church or other large public building can have two purposes: it can reduce the background noise to a level suitable for speech intelligibility and it can improve the quality of the sound by modifying the reverberation time. The latter is very much a subjective matter and there has been much discussion about the desired sound quality of a concert hall, for the proper appreciation of music. A concert hall with a maximum amount of absorptive treatment applied to the walls and ceiling would have a peculiarly dead feel, and so it is generally considered that for each use to which a hall may be put there is an optimum reverberation time.

Speech Intelligibility in Halls

The evaluation of speech intelligibility was previously discussed in the chapter on *Noise in Commercial Buildings*. Some of the aspects to be considered are

> mapping the intelligibility throughout the hall

> investigating the contribution of background noises

> evaluating the public address system with various background levels and different loudspeaker positions

With the use of a Speech Transmission Meter, it is possible to evaluate the RASTI values throughout the hall (or at least in those places where the occupants are likely to be seated). RASTI contours can then be drawn and those areas with a poor intelligibility can be identified. It is desirable to determine the RASTI values with varying amounts of human occupancy, because the presence of an audience can modify considerably the sound absorption properties of the hall. The RASTI contours are first determined without a background noise and then with an artificial noise source representative of the real conditions. The Bruel and Kjaer Speech Transmission Meter (see chapter on *Commercial Buildings*) can correct within the instrument for an imaginary background noise. Figures 1 and 2 give typical contours with and without background noise.

Reverberation Times

The reverberation time is defined as the time taken for a signal to decay by 60 dB (or 10^6) after it has been shut off. It is a measure of the ability of the room to absorb sound. The ideal way of determining the reverberation time is by measurement with a reverberation time meter, or a building acoustics analyser. The basics of such an instrument are: a noise generator with a set of bandpass filters, a sound level meter with the same set of filters and a graphic level recorder with a better than 60 dB range. The average decay rates should be measured rather than the decay

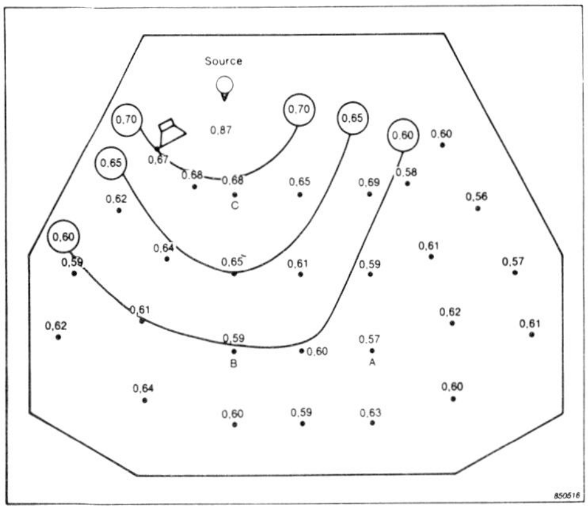

Figure 1 Iso-RASTI contours for an auditorium without public, background noise and PA system

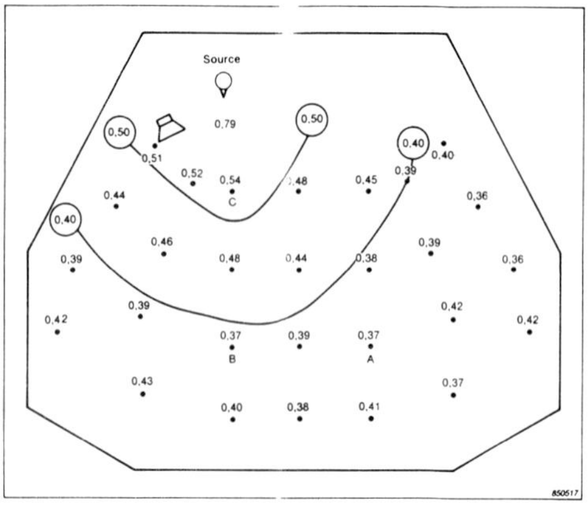

Figure 2 Iso-RASTI contours for the same data points as given in Figure 1 but corrected for an imaginary background noise with an octave band level of 40 dB

times. These functions are all included in a building acoustics analyser, although the equipment can be built up from the separate building blocks.

If the reverberation time cannot be measured, it can be calculated from the empirical Sabine formula

$$T = \frac{0.163V}{S\alpha}$$

where T is the reverberation time in seconds, S is the room surface area in m², V is the room volume in m³, α is the mean absorption coefficient.

It should be noted that α is frequency dependent and therefore the time T is a function of frequency; it also varies with position in the room. This formula is reasonably accurate, provided that the field is diffuse and not excessively absorbent. It will be apparent that if $\alpha = 1$, i.e. if the room is totally absorbent, the reverberation time should be zero, whereas the formula gives a finite time. A more accurate formula, which should be employed for·highly absorbent rooms is due to Eyring:

$$T = \frac{-0.163\, V}{\log_{10}(1-\alpha)}$$

This cannot be used where the value of α is equal to or greater than 1.0 (see the chapter on *Sound Absorption Materials*).

Air Absorption

In very large halls, the absorption due to the air itself may be important. Usually it can be neglected, but at high frequencies, there may be significant loss from this cause. The effect of the air absorption can be catered for by calculating a modified value of the absorption index

$$\alpha_c = \alpha_w + \frac{m\,V}{10^3\,S}$$

where α_w is the absorption coefficient for the walls of the room and α_c is the corrected coefficient, taking into account the absorption of the air. Tables of m in terms of dB per km are to be found in the chapter on *Sound Propagation*, but when calculating α for rooms, the mean distance travelled by a plane wave has to be assumed as 4 V/S, and this leads to the above equation. For convenience, Table 1 gives the values of m, for estimation of the likely significance of the absorption of the air.

TABLE 1 – AIR ABSORPTION IN A ROOM (M/1000) IN METRE UNITS AT A TEMPERATURE OF 20 O° C

Frequency (Hz)	Relative humidity					
	30	40	50	60	70	80
1	4	4	4	4	4	3
2	13	10	9	9	8	8
4	44	32	27	23	21	20
8	150	120	92	77	66	60

Optimum Reverberation Times

The important question which remains, after the reverberation time has been calculated or measured, is whether it is the correct value for the intended purpose.

The sound pressure level depends on the reverberation time, so that each halving of the time reduces the sound level by 3 dB. When listening to speech there is a conflict between volume and clarity: if each syllable is to be clearly heard, the reverberation time must be short with all the sound coming directly to the listener, but then the volume will be low. As always a compromise has to be sought and this may result in a different reverberation time for each purpose for which the hall is to be used.

As an example, it is a common experience that large churches of traditional construction have a noticeable echo (a long reverberation time), because the surfaces are very reflective and the walls are parallel. Organ and church music generally is written with a long reverberation time in mind, so for its proper appreciation, it must be heard in church or in a hall which has been designed to have that characteristic.

One empirical formula which has been suggested for determining the optimum reverberation time, T, at 500 Hz is:

$$T = K\,(0.0118\,V^{1/3} + 0.107)$$

where V is the volume in m^3, K = 4 for speech, 5 for orchestras and 6 for choirs. Another suggestion, giving a broad band of acceptable times, is shown in Figure 3. For low frequencies, down to 125 Hz, the reverberation time should be multiplied by about 1.5 for music. Figure 4 represents this graphically; the lower line should be used for speech and the upper for music.

Because of the differing requirements of multi-purpose halls, it may be necessary to modify the acoustic character by the use of extra curtains or absorbers which can be brought in when necessary. Another option is the use of swivelling or retractable acoustic panels; this offers the

Multiplication Factor for Reverberation Time at 125Hz

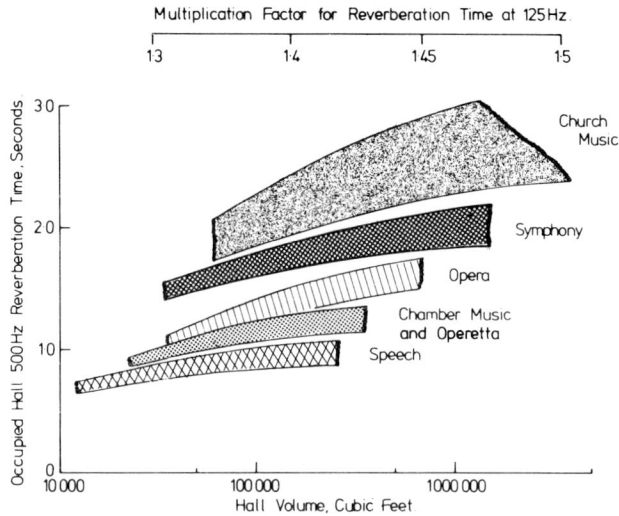

Figure 3 Guide to optimum values of reverberation time

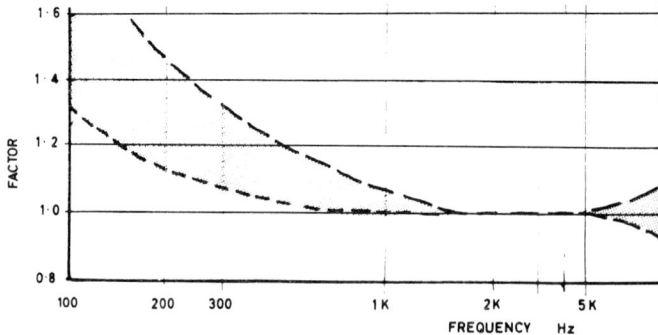

Figure 4 Correction factors for optimum reverberation time dependent
on frequency of sound

possibility of using panels tuned to different frequencies, so that not only can the average absorption be changed, but also modified differentially in the appropriate frequency bands.

General Considerations for Design of Halls

Auditoria

A fuller or richer quality is given where the reverberation time for lower frequencies is greater than that for middle and higher frequencies, although this is more generally noticeable with music rather than speech. However, variable absorption characteristics, giving reverberation times varying with frequency, can interfere with speech to the extent that there can be excessive

TABLE 2 – ACOUSTICAL DEFECTS

Defect	Dependency	Remarks
Echo	(i) size and shape of reflecting surfaces (ii) relative position of sound source and listener (iii) type of sound programme	echo results if time interval between direct and reflected sounds originating from same source is 0.04 to 0.1 s
Short-delay echoes	(i) size and shape of reflecting surfaces (ii) relative position of sound source and listener (iii) type of sound programme	result in blurring or masking of direct sound
Flutter echo	parallelism of reflecting surfaces	sound sources should not be located between parallel reflecting surfaces
Distortion	balance of absorption characteristics of boundary surfaces	sound quality can be damaged by excessive or unbalanced absorption
Shadow	sound shadowing effects by banners or overhangs	reduce sound levels in shadowed areas
Room resonance (colouration)	parallelism of reflecting surfaces at critical distances (ie sub-multiples of the wavelength of sound)	can set standards for room sizes and proportions
Hot spots	sound reflections from concave surfaces	produces local areas of sound concentration. Can also produce 'whispering gallery' effects
Coupled spaces	adjacent rooms with interconnecting air volumes	can modify the reverberation characteristics of the auditorium

absorption of consonant or vowel sounds. For general speech quality, therefore, it is desirable that the absorption coefficients of finishes used in acoustic treatment should be as uniform as possible over the frequency range 250–7000 Hz.

Subtleties of speech are provided by good musical qualities defined as 'presence' or 'intimacy'. As a general rule this is gained by reinforcement of direct sound waves with reflected waves, with a very short initial time delay – e.g. of the order of 5 ms, or a difference in sound path of about 10 m.

Account must be taken of the directional characteristics of typical speech. High frequency sound levels fall off rapidly outside about 140° subtended at the position of the speaker. Hence, ideally, seats should be laid out in a pattern falling within this section (see also Table 3).

Reflecting Surfaces

Without amplification, the acoustic power of human speech lies within the range of 10 to 50 microwatts. For adequate reception sound paths to the audience should be as direct as possible to reduce sound losses in the air. This implies a compact room shape with a low volume per seat (e.g. see Table 4), a raised speaker's platform, and the elimination of all physical obstructions between speaker and audience (e.g. in an open room, a sloping floor rising to the back). A curved

TABLE 3 – SHAPES FOR MUSIC AUDITORIA

Shape	Acoustic Advantage(s)	Acoustic Disadvantage(s)	Remarks
Rectangular	good cross reflections	flutter echo likely. Room resonance may be apparent	traditional for orchestras
Circular	none	echoes, long time-delayed reflections, hot-spots	often associated with domed roof which can further degrade the acoustical properties
Horseshoe	good definition (but not for orchestral music)	tends to be 'dead' (low reverberation times)	traditional for operas
Fan	good balance, good presence and loudness	long time-delayed reflections. echoes and hot-spots	
Irregular	maximum intimacy, definition and brilliance		
Combined Shapes	can include advantages of all types	can include disadvantages of all types	

TABLE 4 – RECOMMENDED ROOM VOLUMES FOR DIFFERENT TYPES OF HALLS
(Room volume per person in m³)

	Minimum	Optimum	Maximum
Rooms for speech, lectures, etc		2.8	5
Cinemas		3.1	4–4.2
Churches	5.7	7–10	12
Concert halls	6.5	7–7.5	10
Opera houses	4	4–5	6

ceiling shape can also help by working as a reflector to provide more uniform distribution of sound energy – Figure 5.

In the case of larger halls or auditoria, even with favourable room acoustics the speech level may still be too low for satisfactory listening. In this case a sound amplification system can be installed to achieve a high degree of speech intelligibility. Recommendations in this respect vary considerably and are also influenced by the type of auditorium covered and the normal background noise level.

It is possible to provide flexible acoustics by large scale versatility in the building fabric (e.g. swivelling and height adjustable acoustic panels), but the cost of this becomes increasingly prohibitive as the size of the hall increases. It is thus more usual to design auditoria specifically for either speech or music.

The basic requirements for good listening are:

(i)   *Absence of background noise* – exclusion or reduction to a suitable level of all extraneous noises.

(ii) *Adequate loudness* – extending to all parts of the auditorium.

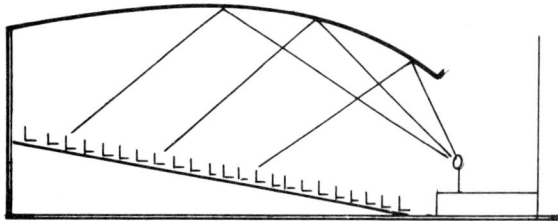

Figure 5 Curved reflective ceiling and
inclined seating levels

(iii)   *Uniform distribution of sound* – ideally sound should be equally loud in all hearing positions.

(iv)   *Optimum reverberation time* – related to size of room, and the type of performance, (e.g. speech or music).

(v)   *Optimum acoustic qualities* – exclusion or reduction to a suitable level of all extraneous noises.

Of these criteria, (i) and (iv) can be directly controlled by sound absorption treatment; and (v) largely controlled by such treatment.

Background noise level is usually rigidly controlled by sound insulation applied to all surfaces of the enclosure, as necessary. The suppression of noise from ventilating ducts and grilles may be particularly important in this respect. Background noise levels may be required to be held down to 40 dB, or better, over all frequencies within the audible frequency range.

Lecture Halls

Lecture halls can vary widely in size. Larger lecture halls can be regarded very much as theatres, posing the same acoustic design problems as listed previously. Provided the acoustic design is good, sound amplification systems should not be required until the size exceeds about 1500 m^2 or the hall accommodates an audience of more than 500. In all cases reverberation time is probably best calculated on the basis of a 2/3 audience attendance (i.e. for average conditions which assume that the lecture hall will be two-thirds full).

Modern lecture halls are normally designed with particular attention to the exclusion of external nose, with consequent exclusion of natural light and ventilation. This can place a premium on the design of ceiling units carrying necessary artificial lighting, and on achieving suitable silencing in ventilation and air conditioning services to the room.

Smaller lecture halls forming part of a building may demand special attention to exclude external noise generated in adjacent rooms, with particular study given to flanking transmissions.

Architectural Materials for Sound Absorption

Materials suitable for a factory or office may not be architecturally pleasing in buildings where the public are admitted. The usual absorbent panels made of foamed plastic or mineral wool, covered with perforated hardboard or perforated steel, however acoustically effective they may be, may not meet with acceptance from the interior designer. Fortunately, there are attractive alternatives. Slatted steel sections backed by mineral wool can be painted in a variety of colours and offer the possibility of very good absorption coefficients, Figure 6 and Table 5.

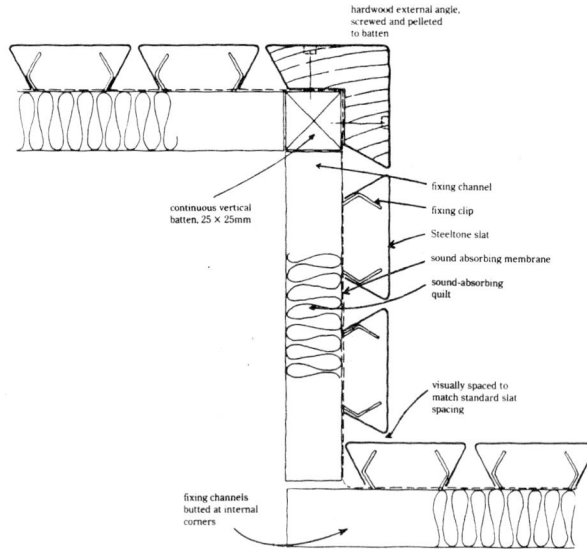

Figure 6 Slatted steel sections supported over
mineral-wool insulation panels

TABLE 5 – SOUND ABSORPTION COEFFICIENTS OF STEEL SLATS ON MINERAL WOOL PANELS AS SHOWN IN FIGURE 6

Frequency (Hz):					
125	250	500	1000	2000	4000
0.13	0.46	0.99	0.85	0.51	0.53

The Helmholtz resonator type of panel, described in the chapter on *Sound Absorption*, can be effective if the size of the cavity is correct for the frequency it is required to control. For comprehensive control over a full frequency range, an array of resonators behind a false ceiling can be used in conjunction with assisted resonance.

Timber linings, either fitted directly on support battens or over an absorbent liner can be particularly attractive. They too can be machined to incorporate continuous Helmholtz resonators. Some sections are given in Figure 7, with the absorption coefficients in Table 6. Sections are also available in MDF (medium density fibre-board).

TABLE 6 – SOUND ABSORPTION COEFFICIENTS FOR HARDWOOD SECTIONS IN FIGURE 7

Profile	Frequency (Hz):					
	125	250	500	1000	2000	4000
Profilia and Tonewood:						
Slots at 200 mm centres	0.10	0.35	0.80	0.40	0.25	0.35
Slots at 400 mm centres	0.15	0.50	0.65	0.35	0.20	0.35
Sylvatone	0.06	0.32	0.24	0.15	0.16	0.33

Profilia

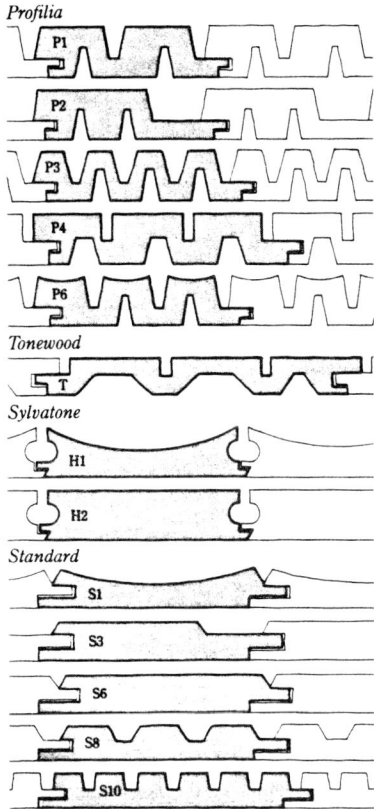

Tonewood

Sylvatone

Standard

Figure 7 Hardwood sections – can be machined with side slots to allow sound to pass through to a backing sound absorption material. Note the resonator cavity or Sylvatone section

Public Address Systems

Amplification systems in concert halls and churches are normally restricted for use with the spoken word. An orchestra or choir is considered to have sufficient volume so as not to need amplification. Pop concerts are quite different; the amplification system used there is part of the equipment of the groups rather than of the hall they use. Faithfulness of tone is rather less important than volume.

Public address systems which are installed in churches to improve audibility of the spoken word are often not effective unless a number of directional loudspeakers are used. For clarity, the sound from the loudspeakers should be heard directly rather than from reflections off the walls or ceiling. It is not usually acceptable to change a traditional building by surface treatment to absorb the reflections.

The design of the loudspeakers is of critical importance in this application. The ideal is to produce a sound field which ensures that the amplified sound is directed towards the congregation rather than to the walls or ceiling. This can be achieved by the use of a line source or column loudspeaker, which consists of several ordinary loudspeakers mounted one above the other with a controlled phase difference between them. This ensures that the loudspeaker radiates strongly downwards. The maximum energy is delivered to the congregation and the minimum upwards into the roof void. Figure 8 shows the directional quality of one system and Figure 9 its installation in a church.

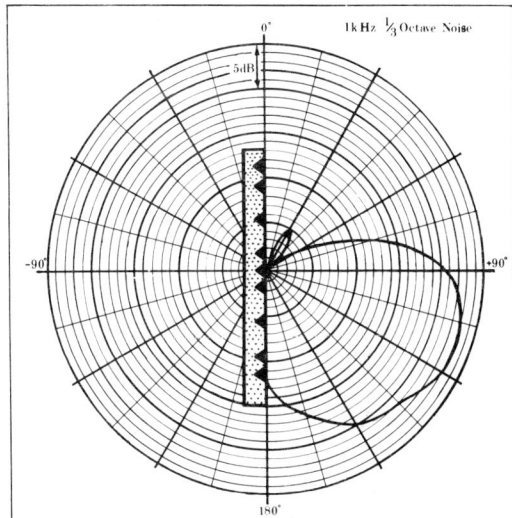

Figure 8 Column loudspeaker vertical polar plot

Figure 9 Column loudspeaker in a church

It can be shown that speech will be intelligible about 10 times as far from a column loudspeaker as from a non-directional speaker, which points to the vital importance of correct selection of equipment.

Assisted Resonance

This is a method of modifying the reverberation characteristics of an auditorium by electronic means. It employs acoustic feedback so as to increase the reverberation time by up to twice its natural value. The principle is to place a loudspeaker at a peak in the acoustic pressure of a room and a microphone at another peak, connect the two with an amplifier and adjust the feedback loop until the desired reverberation time is reached. Eventually, of course, increasing gain will cause feedback howl, and that has to be avoided. If the system is well designed, the listener is not aware of its existence. Assisted resonance cannot reduce the reverberation time – it can only lengthen it, so the original design has to have a low value initially. If assisted resonance is to be considered, acoustic treatment to improve the absorption characteristics may first be necessary. This may consist of wall or roof treatment, but might equally well incorporate an array of Helmholtz resonators.

Figure 10 shows the arrangement with just one channel, but in practice, a large number of channels are used, each for a specific frequency, so that with different gains on each channel, the reverberation time and the sound quality can be matched to the requirements of the hall. This also allows for differing reverberation times for each of the uses to which the hall is put and to cater for variations in the number of occupants. It has been found that up to 100 channels or so are required for ideal response in the frequency range of 63 to 1000 Hz. The system was first used in the Royal Festival Hall in London which, in its original unmodified design, was the subject of

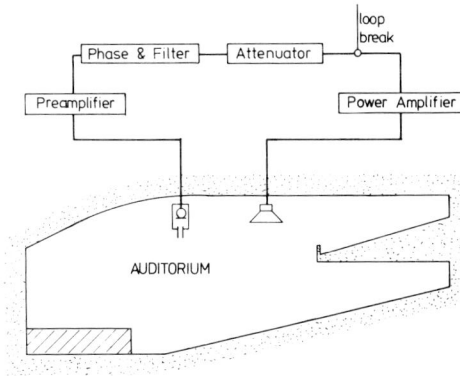

Figure 10 Schematic of a single channel of an
'Assisted Resonance' system

complaints about the sound quality; Figure 11 shows the improvements in reverberation time achieved. Figure 12 shows the improvement for a later design of hall.

It will be apparent that if the reverberation time of an auditorium is changed by architectural means, then for a given volume the energy density will be increased as the sound absorption is reduced. For example, if the reverberation time is doubled by halving the sound absorption, the energy density will be doubled. If the effect is to be achieved through electronic means, the same considerations apply, with the assisted resonance having to compensate for the energy absorbed. This means that the amplifiers have to supply that energy. A channel output of the order of 50 watts maximum has been found satisfactory for the Royal Festival Hall, with 168 channels. The maximum power is seldom needed except for full symphony concerts.

Although assisted resonance has been mainly adopted as a remedial measure in auditoria which have already been built and have proved acoustically unsuitable, it is probably most effective when considered at the design stage, so that the acoustic treatment, the location of the loudspeakers and the amplification equipment can be considered as a whole.

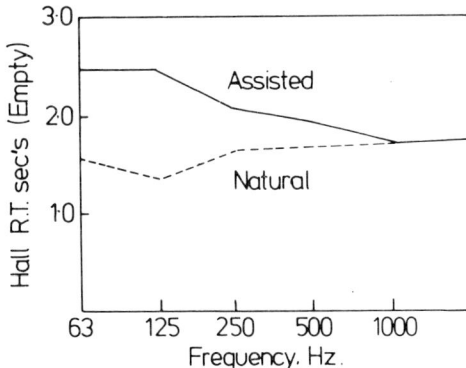

Figure 11 Reverberation enhancement at
the Royal Festival Hall

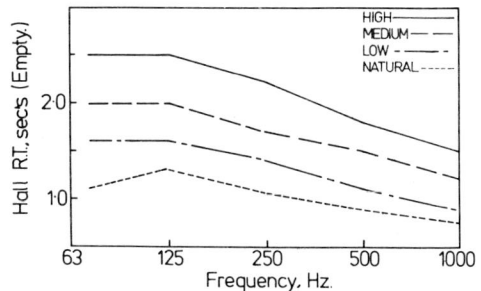

Figure 12 Reverberation enhancement from the
'Assisted Resonance' system at Central Hall, York

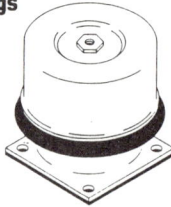

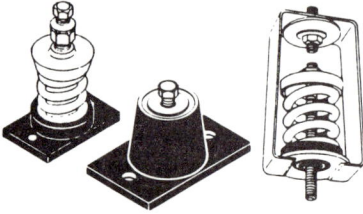

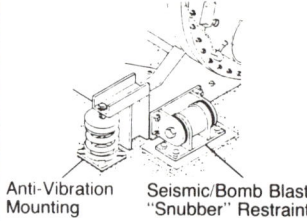

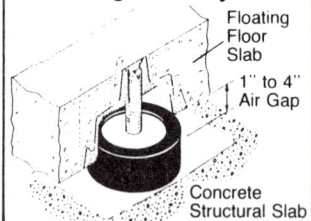

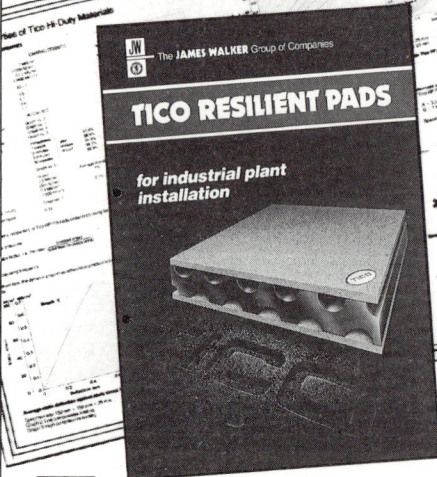

SECTION 7

Chapter 1
Machine Balance and Vibration

Machine Balance

It is important to distinguish between static and dynamic balance. Consider a basic rotating machine element as an equivalent rotor mounted on a shaft; static unbalance will generate vibrations in a plane at right angles to the shaft (A–A in Figure 1). Any dynamic unbalance will generate additional vibrations in other planes which can be represented by resultants in two diagonal planes (B–B and B'–B').

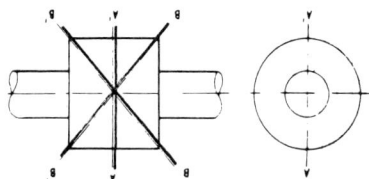

Figure 1

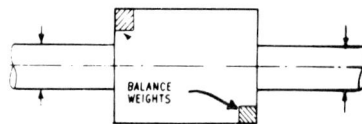

Figure 2

Primary balancing is a process where the primary forces caused by an unbalanced mass are resolved into one plane and balanced by an additional mass placed in that plane; this produces static balance. Secondary balancing is the process whereby the primary forces and secondary couples are resolved into two planes and additional masses placed in those planes; this produces dynamic balance. Static balance can be easily obtained by placing the rotor on knife edges and adding, in one plane, sufficient extra mass by trial and error to ensure that rotor does not turn in whatever position it is placed. Dynamic balance is only apparent when the object is rotating and is less easy to achieve without proper balancing equipment. Dynamic balance implies that static balance is also obtained, but static balance may still leave an unresolved dynamic unbalance. Static balance requires the addition of mass in just the one plane, whereas dynamic balance in general requires the addition of masses in two planes, Figure 2.

In any rotating rotor, an unbalance in the mass generates an accelerating force of the form

$$F = m \omega^2 r$$

where is the angular velocity in rad/s

$$\omega = f/2\pi$$

m is the out of balance mass and r is a vector of the radius at which the mass acts.

It follows from this that the frequency of the vibration which results from this accelerating force is equal to the rotation frequency of the shaft and the force itself is a vector quantity.

The following list gives a number of terms used in balancing. For more precise definitions of these terms and others refer to BS 3851 (ISO 1925).

Unbalance is the condition which exists in a rotor when a vibratory force is imparted to its bearings as a result of centrifugal forces.

The amount of the unbalance is the product m r, usually expressed in gram mm.

The unbalance vector is a vector whose magnitude is m r and whose direction is the angle of unbalance.

Static unbalance is the condition in which unbalance can be corrected in the plane of the centre of gravity of the rotor. Its magnitude is given the symbol U.

The specific unbalance, e, is given by

$$e = U/m$$

where m is the total mass of the rotor. This is also known as the mass eccentricity. e is usually given the units of gram mm per kg.

Couple unbalance is when the unbalance is a simple couple about the centre of gravity.

Dynamic unbalance is when the unbalance consists of two vectors in two specified planes. It is the most general form of unbalance.

Residual unbalance is the unbalance that remains after the balancing procedure has been completed.

Permissible residual unbalance is the permissible unbalance appropriate to the application for which the rotor is to be used, and it follows that the permissible residual specific unbalance is given by the relationship

$$e_{per} = \frac{U_{per}}{m}$$

Where the remaining unbalances in a rotor can be reduced to a single unbalance in the plane of the centre of gravity, e_{per} can be considered as the permissible displacement of the centre of mass from the rotor axis.

The grade of balance quality required for a particular type of rotor depends upon the application. Experience shows that, for rotors of the same type, e_{per} ω varies inversely as the rotor speed over the speed range shown in Figure 3. This leads to the definition of the quality grades shown in Table 1, where the numerical value of the grade is equal to the product e_{per} ω in mm/s, where ω is the angular velocity in rad/s.

Equipment for Machine Balancing

A balancing machine measures the unbalance in a rotor and can be used for adjusting the mass distribution so that the unbalance can be reduced. A balancing machine can be either non-rotating, when it can measure the static unbalance, or rotating, when it is capable of measuring the dynamic unbalance. In practice a rotating type machine is used for both kinds of balancing.

Where the rotor to be balanced can be supported in a standard two plane balancing machine such as the one shown in Figure 4, that is probably the ideal but, in principle, the vibration can be assessed by measurements taken from a rotor supported on its own bearings. A balancing machine for two-plane balancing should satisfy the requirements of BS 3852. Note that although

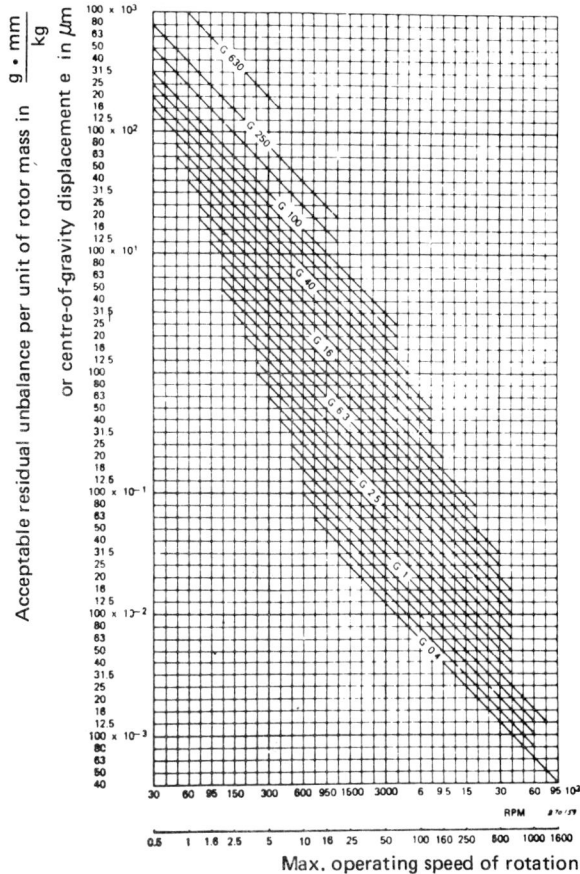

Figure 3 *Acceptable Residual Unbalance for the Rotus Desumbes in Table 1*

BS 3852: Part 1 is based on the International Standard ISO 2953, there are minor differences; it is however possible to satisfy the requirements of both standards in the same machine.

For a basic description of the method adopted by these machines, refer to Figure 5. The procedure is:

> Attach accelerometers near the bearings of the rotor to be balanced. A photo-electric trigger is mounted to give a pulse at a rate of once per revolution.

> Run the rotor and note the vibration amplitude and phase for Plane 1 and repeat for Plane 2. Fix a known mass to the rotor at the radius and in the plane close to Plane 1 where the correction is to be made. Run the rotor and note the amplitude and phase in Plane 1 and Plane 2.

> Remove the test mass and fix another mass close to Plane 2 at the radius and in the plane where the correction is to be made. Note the amplitude and phase in Plane 1 and 2.

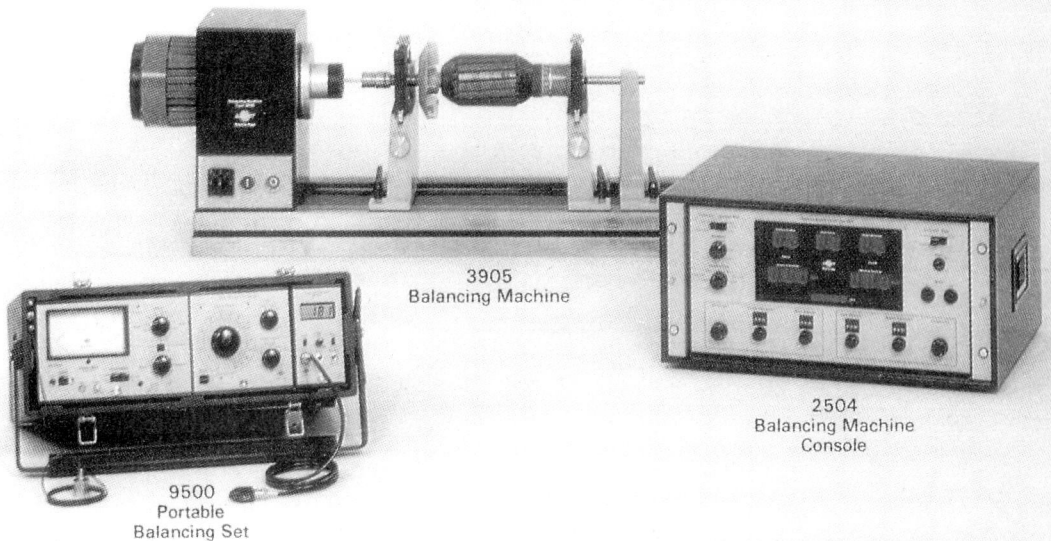

Figure 4 Equipment for Rotor balancing

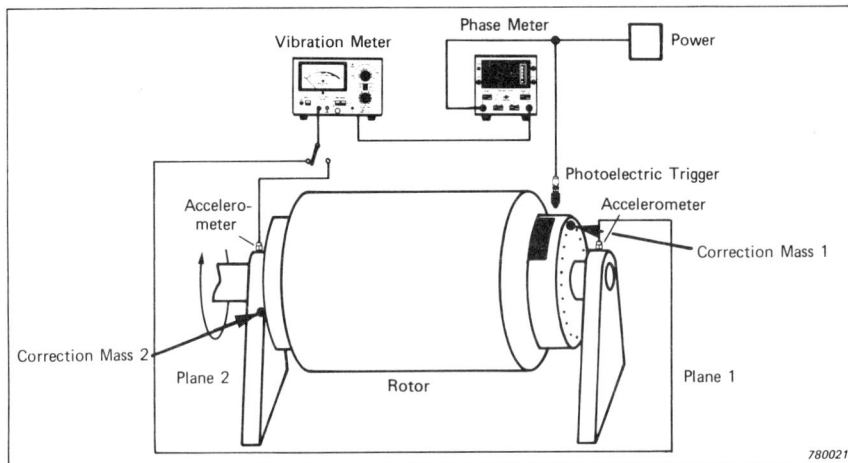

Figure 5 Diagram of a Balancing Machine

Remove the second mass

The six values of phase and amplitude are then entered into a PC or a calculator to determine the final values for the masses and angles where the corrections are to be made. These are then applied to the rotor. If the instrumentation is sufficiently accurate, there should be no further balancing required.

The vibration meter and the phase meter are shown separately in the diagram. Modern balancing equipment contains all the instrumentation in one package.

Dynamic balance can be obtained in a number of ways. The balancing planes can be arbitrarily chosen as can the radii of the added or subtracted masses.

There are several methods of measuring the shaft vibration level. Refer to BS 6749 for some suggestions. Either absolute or relative measurements can be taken. An absolute measurement of velocity or acceleration can be taken either with a shaft-riding probe or a non-contacting transducer (a proximity transducer), combined with a seismic transducer mounted on a bearing housing; the two readings are vectorially summed to give the correct amplitude. Relative measurements can be taken with a non-contacting transducer mounted off the bearing housing, and the relative vibration between the two measured directly.

Balancing of Flexible Rotors

The above discussion is applicable to the balancing of rotors which remain rigid while rotating. The implication of this is that the balancing of such a rotor can be accomplished by a correction applied in any two arbitrary planes and, when that is done, the unbalance does not exceed the prescribed tolerances for unbalance at any speed up the maximum service speed. A flexible rotor is one which deflects when rotating to such an extent that these conditions are not satisfied.

A flexible rotor possesses modal unbalance so that when rotating it can deflect in a number of flexural modes. Each mode is associated with a critical rotational speed and if that speed is one which is within the operating range of the machine, there will be a maximum flexure of the rotor which will be significantly greater than the vibration measured on the support journals.

A complete rotor will have imperfections due to manufacturing, caused by errors in casting, machining, assembly etc., which will produce a distribution of unbalance along the length of the rotor. If the rotor is rigid, the actual distribution of the imperfection is unimportant, since dynamic balance can be obtained by placing two balance masses on arbitrary planes as described above.

A flexible rotor requires a rather different and more complex treatment. An ideal balance would be obtained if, in all the planes in which there was an unbalanced mass, a correction could be applied. Generally these planes are not easily identifiable, although in cases where they are (if for example a rotor is built up of a number of separate components attached to a shaft, with each component having a possible unbalance), balancing should be done in those planes. It is not usually possible to obtain good balance in all the modes of vibration, so usually the procedure adopted is to balance for the first flexural mode and then proceed through the other modes in turn until the mode corresponding to the maximum operating speed of the machine has been reached. It is likely to prove easier to balance a composite rotor (one built up from a number of separate components) if each component is separately balanced before assembly and if particular attention is given to the tolerances of the shaft diameters on which the components are located.

Figure 6 shows a typical rotor and the first three modes of flexure. The modes after the first have nodes on the axis of rotation. When balancing a rotor of this kind, a balance mass placed in the plane of maximum deflection for a mode will have the maximum effect. Conversely, a mass placed in the nodal plane of a mode will not affect the response in that mode. A balance mass placed in plane P_3 will be most effective in the first mode; when placed at point P_2, a mass will not affect the second mode and so on. This gives a guide as to which plane to choose for trial purposes. The flexure modes can be identified before the balancing procedure starts since, although the magnitude of the modal displacement will depend upon the magnitude of the unbalance, the shape and nodal positions of the first few modes will be generally independent of the unbalance.

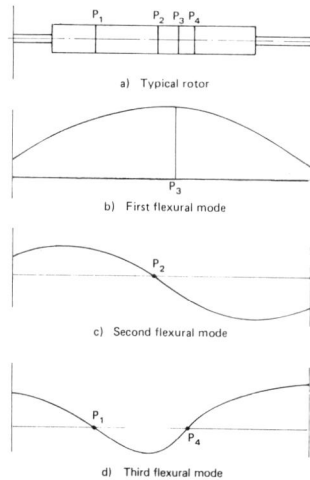

a) Typical rotor

b) First flexural mode

c) Second flexural mode

d) Third flexural mode

*Figure 6 Typical mode shapes for flex-
ible rotors on flexible supports*

Classification of Rotors for Balancing

In considering the methods of rotor balancing, rotors can be classified as follows:

Class 1 – Rigid rotors whose unbalance can be corrected in any two arbitrary planes so that after correction the residual unbalance does not change significantly at any speed up to its maximum. The procedure for dealing with this class is described above. The balance can be assessed and corrected by a conventional low speed balancing machine.

Class 2 – Quasi-rigid rotors which cannot be considered rigid, but can be balanced using modified rigid rotor balancing techniques. Class 2 rotors can be further subdivided, in accordance with the system described in BS 5265: Part 2, into classes 2a to 2h according to how the axial unbalance is distributed. This class of rotors can be balanced in a low speed balancing machine, so that not only is the rotor statically and dynamically balanced but also the modal unbalances are sufficiently small to ensure satisfactory running of the rotor in its final installation. Different balancing techniques are applicable to the various subdivisions of this class, which are thoroughly discussed in BS 5265.

Class 3 – Rotors that cannot be balanced using modified rigid rotor balancing techniques (as in Class 2) but require special high speed balancing methods. Class 3 rotors are further subdivided according to the number of modes of unbalance. A high speed balancing machine is required, preferably with bearings which approximate to the actual bearing support conditions on site, capable of running up to the maximum operating speed in service. Balancing is then carried out at the critical speeds corresponding to the modal shapes. Various procedures are described in the standard.

Class 4 – Rotors which fall basically into one of the previous classes but in addition have a component which is flexibly mounted so that balance may change with a change in speed. Two

TABLE 1 – BALANCE QUALITY GRADES FOR VARIOUS GROUPS OF REPRESENTATIVE RIGID ROTORS

Balance quality grade	Product of the relationship $(e_{per} \times \omega)$ mm/s	Rotor types – General examples
G4 000	4 000	Crankshaft/drives of rigidly mounted slow marine diesel engines with uneven number of cylinders
G1 600	1 600	Crankshaft/drives of rigidly mounted large two-cycle engines
G630	630	Crankshaft/drives of rigidly mounted large four-cycle engines Crankshaft/drives of elastically mounted marine diesel engines
G250	250	Crankshaft/drives of rigidly mounted fast four-cylinder diesel engines
G100	100	Crankshaft/drives of fast diesel engines with six or more cylinders Complete engines (gasoline or diesel) for cars, trucks and locomotives
G40	40	Car wheels, wheel rims, wheel sets, drive shafts Crankshaft/drives of elastically mounted fast four-cycle engines (gasoline or diesel) with six or more cylinders Crankshaft/drives of engines of cars, trucks and locomotives
G16	16	Drive shafts (propeller shafts, cardan shafts) with special requirements Parts of crushing machines Parts of agricultural machinery Individual components of engines (gasoline or diesel) for cars, trucks and locomotives Crankshaft/drives of engines with six or more cylinders under special requirements
G6,3	6,3	Parts of process plant machines Marine main turbine gears (merchant service) Centrifuge drums Paper machinery rolls; print rolls Fans Assembled aircraft gas turbine rotors Flywheels Pump impellers Machine-tool and general machinery parts Medium and large electric armatures (of electric motors having at least 80 mm shaft height) without special requirements Small electric armatures, often mass produced, in vibration insensitive applications and/or with vibration-isolating mountings Individual components of engines under special requirements
G2,5	2,5	Gas and steam turbines, including marine main turbines (merchant service) Rigid turbo-generator rotors Computer memory drums and discs Turbo-compressors Machine-tool drives Medium and large electric armatures with special requirements Small electric armatures not qualifying for one or both of the conditions specified for small electric armatures of balance quality grade G6,3 Turbine-driven pumps
G1	1	Tape recorder and phonograph (gramophone) drives Grinding-machine drives Small electric armatures with special requirements
G0,4	0,4	Spindles, discs, and armatures of precision grinders Gyroscopes

examples are rotors with flexible fan blades and rotors which carry a centrifugal starting switch that comes into operation at a certain speed. These rotors can be balanced initially by simple dynamic methods but then need to be balanced at the operating speed of the rotor.

Class 5 – Rotors which would normally fall into Class 3, but for some reason are only balanced at one operating speed. These are rotors which run at only the one speed and pass through other critical speeds so rarely or so quickly that vibrations in those modes are not important. Usually the reason for putting a rotor into Class 5 is that such a rotor would be unnecessarily expensive to balance by Class 3 methods.

Whirling of Shafts

A rotating shaft can bow out at certain speeds and whirl in a complicated manner. This effect can result from a number of causes, such as mass unbalance or bearing friction. It can occur in a shaft which is dynamically balanced, although the vibration level will be greater if an unbalanced mass is present in one of the planes of maximum response. In general, whirling can be either in the same direction as the rotation or in the reverse direction and the whirling speed need not be the same as the rotation speed. Synchronous whirling is when the whirling speed equals the rotational speed. This is the critical speed of the rotor (or, in the case of a rotor with a distributed mass along the shaft, one of the critical speeds). As discussed in the chapter on *Bearings*, a plain bearing can contribute to the problem of whirl by allowing shaft displacement within the bearing clearance. Elasticity in the bearing support can also affect the magnitude and frequency of whirl.

The critical frequency can be found by treating the vibration of this rotor in the same way as if it were a beam carrying a central mass. The whirling speed is equal to the frequency of transverse vibration of the shaft. For details of some of the standard cases of vibration of beams, refer to the chapter on *Multi-degree of Freedom Systems*. Any known solution for lateral vibration can be applied to whirling calculations.

As an example of whirl, consider a weightless shaft supported on rigid bearings and carrying a rotor at a central position along its length as shown in Figure 7.

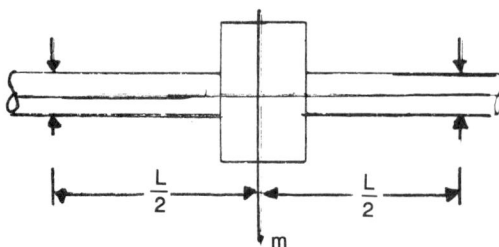

Figure 7 Shaft on simply supported bearings

The frequency of vibration of this as a beam is given by

$$f = \frac{3.46}{\pi} \left(\frac{EI}{m\,L^3} \right)^{1/2}$$

where E is the modulus of elasticity and I is the second moment of area of the shaft. This is also the frequency of whirl of the rotor. Similar relationships can be derived for any combination of masses or any distributed mass. As will be apparent, there are as many critical or whirling speeds as there are modes of vibration.

Shafts which have to operate above the first critical whirling speed should avoid the higher whirling modes, but it is preferable to design the shaft so that it is stiff enough not to reach the first speed, otherwise there will always be the risk of whirling during the run-up. At the critical speed the whirl displacement is theoretically infinite, limited in practice by friction and non-linear effects. At speeds away from the critical one, the deflection due to whirl is given by

$$y = \frac{f^2 \delta}{f_c^2 - f^2} = \frac{\omega^2 \delta}{\omega_c^2 - \omega^2}$$

where f_c is the critical frequency and f is the actual frequency. δ is the static deflection of the mass m, due to the force m g.

Chapter 2
Vibration Isolation

Virtually all dynamic machines generate unwanted forces. Some are of minor importance, such as those caused by unbalanced rotors and machine tools. Others may be much worse, forging hammers for example. The purpose of vibration isolation is to ensure that such vibration as is present in the machine does not affect the structure on which it is mounted, the persons nearby, or adjacent equipment. It is possible by taking extreme measures to install a drop forge next to a precision inspection area, but this would clearly be uneconomic; so one of the first stages in tackling problems of isolation is the intelligent placing of the vibrating machines in relation to the total needs of the factory.

There are two alternative approaches to vibration isolation – either the vibrations generated by the machine itself can be isolated from its support, or the vibrations of the support can be prevented from disturbing the behaviour of a machine. The former would be used where a strongly vibrating machine was liable to disturb much of the adjacent equipment, the latter where one delicate machine was to be installed in a generally vibrating environment.

Basic Principles

The ideal system for vibration isolation is where the vibration generator is separated from its support structure by free space, so that no force transmission is possible; this defines a transmissibility of zero (i.e. no transmission of force from machine to support), a clear impossibility. The concept does however illustrate that the aim of designing for low transmission is to have a very soft support, which usually implies a low rate spring.

From previous chapters dealing with the principles of vibration, it has been established that the force on the support of a vibrating mass is given by

$$F_s = k\,y$$

where F_s is the force on the support, k the spring stiffness and y the displacement of the mass on its spring.

If the force accelerating the mass is F, the force transmissibility, T_F, is given by

$$T_F = F_s/F = \frac{1}{1 - (\omega/\omega_n)^2} = \frac{1}{1 - (f/f_n)^2}$$

Similarly where the support is vibrating, and the aim is to isolate its movement from the supported machine, the displacement transmissibility, T_D, is required:

$$T_D = Y/Y_s = \frac{1}{1 - (\omega/\omega_n)^2}$$

Y is the displacement of the mass and Y_s the displacement of the support.

Figure 1 gives a plot of both force and displacement transmissibility, which are numerically the same. It will be apparent from a study of this curve that the transmissibility is greater than 1 for values of ω/ω_n up to a value 2, which implies that, in that region, there is an amplification rather than a reduction in transmitted vibration. This means that there is a possibility, if the isolating system is incorrectly designed, that the situation is worse than if there were no attempt at isolation at all.

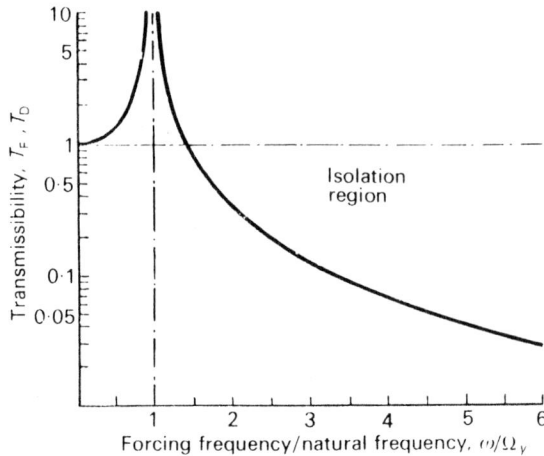

Figure 1 Transmissibility of a single degree of freedom
system in the absence of damping

Any isolating system has to be designed with a natural frequency that is much lower than any frequency present in the excitation. The ratio f/f_n is known as the tuning ratio and should have a value of about 3 as a minimum for isolation purposes. This would result in a transmissibility of 0.125. To achieve a transmissibility better than 0.1, a ratio of 4 is required. Note that if the isolation system is badly designed, so that for certain operating conditions the ratio is less than 2, the force applied to the support is greater than if the mass were rigidly attached to its support.

If $\omega > \omega_n$, the transmissibility is negative, indicating that, in the isolating region, there is a 180° phase difference between the excitation and the response. Although it is mathematically negative it is frequently plotted, as in Figure 1, as an absolute value.

The effectiveness of a vibration isolation system is its isolation efficiency expressed as a percentage:

$$= 100 \left(1 - \frac{1}{1 - (f/f_n)^2}\right)$$

This is plotted in Figure 2.

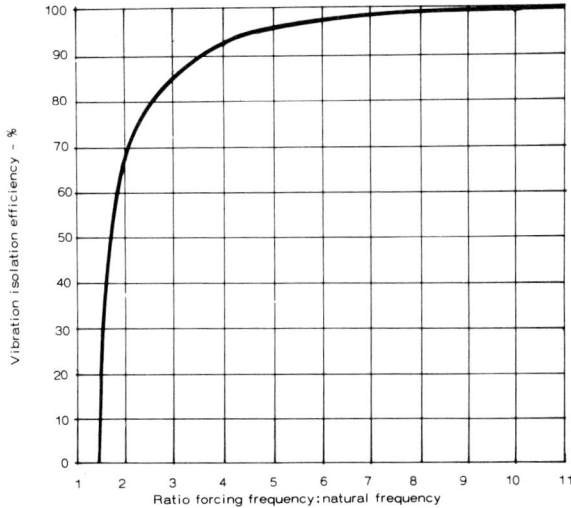

Figure 2 Vibration efficiency

Practically all vibration isolators are springs, and as the natural frequency of a spring is a function of its static deflection under load, the transmissibility can be related to the deflection due to the weight of the machine. This can be done by first expressing the natural frequency in terms of the static deflection, δ mm which can then be inserted in the above expressions.

$$f_n = \frac{1}{2\pi} \left(\frac{q}{\delta} \right)^{1/2} = \frac{15.8}{(\delta)^{1/2}}$$

It is not necessary to know the spring rate in order to apply this relationship. This relationship is plotted in Figure 3.

In a machine with a very low frequency of operation and therefore a low vibration frequency, the suspension may have to be very soft, if the low frequencies need to be isolated. It can be seen from a study of the above equations that this can mean a very large static deflection. For example,

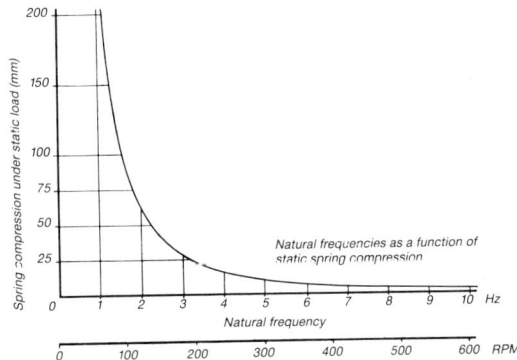

Figure 3

if an isolation efficiency of 99% at 600 rev/min is required, the suspension system will have a static deflection of 250 mm. A helical spring with this deflection would be difficult, if not impossible, to design.

There are several ways of dealing with this problem. One approach is to ensure that the excitation force at the fundamental frequency is dynamically well-balanced, even if this should make matters worse for the higher harmonics. These higher harmonics are more easily dealt with by the isolation system. This is known as supercritical tuning.

A second approach is to use a pneumatic isolator, which is a form of active isolation where the pressure in the support cylinders or bags can be varied to control the height of the machine.

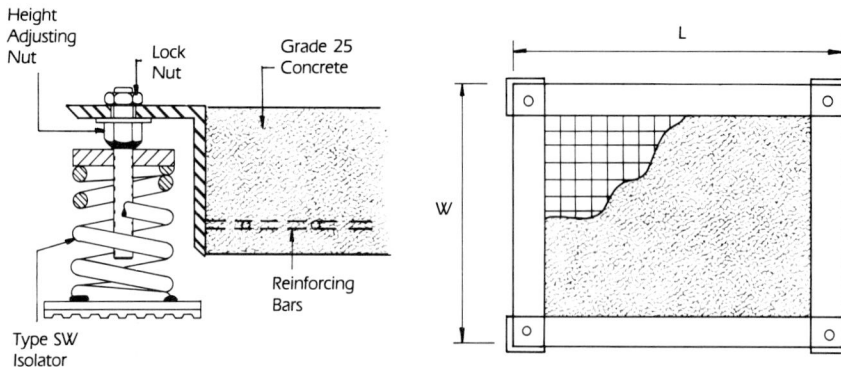

Figure 4 Inertia Base

A third approach is to mount the vibrating machine on a heavy base which reduces the vibration displacement. An inertial base has further advantages which may be important: it lowers the centre of gravity of the machine, improving its stability, and it spreads the base of the machine, making the installation easier. Such a base is usually made of reinforced concrete (Figure 4). In order to understand how the extra mass works, recall the value for the displacement of a mass on its support.

$$y = \frac{F}{k \{1 - (f/f_n)^2\}}$$

where f_n, the natural frequency of vibration, is given by

$$f_n = \frac{1}{2\pi} (k/m)^{1/2}$$

In order to retain a good isolation efficiency, the value of f_n has to remain the same. If the mass, m, is increased by the addition of a support mass, M, the value of k has to change to keep the ratio k/m the same, so:

$$\frac{k_m}{m + M} = \frac{k}{m}$$

The value of y with this new stiffer spring is

$$y_m = \frac{m\,F}{(m + M)\{1 - (f/f_n)2\}} = y\,\frac{m}{m + M}$$

Design of Foundations for Shock Loading

There is one special application for isolators – to support machines which are subject to intermittent shock loading. A typical example is a forging hammer. These machines, if not properly installed, can cause considerable annoyance to the neighbourhood and may result in damage to nearby equipment. Furthermore, if the soil on which the machine is resting is weak, the dynamic forces, being several times greater than the static forces, can cause irregular settlement beneath the machine which can hardly ever be corrected. The traditional way of overcoming this problem is by the use of a massive concrete foundation, which can involve a large and expensive excavation. An improvement on this method is the introduction of spring isolators and an increase in the stabilising mass as described above. A further improvement, which can be very effective, is the use of correctly designed viscous dampers in parallel with the spring isolators. The isolators are attached directly to the machine and a comparatively lighter foundation mass is used. The three stages are shown in Figure 5. The third method is of value in those applications where the foundations are subject to intermittent shock, and all frequencies are excited. It is of lesser use where the excitation frequency is fairly constant.

The natural frequency of the soil itself is relevant to the efficiency of the suspension. Some typical structure and soil natural frequencies are given in Table 1.

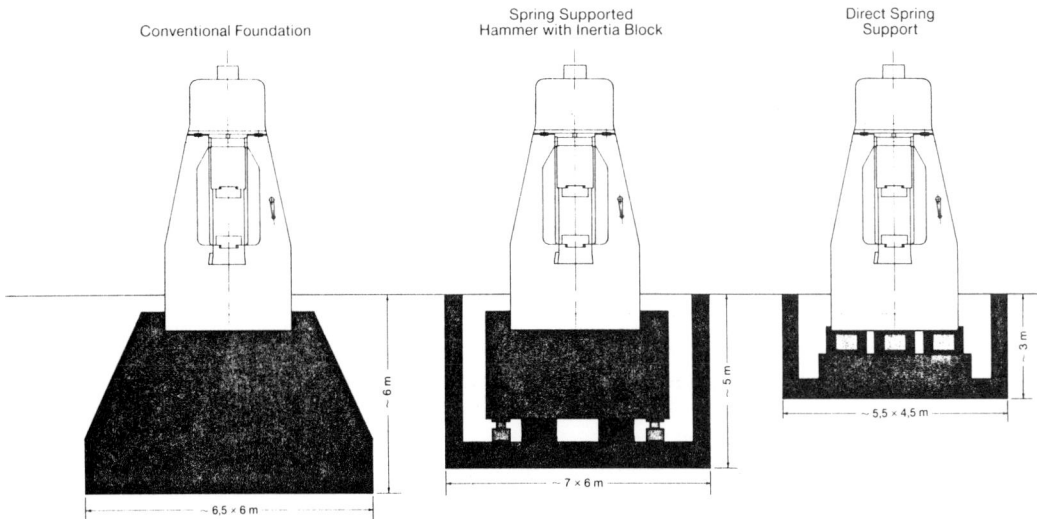

Figure 5 Alternative foundation methods for a Forging Hammer

TABLE 1 – NATURAL FREQUENCIES OF SOILS

Ground or structure	Frequency (Hz)
Suspended concrete floor	10 – 15
Ground floor	12 – 34
Soft clay	12
Medium clay	15
Stiff clay	19
Loose fill	19
Very dense mixed grain sand	24
Limestone	30
Hard sandstone	34

To find the isolation efficiency, determine the natural frequency of the ground f_e and that of the isolation material f_n. Where the damping factor, C/C_c is known, refer to Table 2. Thus in the example shown in the Table by an arrow, the ground frequency is 24 Hz, the isolator material has a damping factor 0.08 and a natural frequency of 8 Hz. The isolator efficiency is therefore 85% approximately.

TABLE 2 – ESTIMATION OF ISOLATOR EFFICIENCY

Damping Factor C/C_c	Frequency ratio R – fe : fn							
	1.5	2.0	2.5	3.0	3.5	4.0	4.5	5.0
0.05	20	66	80	87	91	93	94	95
0.10	19	64	79	85	89	91	93	94
0.15	17	62	76	83	87	90	91	93
0.20	16	59	74	81	85	87	89	91
0.30	12	52	67	75	80	83	85	87

▲
% ISOLATION EFFICIENCY

The Use of Damping

A machine with a correctly designed suspension system, to cater for the excitation forces when operating at its design speed, may nevertheless have to run through a resonance region every time it is started or stopped. If this is likely to be a problem, the suspension can be fitted with a snubber, which is a mechanical stop preventing excessive movement during the time the machine passes through resonance. A rigid stop is not desirable because there is a risk that force applied to it may cause it to fracture so, if a snubbing device is employed, it should be made of a resilient material such as rubber. Progressively increasing damping towards the extremes of isolator travel can be of value in such a case.

Damping can also be of value in depressing the amplitude of machine vibrations and in forcing the decay of vibrations caused by shock. The following analysis of damping should be read in conjunction with the previous chapters on the principles of vibration. In the above treatment of spring isolators, the analysis has implicitly incorporated some simplifying assumptions which need to be understood. Damping has been neglected; this can have a significant effect on the vibration isolation efficiency (in many cases it can be beneficial). Springs have been assumed to have a linear stiffness (force proportional to deflection). For conventional metal helical springs this is a fair assumption, but for non-metals it may be far from the truth; air springs, rubber and other elastomers, cork and composite materials have far from linear properties. Any isolator with a calculated spring stiffness based on static considerations may not behave in the predicted way because the dynamic stiffness differs from the static stiffness.

When calculating the effect of viscous damping, it is usual to introduce the concept of critical damping, which is the amount of damping just sufficient to suppress vibratory motion. It is rare for damping in an isolator to exceed the critical value. The value of this critical damping is given by:

$$C_c = 2 \, (k \, m)^{1/2}$$

If the actual value of damping in the system is C, the damping ratio is C/C_c. the presence of damping changes the natural frequency, so that

$$f/f_n = \{1 - (C/C_c)^2\}^{1/2}$$

In order to appreciate the effect of added damping, refer to Figure 6. It can be seen that with increased damping, the curve of transmissibility is progressively flattened, so that in the region near to resonance it is reduced, but in the region where greater isolation is required it is increased. The curves show that, if there is a significant amount of damping in an isolator, its spring stiffness has to be reduced to compensate, if one wishes to retain the same degree of isolation.

Damping in the isolator has a beneficial effect on the machine because it helps to suppress its vibration, but it also leads to a loss of isolation efficiency, according to the curves of Figure 6. A compromise has to be reached in practical cases.

An ideal isolator would have as little damping as possible in the operating range and as much as possible near to resonance. A conventional damper does not have properties like that, but devices are commercially available which give viscous damping for low frequency/high amplitude motion and combine that with material damping for high frequency/small amplitude motion. Spring isolators can be combined with viscous dampers to use the beneficial advantages of both, Figure 7.

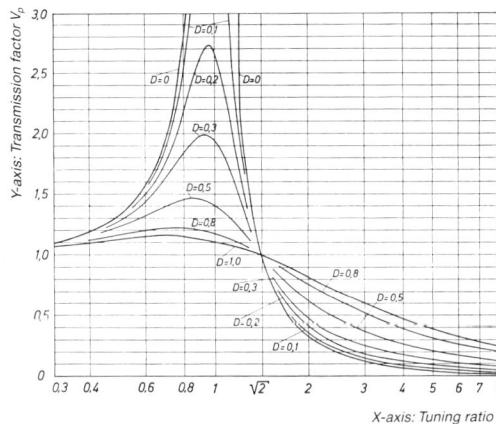

Figure 6 Effect of Damping – note that D = C/C_c

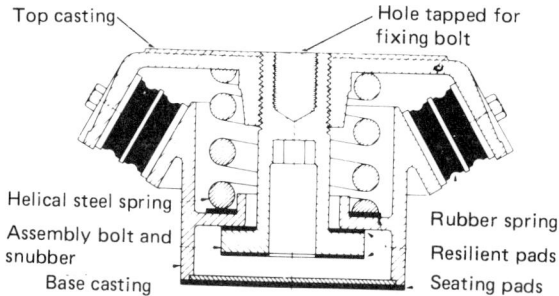

*Figure 7 Combined spring and rubber
in shear/compression mount*

Top casting

Hole tapped for
fixing bolt

Helical steel spring

Assembly bolt and
snubber

Base casting

Rubber spring

Resilient pads

Seating pads

Many materials are used for vibration isolation: air springs (active and passive), metal springs, wire mesh pads, rubber both natural and artificial, felt, cork, glassfibre and various composite materials combining rubber with fabric reinforcements. Rubber blocks can be used in compression or shear.

Degrees of Freedom

The minimum number of point mountings of an isolator system is three (this excludes the mat type isolator which can be considered to be a single mounting distributed over the whole base), and this introduces the possibility of complex modes of vibration. In general there are six degrees of freedom of a body supported resiliently in all directions:

vertical translation in z direction

horizontal translation in x and y directions

rotation in a horizontal plane about the centre of gravity

rotation in the x and y planes about the centre of gravity (rocking modes)

A vibrating machine can, in general, vibrate in any of these modes or a combination of them, provided that it is not constrained. Whether it actually does so depends on the direction of the excitation forces. Usually in well designed isolator systems, the likely modes to be excited are vertical translation and the two rocking modes. The other three motions will only occur if there is lateral flexibility of the spring supports, which is usually suppressed in commercial isolators. If horizontal excitation is known to be present, it is better to use horizontal springs than to rely upon lateral flexibility of the vertical springs, whose stiffness is difficult to calculate. It is always desirable, when designing a support system, to check the possible modes of vibration and calculate the natural frequencies.

Anti-vibration Mounts

Table 3 illustrates a variety of machine mounts that can be used. As a general guide, Table 4 can be used to identify the working range in terms of amplitude and natural frequency.

Metal Springs

As can be seen, metal springs are capable of accommodating very large deflections so they are preferred for isolating low frequencies. In this respect, they are superior to any other isolating material, except for air springs. One point that should be borne in mind is that a metal spring can transmit high frequency vibrations well away from its designed natural frequency. In this case it may be necessary to combine it with another form of isolator. In particular, acoustic noise can be readily transmitted from the machine to the support. If a noise barrier needs to be introduced, an elastomer pad or something similar needs to be incorporated.

TABLE 3 – ANTI-VIBRATION MOUNTS

Diagrammatic	Type	Construction	Features and Applications
	Simple pad	Rubber, rubberized fabric, rubberized cork, felt.	Simple, inexpensive mounts for stationary machines – may or may not be bonded in place.
	Area mount	Rubber, rubberized fabric, rubberized cork, cork, felt, ridged or contoured rubber mats	Carpet type mount laid under stationary machines over concrete floor or concrete blocks
	Spring	Helical steel spring	Excellent vibration isolation and can be tuned over a wide range of natural frequencies. Also good as a shock absorber, but has very little damping. Can be designed for very heavy loads. Suitable for high temperature surroundings.
	Damped spring	Helical steel spring with integral damping device	All the advantages of a spring for vibration isolation and steel absorption with large damping. Damper may be of friction, viscous or elastometic (hysteresis) type
	Miniature rubber	Basically a rubber grommet with or without metal mounting plates	Typical natural frequency 15 Hz. Typical maximum load 50 N. Available in a variety of rubbers with different damping characteristics.
	Simple rubber	Cylindrical, square or contoured rubber block with bonded-in studs	Simple, inexpensive vibration isolation mount. Typical natural frequency 5–20 Hz. Maximum load about 1.5 kN. Moderate damping characteristics.
	'Captive' rubber	Rubber block with bonded-in stud, captive in metal base	Excellent general-purpose flexible mount for reciprocating engines and machines up to about 15 kN. Typical natural frequency 9–12 Hz. Large variety of proprietary designs.
	Suspension systems	Helical coil or leaf springs	Used for the isolation of concrete foundation blocks for stationary machines

**TABLE 3 – ANTI-VIBRATION MOUNTS
(contd.)**

Diagrammatic	Type	Construction	Features and Applications
	Hydraulic	Load damping rubber bellows with interconnecting orifice	Superior damping and shock absorbing performance to captive rubber mounts. Recently developed for automobile engine mounting.
	Pneumatic (air spring)	Heavy duty sealed bellows internally pressurized at 20–100 bar	Typical natural frequency 2–5 Hz. Loads up to 100 kN Excellent performance as isolation mount and shock absorber
	Pneumatic (self-levelling)	Externally pressurized air spring	For vibration-free fixed level isolation of precision machine tools, etc. Natural frequency may be as low as 1–1.5 Hz.
	Free-standing mounts	Various, but commonly metal spring with elasto-meric damping and incor-porating height adjustment	Mounts for stationary machines and attached only to the machines
Proprietary	Isolating mounts	Numerous, including designs and constructions for bolted down and free-standing mounts	Designed as vibration isolation mounts, or combined isolation mounts and shock absorbers
	Woven mesh (cushions)	Crimped stainless steel mesh rolled into a cylindrical cushion	Particularly suitable as a damping material used in conjunction with metal spring isolators for high temperature applications. Also useable on its own as an isolator.

TABLE 4

Type of Mount	Amplitude (mm)	Frequency (Hz)
Air	20 – 200	1 – 15
Metal spring	3 – 30	3 – 10
Glassfibre	3 – 30	7 – 15
Rubber	0.2 – 5	10 – 30
Cork and felt composites	0.1 – 0.5	20 – 50

Of the metal springs in common use, helical springs are the most usual, although leaf springs are also to be found in special applications, such as road vehicles, where they have an advantage in that they provide damping through interleaf friction. Conical compression springs are also used; these have a particular advantage in that the spring stiffness increases with load.

Helical springs can be used in tension or compression. For heavy loads, compression springs are better, mainly because of the difficulty of attaching the end connection of a tension spring. Even when the application is one in which the mass is supported overhead, such as in pipe hangers, the spring is usually placed in compression, Figure 8.

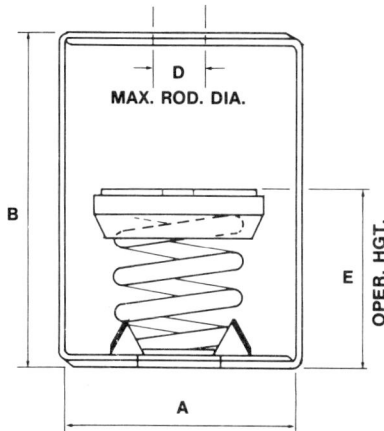

Figure 8 Pipe Hanger isolator, with elastomer cushion

When considering the use of a simple helical spring, it is important to determine (or obtain from the manufacturer) the lateral stiffness as well as the axial stiffness. For calculation of spring stiffnesses it is best to refer to a specialist text, but in summary, the axial stiffness is given by:

$$k_a = \frac{G\,d^4}{8\,n\,D^3}$$

where G is the shear modulus of the material of the spring, d is the diameter of the spring wire, n is the number of active coils, and D is the mean coil diameter. The lateral stiffness is a more complicated calculation, and it is better to rely on data supplied by the spring manufacturers.

A check should be made on the spring stability. A helical spring is stable if

$$\frac{k_l}{k_a} \geqslant 1.2\,\frac{\delta}{L}$$

K_a is the axial stiffness, k_l is the lateral stiffness, L is the working height and δ is the axial displacement of the spring.

This is an approximate relationship, and it may be necessary to do a more careful analysis if the stability is marginal.

Air Springs

A full description of the use of air springs as isolators is given in the Pneumatic Handbook, published by Elsevier, to which reference should be made.

Elastomeric Mounts

These have advantages over metal springs, when the isolating frequency is not too low. In general they are easily mounted, occupy very little space and possess a degree of internal damping; they are relatively cheap. They can be used as specially made units, as pads which can be cut from the solid or as carpets occupying the whole base area of the machine; this is sometimes done where inertia blocks are used.

The stiffness of rubber as a spring can be determined in terms of the dynamic modulus of the material and its geometry. Through the skill of the rubber technologist, it is possible to obtain any desired stiffness. It is the dynamic modulus which has to be used in calculations, rather than the static modulus. There can be a considerable difference between the two values. Natural rubber tends to have more resilience than synthetic rubber and so can provide for lower frequencies with the same geometry.

All rubbers have properties which vary with temperature. In a high damping rubber, the temperature can increase due to the internal absorption of energy, so the properties need to be established at the actual temperature reached in practice. Some rubbers, particularly natural rubber, are affected by the presence of oil or other hydrocarbons.

Rubbers are generally incompressible; under a compressive load, they must be allowed to expand laterally, so sufficient clearance must be provided. Anything which prevents lateral movement such as bonding to a metal plate, will increase the effective stiffness. Even if a rubber block is not bonded, the friction between the face of the rubber and the surface on which it rests acts as a restraint, so it is almost impossible to develop pure compression without surface lubrication.

The incompressibility of rubber limits the use of large area mats, where lateral expansion can only take place at the periphery of the mat, unless some form of ribbing or grooving is incorporated, as in Figure 9.

Figure 9 Two types of carpet foundation isolators made from natural or nitrile rubber

Foam rubber behaves in a different way: the air in the pockets allows compression so large area mats of this material can be used without resorting to ribbing or grooving.

Elastomeric Unit Mounts

Rubber can be stressed in compression, tension, shear or a combination of these as shown in Figures 10 and 11. Torsion can be considered as a form of shear, since the strains other than shear in a torsion component can be neglected in most cases.

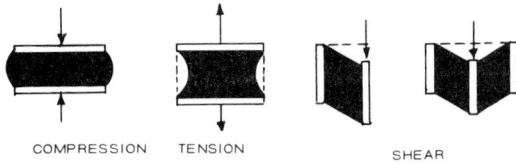

COMPRESSION   TENSION

SHEAR

Figure 10

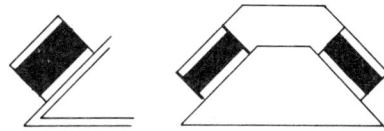

Figure 11

For a given load and volume of rubber, deflection is least in compression and greatest in shear. Tension loading is seldom used except for very light applications, since the integrity of the mount depends upon the strength of the rubber-to-metal bonding, which is quite low (typically 300 to 500 kPa). A parallelogram block, loaded as in Figure 10, stresses the rubber uniformly. Torsion is not as efficient in its use of material as pure shear because the stress across a torsion bar is not uniform.

The calculation of the stiffness of a rubber section is not as straightforward as it would be for a metallic section, because the relatively large deformations cause changes in the geometry of the section which cannot be ignored. As mentioned previously, the properties of rubber also vary with rate of change of stress. When calculating the behaviour of a spring mount, refer to the manufacturer's literature, which usually quotes the appropriate stiffness values. The following summarises some of the applicable formulae.

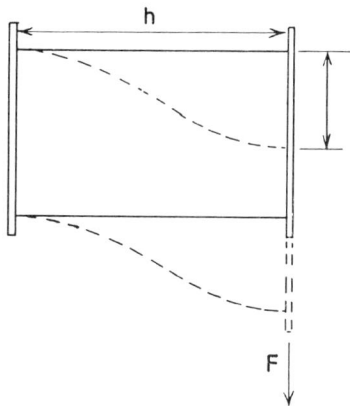

Figure 12 Block of rubber in 'pure' shear

When strained in shear, a rectangular rubber pad will assume a parallelogram shape. The deflection, δ, Figure 12 is given by

$$\delta = \frac{F\,h\,\{1 + h^2/(36\,k^2)\}}{G\,A}$$

where F is the force, h the thickness, k the radius of gyration of the section about the axis of bending, G the shear modulus and A the cross-sectional area of the block. This is an extension of the simple theory which ignores the bending stiffness:

$$\delta = \frac{F\,h}{G\,A}$$

It can be shown that the simple theory is a sufficiently good approximation when the height/diameter ratio is less than 0.5. It is often more convenient, when considering the compression of rubber, to employ the shear modulus, G, rather than the elastic or Young's modulus, E. For an incompressible material such as rubber

$$E = 3\,G$$

For rubber in compression with bonded end faces, the relationship for compressive stress is

$$\frac{F}{E\,A} = \frac{1}{3}\left(\frac{1}{\lambda^2} - \lambda\right)\,s$$

F is the compressive load, A is the cross-section in compression, λ is the ratio of the strained to the unstrained length and s is the shape function which depends on the shape of the block. s is equal to 1 where there is no end restraint, so it represents a correction to take account of restraint. This function has been derived for most shapes in current use.

The stiffness for calculation of frequency of isolation (provided that the amplitude is small) is:

$$k = \frac{E\,A}{3}\left(\frac{2}{\lambda^3} + 1\right)\,s$$

For λ close to 1, that is with a small deformation, this simplifies to

$$k = E\,A\,s$$

For a square cross-section, side a:

$$s = 1 + B\,(a/h)^2$$

For a round cross-section, diameter d:

$$s = 1 + B\,(d/h)^2$$

h is the thickness in the direction of compression. B varies with the modulus according to Table 5.

For shape functions other than square or round, it is best to refer to a specialist text.

It can be seen that the shape factor is of importance for pads where the thickness is much smaller than the width.

Rubber can be combined with helical springs to give the advantages of both types. The combination shown in Figure 13 adds damping to the unit in a more economical way than by using a viscous damper.

TABLE 5 – VARIATION OF FACTOR B WITH ELASTIC MODULUS

E (M Pa)	<2	2 to 3	3 to 4	4 to 6	>6
B	0.12	0.103	0.08	0.063	0.056

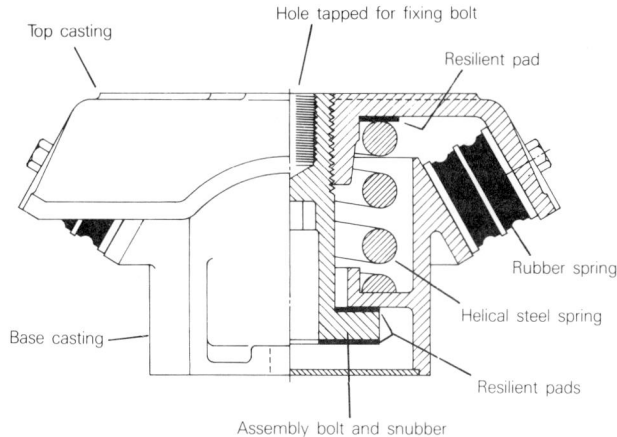

Figure 13 Combined spring and rubber mount

Pad and Area Mounts

Materials used for pad and area mounts include:

Configured rubber – studded pads
– ribbed pads
– honeycomb sections with band facings

Reinforced rubber – metallic reinforcement
– non metallic reinforcement

Cork – agglomerated cork
– rubber bonded cork
– reinforced

Felt
Foamed materials
Glass fibre
Steel mesh

Configured rubber may be adopted for mounting pads, the object being to increase the static deflection of the material. Configuration can be studs or ribbing of both surfaces. If concrete has to be poured over it, then a rigid sheet has to be laid over it to prevent filling of the open spaces.

Reinforced rubber can provide very high load bearing capacity while retaining good resilience. Rubberised fabric is a form of reinforced rubber. Asbestos has been a traditional reinforcing material with good properties, now out of fashion for safety reasons. Various fabrics are now used. They can carry loads of up to 10 M Pa and still retain useful elasticity.

Natural Cork

Cork sheet consists of particles bonded under pressure into a sheet material. It is a compressible material, unlike rubber. It can accommodate up to 30% compression without lateral expansion. Care should be taken that the maximum load without permanent set is not exceeded. Its main advantage is cheapness, but its disadvantages are its tendency to crumble under repeated loading and its deteriorating performance when waterlogged.

Steel reinforced cork pads are used for isolating large machines, where spring mounting is too expensive. The strips of cork are assembled in a mosaic and set into a frame, stiffened with wire struts. These pads are supplied in standard thicknesses, according to the isolation frequency.

TABLE 6 – TYPICAL PROPERTIES OF REINFORCED CORK PLATE – 60 MM THICK

Static Load kPa	Modulus E M Pa	Static Deflection mm	Natural Frequency Hz
15	19.5	0.46	23.3
18	22	0.49	22.5
20	23.2	0.52	22.0

Foamed materials are used for mounting pads but they are limited in their load carrying capacity. For light loads and low frequency isolation they can be effective. Their main use is as acoustic absorbers (see the chapter on *Sound Absorption*).

Steel mesh can be used when moulded into pads, Figure 14, but they have very high hysteresis, and for this reason are often used in conjunction with a steel spring to promote the return action. They are mainly used for shock absorption, rather than for continuous vibration.

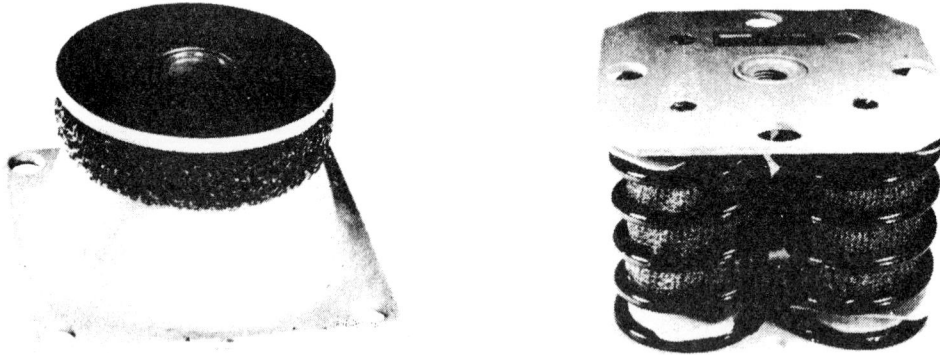

Figure 14 Two types of mounting employing knitted stainless steel mesh
On the left a single unit has a natural frequency 13 to 22 Hz
On the right, the unit has a natural frequency of 4 to 4.5 Hz damping factor
0.15 to 0.2

Glass Fibre

This is claimed to be superior to other pad type isolation materials. Due to the precise manufacturing techniques it has predictable properties. It consists of a high density matrix of precompressed moulded glass fibres with bonding at the fibre intersections. When used as an isolation pad, it is coated with a flexible moisture-impervious membrane, which allows air movement between the fibres. The effect of this is to introduce viscous damping. It is available in a range of densities to provide load bearing capacities from 7 k Pa to 3.5 M Pa. Glass fibre is resistant to moisture and oil and has constant characteristics from $-15\,°C$ to $121\,°C$.

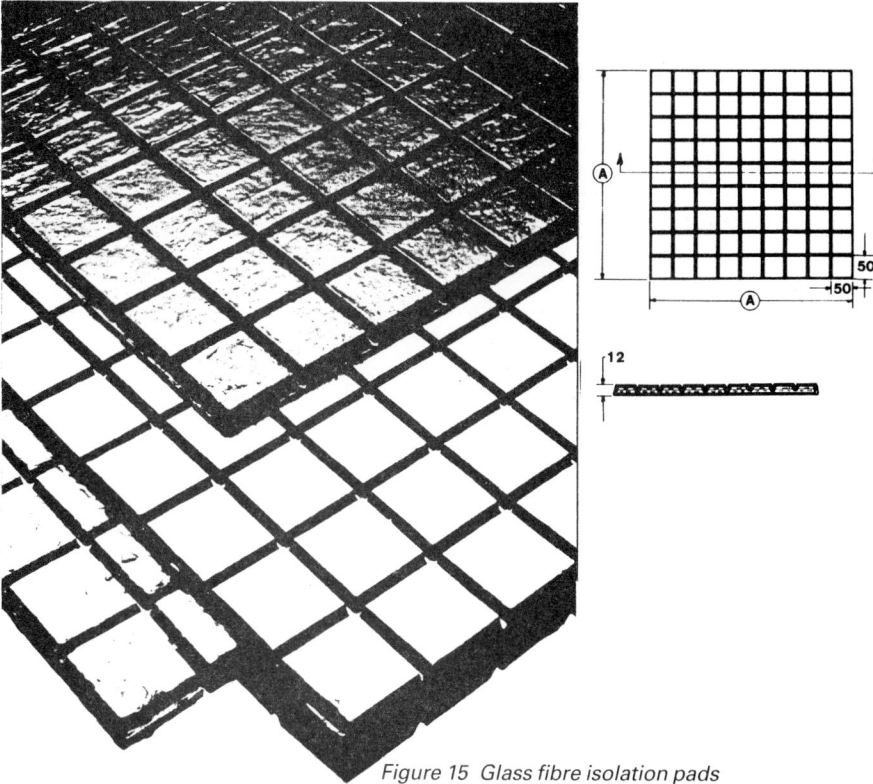

Figure 15 Glass fibre isolation pads
Glass fibre pads can also be incorporated into individual spring mounts.

When used as an isolation pad as shown in Figure 15, each individual pad, size 50 × 50 × 12 mm, can sustain a static load of 90 to 1800 N. Individual pads of varying thicknesses up to 100 mm are also available, with natural frequencies less than 15 Hz.

Commercial Isolators

Figures 16 to 19 show a range of commercially available isolators, for a variety of purposes. It will be found that in many cases, a problem of vibration isolation can be solved by one of the many proprietary isolators available on the market.

Figure 16 Neoprene isolator incorporates cast-in steel plate. Loads from 250 N to 5 kN, static deflection 3 to 12 mm

Figure 17 Range of open spring mountings with
integral rubber end fixings

Figure 18 Leaf spring vibration mount made
from laminated stainless steel leaves with
epoxy resin. Resonance frequency in the
range 3–9 Hz with low damping

Figure 19 Glass fibre isolator, suitable for
operating frequencies in excess of 1440 rpm

Chapter 3
Resilient Mounting of Structures

The resilient mounting of structures has come into prominence in recent years with the recognition that vibration can affect the ability of people to function efficiently when the building they are in has a certain level of vibration.

While severe earthquakes are uncommon in UK, the design of buildings to resist earthquakes and seismic motion generally is important for critical buildings such as nuclear power stations, and there is an increasing requirement for seismic qualification of buildings. This subject is too large to cover in a single chapter, but the future will see a greater emphasis on the subject.

This chapter will be restricted, in the main, to considerations of resilient bearings to resist the vibrations that are caused by vehicles and factory generated vibrations. The resilient mounting of structures is, in principle, no different from any problem of isolator design, but there are added features, in particular the internal vibration modes of the structure need to be considered as part of the whole analysis.

Of the standards that are available, BS 6177 is of considerable importance, since it describes the design of resilient bearings and the conditions of their use.

BS 7385 is helpful as a guide to the techniques that can be used for the measurement of vibrations in buildings, with recommendations as to the appropriate kinds of instrumentation. It contains a classification scheme for buildings according to their resistance to vibration and categorises the varieties of foundations and types of soil.

BS 7527, Section 2.6, should be studied for descriptions of earthquake intensity levels and give a classification based on a modified Mercalli scale. Section 1 gives the environmental parameters and their severity.

ISO 6258 (not a BS) gives guidance for the design of nuclear power plants against seismic hazards.

Types of Bearings
Resilient bearings fall into five main types, as designated in BS 6177:1982 – see Table 1. All types are usually rectangular in shape, but can have detail variations (e.g. some may incorporate corrugations or other forms of voids).

For most applications it is sufficient for the mechanical properties of the material to be described by:

> Static load-deflection curves under normal loading (generally compression).

> Long term creep behaviour – e.g. pad deflection against time under constant static (and if necessary dynamic) loading.

TABLE 1 – TYPES OF RESILIENT BEARINGS (BS 6177: 1982)

Type	Construction	Remarks
Plain elastomer bearing	Homogeneous vulcanised natural or synthetic rubber.	Very limited application. Needs location to prevent lateral spread.
Plain composite bearing	Rubber with homogeneously dispersed cellular particles.	Many possible combinations and physical properties.
Elastomeric sandwich bearings	Plain elastomeric material bonded between two parallel steel plates.	Many possible combinations and physical properties.
Multiple elastomeric sandwich bearing	Multiple layers of elastomeric material alternating with metallic or non-metallic reinforcement layers.	Many possible combinations and physical properties.
Multiple composite sandwich bearing	Multiple layers of composite elastomeric material with metallic or non-metallic reinforcement layers.	Many possible combinations and physical properties.

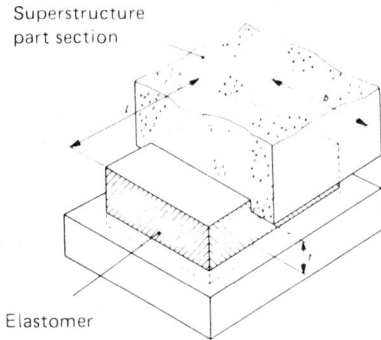

Superstructure part section

Elastomer

(a) Plain elastomer pad

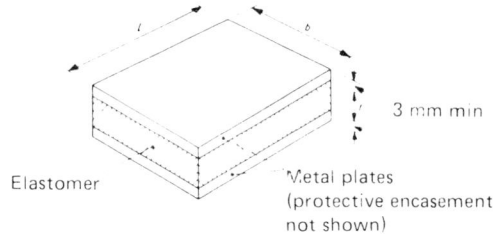

Elastomer

Metal plates (protective encasement not shown)

(b) Sandwich mounting

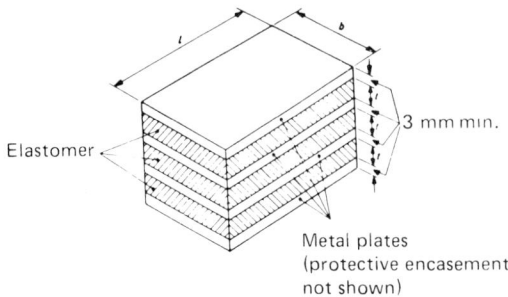

3 mm min

Note:

l = length of elastomer layer
b = breadth of elastomer layer
t = thickness of elastomer layer

$$\text{shape factor} = \frac{\text{load area}}{\text{bulge area}} = \frac{lb}{2\,(l+b)\,t}$$

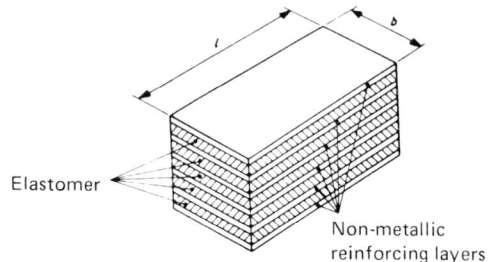

Elastomer

3 mm min.

Metal plates (protective encasement not shown)

(c) Multiple sandwich mounting

Elastomer

Non-metallic reinforcing layers

(d) Composite mounting

Figure 1

Dynamic modulus and loss factor at different frequencies.

Here it is important to appreciate that the dynamic properties of resilient bearings can differ under varying operating conditions and are affected by:

static pressure;

shape factor;

operating frequency;

strain amplitude;

operating temperature;

flexibility of the bearing supports and back-up structure;

age and previous history of the bearing.

It is essential that the installation of a resilient bearing has to be considered at an early design stage to enable an economic, efficient and safe system to be engineered.

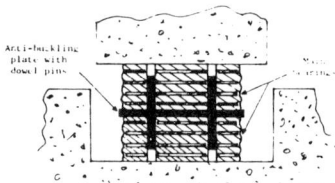

Double unit rubber/steel sandwich bearing with dowel anti-buckling device.

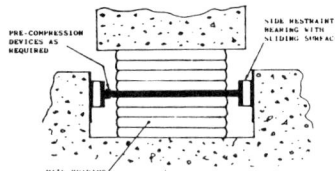

Side restraint anti-buckling system for use with multiple composite sandwich bearings.

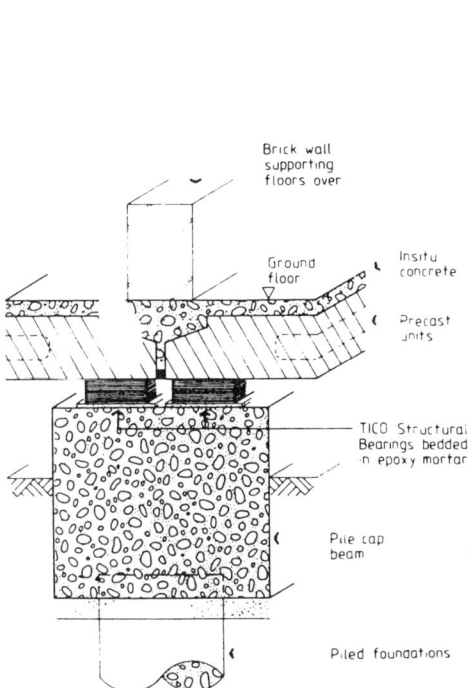

Fire protection fibre coils surrounding bearings, with isolated cover plate.

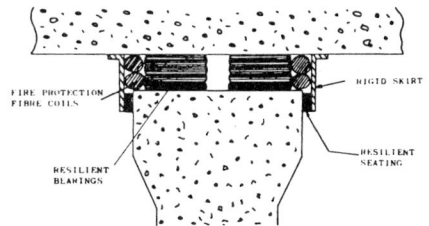

Diagrammatic view showing the ground floor construction at West End Sidings

Irish Centre Community Hall is situated directly over the main railway line from St. Pancras Station – a classic example of when resilient bearings can provide noise and vibration isolation under extreme site conditions

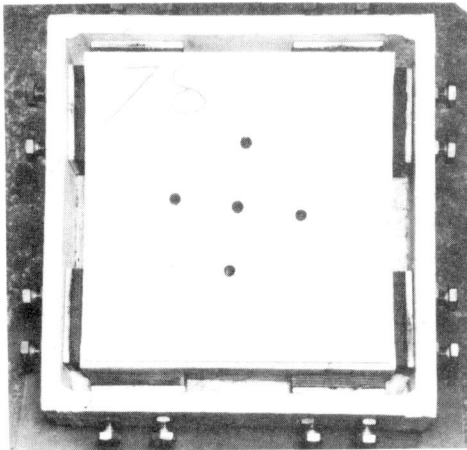

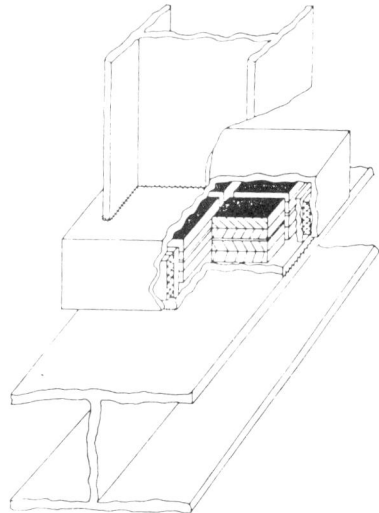

Typical example of module bearing housing

Example of a small accomodation module mounting with wind/shear restraint, (viewed from underside)

Resilient bearings at base of column with steel location dowels and capped resilient isolating collars

Design Procedure

The following stages give a good idea of the design and supervision procedure:

1. Site vibration survey including one for all types of trains – high speed passenger, local, goods, heavy freight, *etc.*, recording several examples in each case. The recordings must be vibrograms of velocity or displacement and the natural frequency of the pick-up must be lower than the lowest frequency of ground wave, or be D.C. Meter readings from 'high frequency' accelerometers do not provide information about the low frequency waves, nor how many cycles they contain; neither do they provide adequate information about peak transients, all of which are required for design of insulation.

2. The quantity of vibration reduction needed (for living units it would be to about the 'boundary of perception') is next calculated, and the natural frequency of the sprung building is determined, based on the dynamic, not the static, elasticity of the bearings. The wind sway calculations of the building are based on the static stiffness.

3. The accurate dead weight of the building is found, and its mass centre. If the latter is eccentric for a rigid building the number of bearings must be adjusted to keep the structure upright and to prevent cracking, etc.

4. As the springs are quite resilient compared with the much more rigid ordinary foundations, the structure must often be made stiffer to accept wind loads without undue distortion, which might otherwise seriously damage lightweight partitions, external cladding or windows.

5. All rubber bearings must be protected against fire in vulnerable places.

6. A structure that is carried on resilient bearings should be designed on fail-safe principles so that in the event of a failure, partial failure or damage to the bearings, it will remain adequately supported and retain its safety and general serviceability under the design load.

7. Long term side stability of the sprung building is maintained by side acting resilient members which also take over the function of wind braces (Figure 2).

8. Finally, all variations of stiffness in the springs shown by these tests are compared with the designed stiffness of the building members, so that excess variations can be adjusted, see Figures 3 and 4.

    Specific design requirements are detailed in BS 6177:1982.

The bearing characteristics can be 'tailored' over a very wide range. Compounding the rubber mix, fabrication, moulding pressure and vulcanisation, plus bearing sizes, can be adjusted to meet particular requirements of static deflection, shear deflection and stress. Where large column loads and areas of bearings are required it is advisable to employ a number of smaller bearing units in modular form to ensure equal dynamic and static properties throughout the structure.

The significant parameters of dynamic behaviour can be stated as:

Dynamic stiffness – a function of the dynamic modulus of elasticity, pad area and thickness. This can be very different from the static stiffness (Figure 5).

Shape factor – laminated construction makes the bearing far less susceptible to excess internal stresses, long term fatigue, and lateral spread compared with homogeneous rubber bearings of the same overall dimensions.

Snubber

Wind

Wind Spring

Figure 2

Varying spring
stiffness ± 15%

induced bending
moment

Column
bending

hard spring   soft spring

Figure 3

Usual B.M.D,

Equally balanced
spring groups

hard
spring   cracks   very soft spring   cracks   hard
spring

increased B.M.D.

cracks

hard
spring   very soft spring   hard
spring

BM reversal   increased BM

Figure 4

Load

curves from tests

Working load

Spring rates
A = apparent
B = static
D = dynamic

deflection   *Figure 5*

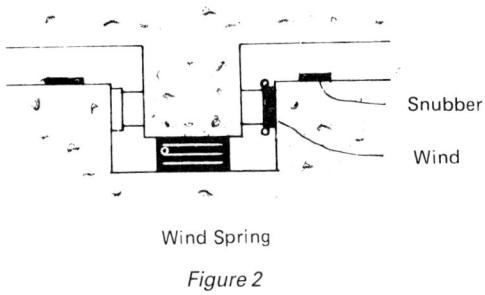

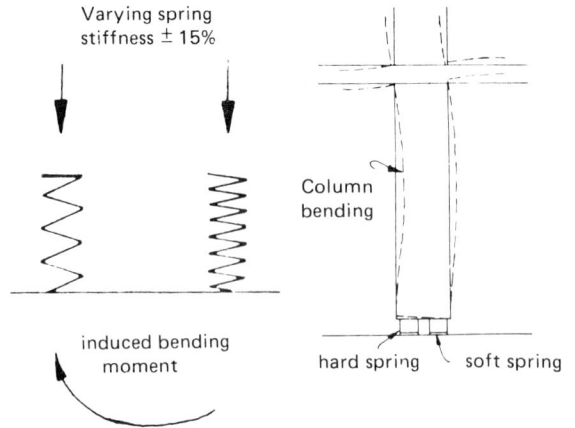

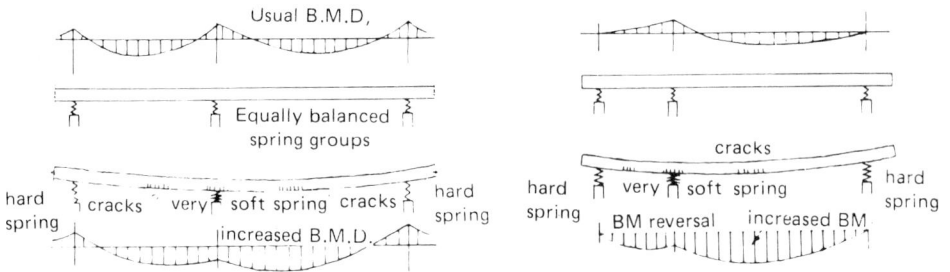

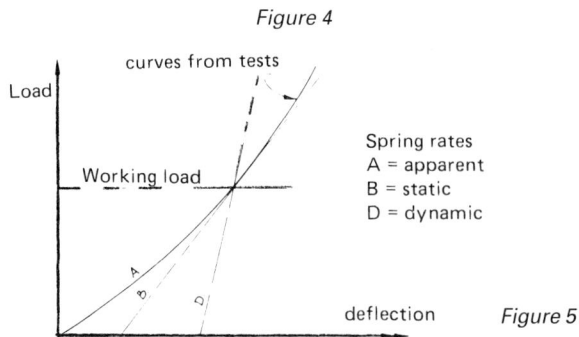

Static and Dynamic Stiffness

Loss factor – a measure of the damping ratio of the pad, defined as the tangent of the phase angle by which the damping force produced by the pad lags behind the stiffness or spring force.

Model factor – model springs are substantially less stiff dynamically than the dimensionally larger, though similarly proportioned, civil engineering resilient bearings.

Leak Paths

It is essential to appreciate that no form of isolation mounting remains effective if it is 'short-circuited' by a rigid connection between the resiliently mounted structure and its surrounds, thus allowing vibration to 'leak' across this path. Also, clearances around a bearing should be such as to permit movement to its loaded profile without restriction. Care must be taken to ensure that restraint to a bearing is not caused through interference either by structural parts or by accumulation of surplus grout or other debris in its vicinity.

Service connections to a mounted structure should either be sufficiently flexible to act as isolators themselves, or incorporate flexible couplings to give the same effect (e.g. elastomeric bellows fitted to drainpipes).

Where it is necessary to seal any small gaps betwen the mounted structure and its surrounds (i.e. its foundation or an adjacent unmounted structure) the sealing must be done with a material of adequate flexibility to accept any probable movement. Due allowance must also be made in the design for any consequent effects on the efficiency of the mounted structure.

To ensure that a mounted structure remains completely isolated from any adjoining structure, sufficient space should be left between them to allow for lateral movement. Such movements will include those from temperature, ordinary deflection and sway under load, foundation settlement, long term creep and deflections of mountings, including possible damage caused by fire or structural overload. Normally building tolerances and clearances should also be taken into account.

Examples of Resilient Bearings

TICO CV/CA is a high stress resilient bearing material of vulcanised laminate construction comprising a reinforcement of alternating layers of high tensile strength fabric bonded to plies of polychloroprene base rubber modified by the inclusion of cellular particles. The face layers are of slightly different formulation to facilitate bonding. It is dimensionally stable under widely varying atmospheric conditions and is rated for a maximum working stress of 7 000 kPa and can achieve natural frequencies down to 10 Hz without the need for additional horizontal stability.

Principles applications are the isolation from outside sources of vibration of tower office blocks, flats and theatres, and heavy duty oil rig accommodation modules, *etc.*

TICO CV/CA bearings with rigid fail safe support with vertical load capacity equal to three resilient bearings

Luxury flat in Ebury Street, London.
This 24 000 ton structure is isolated at first floor
level on 1410 TICO CV/CA bearings 103 m thick,
with basement flats constructed on 'box-within-
box' principle, separated from the foundation by
351 TICO CV/M bearings.
It is Europe's largest resiliently mounted building

TICO CV/M is a medium load structural bearing material formulated from highest quality neoprene modified by the inclusion of cellular particles. It is capable of being used in a variety of ways from modular form bearings on pile caps to continuous strip footings; and may be employed in any thickness (in multiples of 25 mm) necessary to provide the required natural frequency. Maximum recommended stress is 1400 kPa.

Particular applications are for isolating partial or total structures from their foundations on sites subject to medium and low frequency noise and vibration (*e.g.* near railway lines, isolation of domestic housing, studio control rooms, floating floors *etc.*, and also lightweight oil rig accommodation and laboratory modules).

TICO CV/LF material incorporates a moulded configurated rubber layer bonded between flat face plies of rubber composite material. It has been developed as a low load, low frequency structural bearing capable of a relatively high spring rate. It is produced in three grades

Sections of the bearings employed during construction of isolated barrier block at
West End Sidings in close proximity to main line trains; a total of 1950 TICO
CV/M bearings were used.

Fully floating isolated anechoic chamber carried on five layer TICO CV/LF/K bearings providing a natural frequency of approximately 62 Hz at 350 kPa stress.

Example of TICO CV/LF low frequency bearing using five layers.

catering for loads up to a maximum of 700 kPa with natural frequencies down to 7 Hz (two layers), or lower depending on the number of layers employed.

CV/LF materials are designed for total or partial support of all types of internal structures such as anechoic chambers, recording studios, floating floors, theatre stages, false roofs, etc.

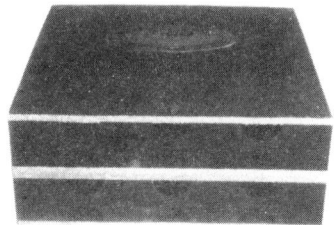

Figure 6 shows recommended working stress ranges and natural frequencies of these materials. All can be bonded to concrete, steel or wood, using the appropriate adhesive.

Cellular Polymers used in Building Construction

If a cellular material is chosen, it is volume compressible and the shape factor has very little significance so large sheets can be used effectively in foundations. One material used for this purpose is a polyetherurethane, with properties as shown in Figure 7. The properties can be varied to meet the required stiffness and load. When this material is dynamically loaded, the inherent damping converts some of the energy into heat. The damping property of sheet or pad materials of this kind is usually expressed as the loss factor (or dynamic loss factor). This is similar to the damping ratio, but is more convenient to use because it relates directly to the pad material rather than to the complete vibrating system. It is defined as

$$\eta = C/\omega_0 m$$

where η is the loss factor, C is the damping, ω_0 is the undamped natural frequency and m is the vibrating mass.

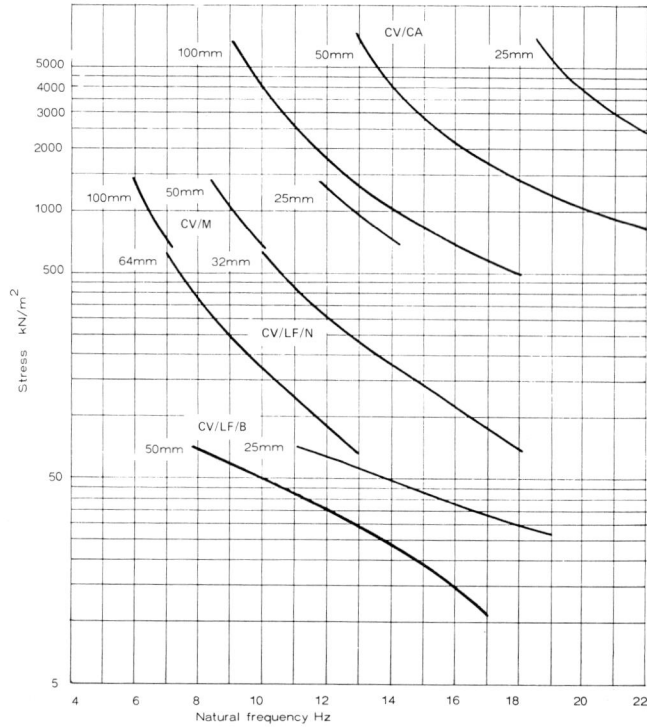

Figure 6 Properties of structural bearings

Remembering that

$$\omega_o = (k/m)^{1/2}$$

the loss factor can also be expressed as

$$\eta = C/(k\,m)^{1/2}$$

The critical damping, C_o, is given by

$$C_o = 2\,(k\,m)^{1/2}$$

From this it can be seen that

$$\eta = 2 \times C/C_o$$

The material in Figure 7 has a dynamic loss factor of 0.1 to 0.3, according to the density.

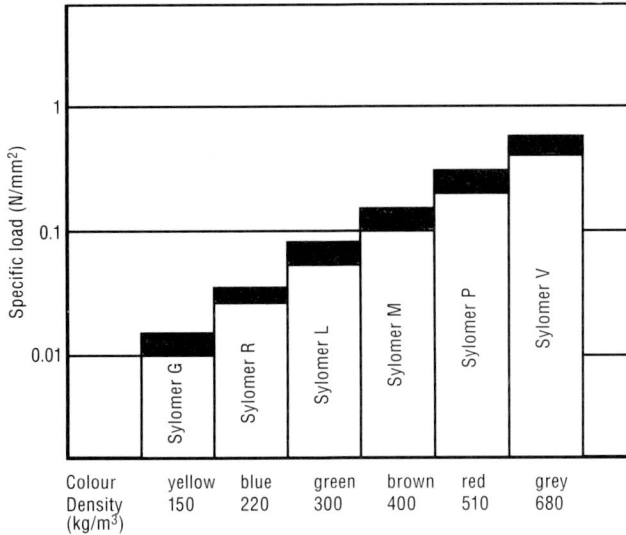

Figure 7 Sylomer-Standard Types: Load Ranges, Identification Colours and Densities

Resilient Mounting of Railway Track

Buildings often need protection against vibrations caused by railway traffic. It may be more satisfactory to prevent the railway vibrations from being transmitted into the ground than by treating the building. The method is shown in Figure 8, which uses a ballast-free track and elastic pad foundations.

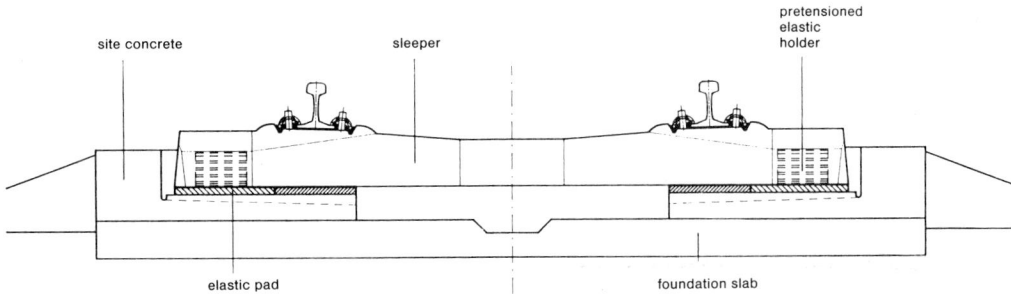

Figure 8 Railway track supporter on Sylomer pads

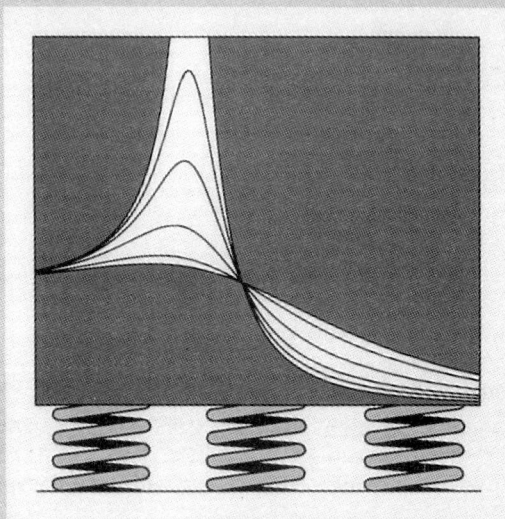

Chapter 4
Vibration and Noise in Tools

Vibrating hand tools are widely used in the construction and manufacturing industries for drilling and breaking materials ranging from rock to steel. Pneumatic operation is still the preferred medium for drilling rock and for breaking concrete on construction sites. In factories, tools such as riveting hammers, chippers and impact wrenches are still mainly operated by compressed air. For a discussion of the various tools that are available, refer to the Pneumatic Handbook, published by Elsevier.

Vibration problems are not restricted to tools operated by compressed air. Among the tools which have been associated with vibration white fingers (VWF) are riveters, caulking hammers, chipping hammers, rock drills, pedestal grinders, portable grinders, road and concrete breakers, swaging tools, chainsaws and shoe pounding-up machines. They are not restricted to reciprocating or impact tools; rotary tools have also been implicated in the occurrence of VWF.

Any tool which acts by impact necessarily employs intermittent application of a force to a piston, whether that force derives from compressed air, hydraulic fluid or electric energy. The force has to be resisted by the person holding the tool, with the possibility that such a vibrating force may cause some form of vibration disease. This is dealt with in some detail in the chapter on *Human Response to Vibrations*. It is now generally accepted that any level of vibration has a statistical probability of causing vibration white fingers in some individuals at least, but until the recommendations contained in BS 6842 (ISO 5349) were published, there was little guidance as to the levels that should be aimed at and the recommended exposure times. Recently manufacturers have made some considerable attempts to design tools which have a reduced vibration level on the handles, and we can expect legislation to be introduced over the next few years, requiring the labelling of vibrating hand tools.

Noise and Vibration from Pneumatic Tools

These tools not only cause vibration but also generate noise from the air exhaust. It would be wrong, however, to assume that another form of power operation would be less offensive merely on the grounds that it did not have an exhaust. A considerable part of the noise that is created by an impact tool derives from the impact mechanism itself, which is present whatever the fluid used to produce the force. Although much attention has been given by the legislators to the noise from pneumatic tools, that merely reflects their widespread use. There is some limited evidence which appears to indicate that if the noise is combined with vibration it has a significantly greater effect than if either were operating alone. For the legislative background to noise from construction sites refer to the chapter on *Noise on Construction Sites*.

Estimating Permissible Exposure Times

The calculation basis of vibration exposure times is a statistical one of estimating the likelihood of showing symptoms of a disease over a typical working life. One such suggestion is shown in Figure 1, derived from BS 6842. The example shown means that a person exposed to 3 m/s^2 for a working day of 4 hours is likely to display symptoms of white fingers after 25 years. Whatever value of permissible exposure that will emerge as a 'safe' level will be based on considerations of this kind.

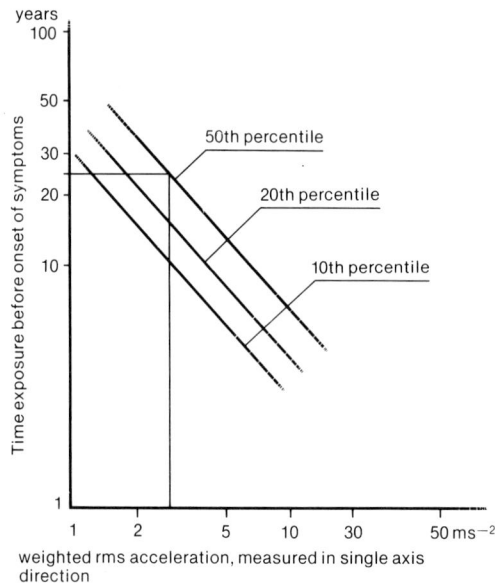

Figure 1   *Dose – response relationship of vibrations in handles*

The weighted acceleration for a period of T hours when the permissible level is known can be found from:

$$a_{wT} = a_{w4} \left(\frac{4}{T} \right)^{1/2}$$

a_{wT} is the permissible exposure for a working day of T hours, when the 4 hour exposure, a_{w4}, is prescribed.

on the assumption of 4 hour limit of 3 m/s^2,

$$a_{w4} = 3 \text{ m/s}^2$$

$$a_{wT} = 3 \left(\frac{4}{T} \right)^{1/2}$$

It is on this basis that the values in Figure 3 are quoted.

Design of Vibration-reduced Equipment

When an operator works with a vibrating tool, he has to apply a steady force to keep the cutting point applied to the surface and has to withstand the vibrations of the handle. When the tool is held pointing downwards (as for example in the case of a road breaker), some of the steady force is catered for by its weight, but it is usual to require the addition of a further applied force for efficient operation. Any techniques of vibration suppression that are adopted must take account of this requirement to apply the constant force. It might be thought that one solution would be to increase the weight of the body of the tool so as to reduce the need for further force, and at the same time to reduce the vibration level through the extra mass to be accelerated. In practice, such a tool would be so unwieldy as to be impractical to use.

The user of a tool such as a chipping hammer has to suffer a vibration of the hand holding the tool (usually the right hand for a right handed operator) and may also be subject to vibration in the hand guiding the bit or chisel. The latter is characterised by a shock wave passing up and down the length of the chisel, repeated at the blow rate of the tool. The operator needs protection from both these kinds of vibration. The former can be tackled by conventional methods of vibration reduction, the latter by some kind of protective sleeve or by wearing of gloves.

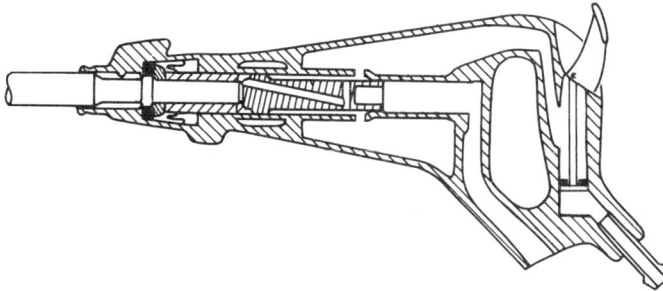

Figure 2 Low vibration chipping hammer

One feature of fluid-operated percussive tools that needs to be appreciated is that, although the frequency of operation is reasonably constant for a given pressure, there is very little inertia in the system to keep the motion constant: each cycle is a single event with all the energy put in during the cycle being extracted at the time of impact. So a single cycle is not affected by the previous one nor does it affect the subsequent ones. The consequence of this is that any form of rotary balancing that is contemplated will fail through the impossibility of synchronising the balance system with the cycle of the tool. A similar argument applies to the suggestion that a measure of vibration balance can be achieved by the use of a counter piston, moving in opposition to the impact piston: it would be impossible to synchronise such an arrangement.

Another principle that needs to be remembered when designing a percussive tool for vibration reduction is that the excitation forces are far from being sinusoidal: there is a much larger force on the power or down stroke and there is an instantaneous impact force at the end of the power stroke. Reducing the vibration level at the operating frequency may still leave a considerable level of vibration at a higher harmonic.

Some simple methods of vibration isolation are available on most modern percussive tools. They have comfortably shaped handles covered with rubber or polyurethane plastic grips, which are effective in suppressing the high frequency 'sting' which is experienced in some tools. These grips need to be replaced regularly, particularly by when there is any sign of wear.

Attenuation of the vibration of the handle is theoretically possible by use of a conventional spring mass isolator. Take for example a typical rivetting hammer which requires a static feed force of 150 N and operates at a frequency of 30 Hz; the handle and hand together weigh 0.5 kg. With a spring giving a natural frequency equal to one half the frequency of the disturbing force, the static deflection will be 30 mm, which is too large for precise control. The spring could be preloaded so that a much lower static deflection is obtained at the nominal applied load, but in that case, the vibration would not be suppressed at lower feed forces. A further problem arises with some tools such as road breakers where the bit can get stuck in the ground, requiring the application of a reverse pull. In that case the full vibration would be felt by the operator.

There are some tools which make use of this principle of passive vibration reduction, but there are more promising approaches which, through fundamental changes in design, can reduce the vibration level. One such, a chipping hammer used for the dressing of steel plates and castings, is shown in Figure 3. In this tool a differential piston moves in the cylinder with the supply pressure constantly acting on its rear surface. When the piston strikes the chisel, a volume in the front of the piston is supplied with air pressure, and because of the larger front area, the piston then moves backwards. The supply pressure is constant, which means that the reaction force on the handle is also constant, which cannot give rise to vibrations. Shock reflections from the chisel are absorbed by a collet. The benefits of this design are a reduction in vibration and a simpler design without the normal valve system usually incorporated in such a tool. The vibration reduction is impressive, with the spectrum as shown in Figure 3 for two tools in this range.

Vibration Level

dB re 1E-6 m/s2

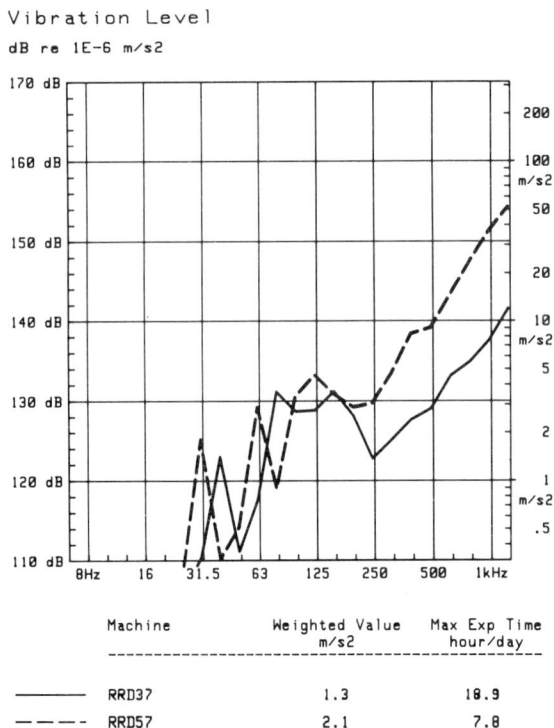

Figure 3 Vibration spectrum for chipping hammers of the type shown in Figure 1. The maximum exposure is based on a permissible 4 hours exposure of $3m/s^2$

Machine	Weighted Value m/s2	Max Exp Time hour/day
RRD37	1.3	18.9
RRD57	2.1	7.8

An example of passive attenuation is shown in Figure 4 on a scaling hammer. This is a tool used for removing scale from a steel sheet by the application of high frequency hammer action through a chisel attached to its head. Vibration reduction is secured by a springmass attached to its handle. This is such a light tool that very little force is required to operate it and so the static deflection problem is not important.

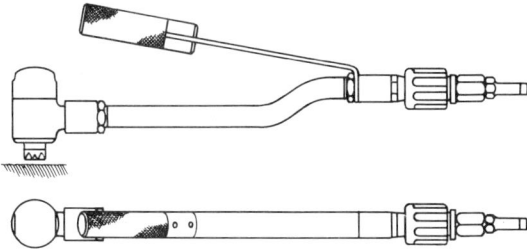

Figure 4 Mass balance in a scaling hammer

An example of a road breaker with a passive attenuation system is shown in Figure 5. A reduction of 90% is claimed in this range of tools as illustrated in Figure 6. The handles of this tool are spring loaded, with a helical spring mounted cross-wise in the breaker head. In conjunction with the spring, a rubber block comes into play when the tool is required to extract a stuck chisel. Another design, shown in Figure 7, has two longitudinal springs.

Figure 5 Road breaker with vibration attenuation incorporated in the handle. A noise muffler is part of the design

There have been recent attempts to design road breakers with the cushioning provided by pneumatic means, which offer some theoretical advantages in that they do not suffer from the problem of a large static deflection mentioned above. They work on the principle of isolating the handle from the body of the tool by means of a pneumatic cylinder, which is charged by air at a pressure equivalent to the optimum down force. Tools based on this principle have to be bulkier to accommodate the extra pneumatic cushions, and so far have had only had a limited degree of commercial success.

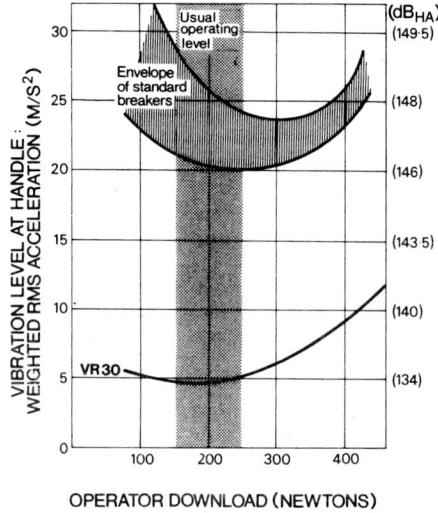

Figure 6 Vibration levels of the breaker of
Figure 5

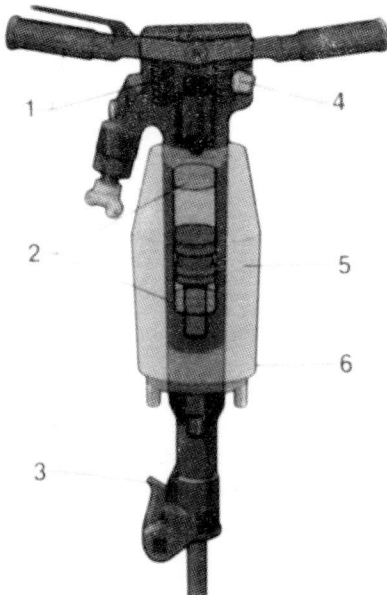

Figure 7 Ergonomically design breaker
1 – spring suspension
2 – cushioning behind piston
3 – latch retainer
4 – built in lubricator
5 – silencer reduces exhaust noise by 15 dB
6 – smooth external design

Grinding Machines

One does not normally associate rotary tools with vibration problems, however they have been implicated in the past. A typical example is a hand held grinder. The vibrating force can come from either an unbalance of the grinding wheel or the contact forces between the wheel and the material being worked. A grinding wheel is not perfectly homogeneous, so it is possible for a wheel to be both geometrically accurate and precisely located on the spindle of the tool and yet give rise to out-of- balance vibrations. It is recommended that grinding wheels should be regularly dressed so as to reduce the possibility of out-of-balance forces, but there may still be a residual vibration which needs to be dealt with by design of the grinding machine. The dynamic contact forces between the grinding tool and the material are normally of the same magnitude as the static feed force imparted by the operator. They are necessary to control the stock removal rate of the grinding operation. If the grinding tool is out of true, the dynamic contact forces may become greater than the feed force which means that the grinding tool jumps on the workpiece, resulting in high vibrations of the machine.

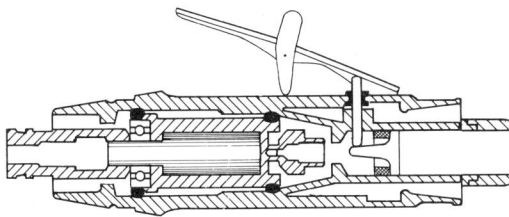

Figure 8 Die grinder showing elastic motor supports

One approach to this problem is illustrated in Figure 8 – a die grinder. The design required a rigid axial attachment in order to secure the motor to the wheel (not shown) and a weak radial attachment to accomplish vibration isolation. This was achieved by allowing the motor assembly to rest against a spherical bearing at the front, where the balls prevent axial motion but roll with low friction during radial movements. These are absorbed by a soft O-ring at the front and a stiffer one at the back. The motor can thus have a circular pendulum-type motion, yet remain stiff in an axial direction. The levels of vibration are small, which to some extent reflects the low levels initially rather than effectiveness of design.

Riveting Hammers

These can cause vibration injuries to the hands not only of the person holding the gun but also to the one holding the reaction block or bucking bar. Riveters may have have to work for prolonged periods through the whole day just doing the one job of riveting, so this can be serious problem.

Figure 9 presents one solution. In this design, the application of a feed force to the riveting hammer opens a port which admits air to the air cushion behind the impact mechanism and in the volume contained in the handle. The impact mechanism is effectively isolated from the operator by an air cushion, the pressure of which is proportional to the applied force. The extra volume in the handle helps to keep the feed pressure variations small to reduce still further the handle vibrations.

The bucking bar can also have a pneumatically cushioned cylinder to relieve the vibrations of the person holding it.

A skilled operator, previously accustomed to using a conventional riveting hammer, may take some time to be familiarise himself with the new tools. Much of his skill comes from an ability to recognise a well set rivet by the feel of the tool; this feel will be lacking in a cushioned tool, so proper instruction needs to be given him to make sure that he does not overdo the feed force.

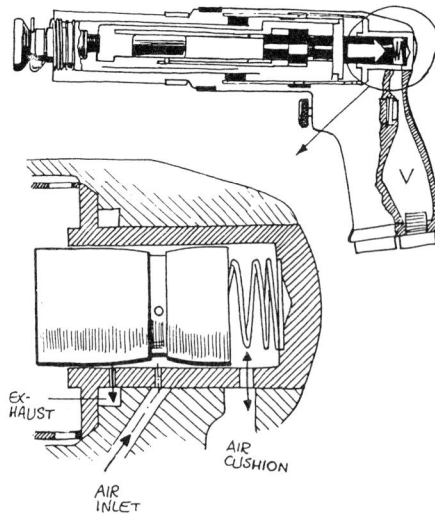

Figure 9 Operation of Riveting Hammer cushioning system

Test Methods

In evaluating the effectiveness of any method of vibration suppression in tools, one is faced with the difficulty of specifying a representative test regime and then of taking measurements. This will be dealt with in a series of International standards under the general number ISO 8662. At the time of writing only a general method has been issued (ISO 8662–1), which has not yet become a British Standard. Eventually all the hand-held power tools will be covered by separate sections of this standard – see the chapter on *Standards*.

General Guidance in the Use of Tools for Minimum Vibration

The chapter on *Human Response to Vibrations* gives some suggestions as to the precautions that should be taken when using vibrating equipment. These guidelines should always be followed, irrespective of whether special low vibration tools are available. Technology is still unable to meet the recommended limits of exposure of 2.8 m/s^2 for all tools in use, and there will always be some workers for whom even that level is too high. We are likely to see human vibration tolerance monitoring along similar lines to audiometry in the future, once a reliable diagnostic technique has been developed. Until then employers and employees alike have a duty to ensure that tools are used sensibly.

SECTION 8

Chapter 1
Legal Aspects of Noise and Vibration

It is only possible to give here a broad indication of the law relating to noise and vibration. English law will be considered in the main, although where European Community law is considered, the application is wider than that of the law of England. Noise as a nuisance can be either statutory or common law. The distinction for the layman is that common law is made by the courts and statutory law is made by Parliament. Whether one takes action under the rules of common law or under statute depends upon the circumstances. In some instances, statute has codified common law. In other instances, statute has created new offences.

Common Law

Noise has been recognised as a nuisance (in the legal sense) for many years. Nuisance can be either public or private. A private nuisance is commonly defined as an unlawful interference with a person's use or enjoyment of land or of some right over it or in connection with it. A public nuisance is an unlawful act or omission causing interference with the health and safety of the public at large. A noisy operation can be both a public and a private nuisance. It is not always easy to answer the question of whether the nuisance complained of is a public or a private one. If the noise nuisance affects just one person or a household, it is clearly a private nuisance. If it affects a class of the public, such as the residents of a community or of the passers-by of a noisy factory, it is likely to be a public nuisance.

The distinction between a public and a private nuisance is important, because it governs the kind of legal action that can be taken. A private action can only be brought by the occupier of the land or by someone who has a legal interest in the land, and he can only take action against the person who cause the nuisance. The remedies available to the plaintiff are damages and an injunction. If the injury is a minor one, which can be compensated for by the payment of money, damages can be awarded. A court also has the discretion to issue an injunction, which is an order to discontinue the nuisance.

It is also open to the injured party to take action to abate the nuisance by self help without recourse to the courts. Thus, for example, a resident who is continually annoyed by a faulty burglar alarm could disconnect it to ensure he has a good night's sleep. Usually such an action is not advised, as it could itself be considered unlawful and bars the plaintiff from recovering damages. He must not do more damage than necessary, and if the action requires him to enter upon another's property, he may have to give notice.

A public nuisance is a crime, and proceedings can be taken by public bodies such as the local authority or the Attorney-General or both. All the criminal sanctions are available to the court, and in addition it has a power to issue an abatement order, which may take the form of restricting the noise to certain hours, or levels, or to cease entirely.

Some of the common law rules that have been established are:

It is no defence to show that all possible care has been taken to minimise the noise.

Noise need not be injurious to one's health. It can be sufficient to show that there is an interference with ordinary comfort.

There is no fixed standard of noise. It depends upon whether a reasonably minded resident, accustomed to the noise of the area, would take the same view as the plaintiff.

It is no defence in a private nuisance to show that the plaintiff came to the nuisance or that the nuisance is beneficial to the plaintiff.

The noise must not be purely of a temporary nature. Thus noise from a building site is not a nuisance, provided that reasonable care has been taken to reduce the noise as far as is practicable.

If the defendant's acts are malicious, the courts are likely to treat them as not reasonable.

There are unsatisfactory features in treating noise as a nuisance, the main one being that the noise has to be made before any action can be taken to stop it. A local authority, through planning regulations, may make it possible to ensure that a noisy operation is not allowed to be established. Controls on noise from construction sites, even though they are of a temporary nature, are also subject to local authority control.

Statute Law

The first attempt to control environmental noise dates from the Noise Abatement Act 1960, which has now been repealed by the Control of Pollution Act 1974. An important UK Government Report, known as the Wilson Report 1963 (Cmnd. 2056), initiated much subsequent legislation in respect of noise in towns, within buildings, from motor vehicles, aircraft, industry and construction, and in setting limits to occupational exposure. The Environmental Protection Act 1990 is the most recent act in the area of environmental noise.

Neighbourhood Noise: Current Legislative Controls in the UK
The Environmental Protection Act 1990

This act (the EPA) came into force on 1 January 1991. Part III restates the law defining various statutory nuisances, and improves summary procedures for dealing with them. The 'Noise Nuisance' provisions of the Control of Pollution Act 1974 (the COPA) have been repealed.

Previously, under the COPA, local authorities were required to inspect their area to detect statutory nuisances. However, the EPA also requires local authorities "to take such steps as are reasonably practicable" to investigate complaints. No guidance has been issued on this new requirement, but a report at the committee stage of the bill provided the Government's view that:

"There will always be constraints on expenditure under any government, so we must be cautious before we oblige local authorities to follow up each and every complaint."

Clearly the level of response to complaints will vary between authorities, and will depend on different perception of the problem of noise and the priority given to noise control in resource allocation. For example, some kinds of domestic noise nuisance will be given an immediate response in one district, whereast in another area complainants will be advised to pursue their own action if there is no 'public' element in the alleged nuisance.

'Noise' includes vibration in Part III of the EPA, and the term 'nuisance' must be understood in its common law sense.

The basic steps in dealing with a complaint and the abatement of a noise nuisance by a local authority are given in Figure 1. Local authorities must serve abatement notices on the 'person responsible', and in some cases on the owner or occupier of the premises.

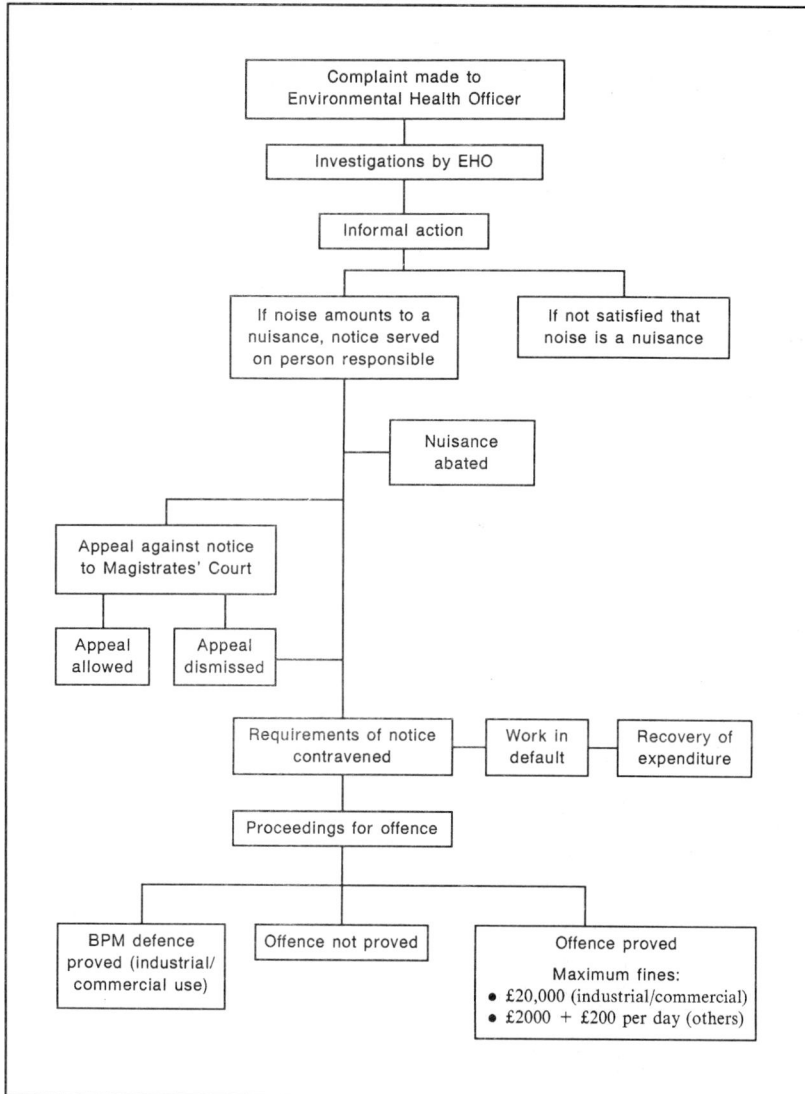

Figure 1 Action by local authority under the Environment Protection Act 1990 (Sections 79, 80, 81): summary proceedings for noise nuisances

Notices may be served in anticipation of a noise nuisance. Appeals must be lodged within 21 days from service, and grounds of appeal are detailed in separate regulations, The Statutory Nuisance (Appeals) Regulations SI 1999:1276. These are similar to regulations issued under the COPA, and stipulate where notices are to be suspended in the event of an appeal.

The EPA (Section 82) also allows an individual to take a private action in respect of noise nuisance. Again the process is similar to that of the COPA, which it replaces (see Figure 2). There are however, some new features – the aggrieved person is required to give at least three days' notice to the person responsible for the nuisance (or owner or occupier as appropriate), of intention to bring proceedings. Also, where the court is satisfied of the existence of a nuisance, it can order the payment of reasonable compensation to the aggrieved person, for the cost of bringing the proceedings.

The noise nuisance provisions of the EPA are largely a re-anactment of earlier powers given to local authorities. It seems therefore that in many cases the abatement procedure will be a lengthy one.

Assessment of Noise Nuisance

The basis for a decision as to whether noise is a statutory nuisance is to consider whether or not the noise would constitute a nuisance at common law. Case law has shown that various factors including sound character, time of day, interference with living arrangements etc., are involved in the establishment of nuisance.

Thus, the test is largely a subjective matter. Such assessments demand a large input of the staffing resource of local authorities. During investigations supplementary evidence of noise impact will be assembled by the Environmental Health Officer by carrying out sound level measurements. In the case of industrial and commercial noise, the most commonly referred to standard method of assessing sound level has been BS 4142, which was first issued in 1967.

The recently revised edition of BS 4142:1990 follows the original approach of measurements to determine whether a given noise (new or existing) is likely to give rise to complaints. Significant new feature of the standard include:

> The use of the equivalent continuous A-weighted sound pressure level.
>
> Tighter specifications for measurement equipment.
>
> Traceable calibration of meters.
>
> Determination of the specific noise level, background noise level, rating noise level and assessment level.

While the standard declares that the assessment of *nuisance* is beyond its scope, previous experience of its use is that BS 4142 assessments are much liked by those in control of court proceedings or appeal hearings.

It is noted that the National Physical Laboratory (NPL) and the Building Research Establishment have produced a data sheet for use with the new standard. This will assist investigating officers in the application of the standard. NPL has proposed that completed data sheets be sent back to it, to contribute to the understanding of the standard's performance, and to identify the need for further refinement.

The Control of Pollution Act 1974

The Control of Pollution Act 1974, Part III (Noise) remains in force to deal with:

> Construction noise
>
> Noise abatement zones

Codes of practice

Noise in streets.

Construction Noise

The COPA enables local authorities to specify the way in which construction, demolition and similar works must be carried out so as to minimise noise. Requirements may be specified before or

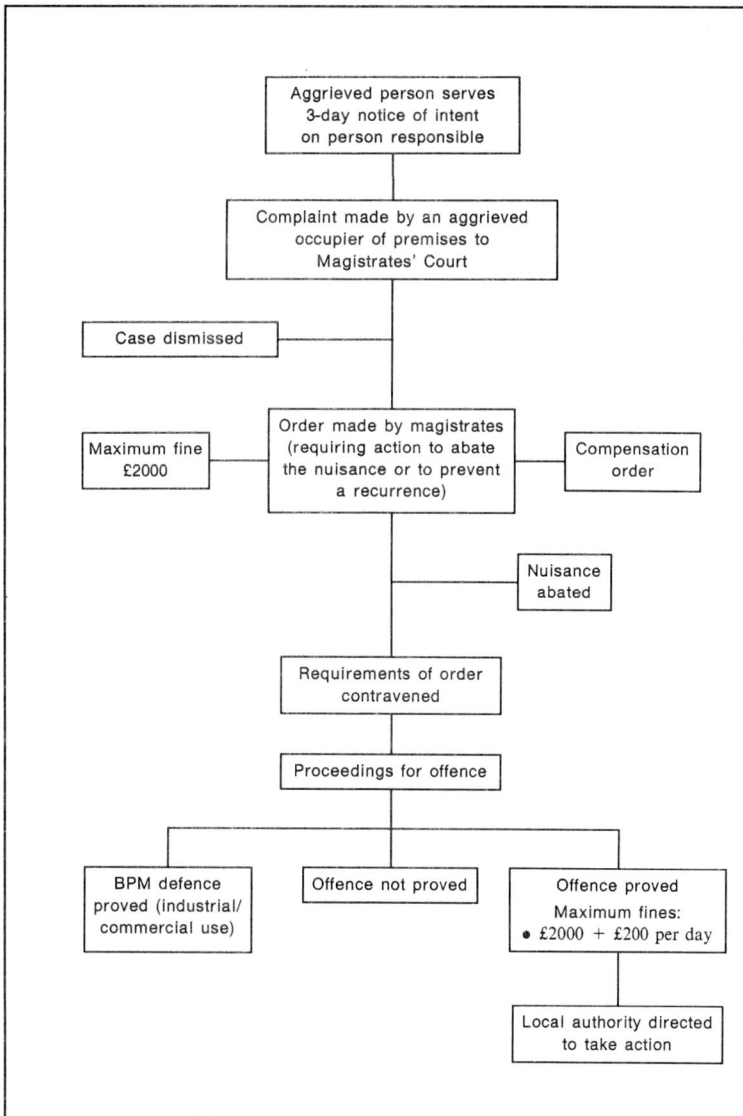

*Figure 2 Action by aggrieved person under the Environmental Protection Act
1990 (Section 82): summary proceedings for noise nuisances*

after work has begun. Control is exercised by the issue of a Notice which may specify the plant or machinery which is, or is not, to be used; the hours of operation; and permitted noise levels from particular types of machinery.

The Act also allows the contractor, or other person arranging for the works to be carried out, to ask the local authority to make its requirements known in advance. The local authority has to reply to an application within 28 days, otherwise an unconditional consent is the result. In general the principle of noise controls must relate to British Standard BS 5228, which provides detailed guidance. See also the chapter on *Noise on Construction Sites*.

Noise in Streets

The COPA bans the use of loudspeakers in streets between 9.00 pm and 8.00 am. There are a few exceptions to this rule, notably for public emergency services such as police, ambulance and fire brigades. During the daytime it is also illegal to use loudspeakers in streets for advertising entertainments, trades and businesses. There is a specific exemption for vehicles selling ice cream etc., but this relates only to the period between noon and 7.00 pm. These controls date back to the Noise Abatement Act 1960, and do not cater for noise sources such as car alarms.

Noise Abatement Zones (NAZ)

The COPA enables a local authority to designate parts of its area as a Noise Abatement Zone. An order specifies the classes of premises to be controlled. Subsequently measurements are made of noise emanating from classified premises in the zone. Methods of measurement are given in regulations, The Control of Noise (Measurement and Register Regulations 1976 SI 1976:37.

These measurements are then registered, and it is an offence for registered levels to be exceeded, unless the local authority gives its written consent. In some circumstances a local authority can require a reduction of the registered noise level. A preventive element of the NAZ procedures is that local authorities can determine an acceptable level of noise for new or converted premises in an area. Most local authorities have avoided NAZ programmes, considering them to be largely impractical and too demanding of resources.

Codes of Practice

The COPA enables the Secretary of State to prepare and/or approve codes of practice for minimising noise. Codes of practice have been issued in respect of:

> Audible intruder alarms
>
> Model aircraft
>
> Ice cream van chimes

The Secretary of State has also approved a code of practice relating to construction noise – The Control of Noise (Code of Practice on Noise from Construction and Open Sites) Order 1984 SI 1984:1992.

Other Noise Controls

There are other opportunities available to local authorities to control noise, including:

> Local Government (Miscellaneous Provisions) Act 1982 – issue etc. of public entertainment licences.
>
> Licensing Act – objection to granting/renewal of liquor licences.
>
> Road Traffic Act – objection to granting etc. of heavy goods vehicle (HGV) operator's licences.

Local byelaws – enforcement of controls (for example noisy music, animals etc.) specific to a particular local authority area or part of a local authority area.

Conclusion

In the last 15 years complaints to local authorities about noise have increased significantly. Most of the law available to local authorities to control neighbourhood noise originated over 15 years ago.

Currently, many complaints relate to newer kinds of noise source which are not covered, or are inadequately covered by present law. It is suggested that the staffing resource of local authorities that is available for noise control work, has not kept pace with the trend in complaints. It seems therefore that a large amount of noice nuisance and annoyance cannot be effectively controlled at present.

The Government has welcomed the wide range of recommendations of the Noise Review Working Party (Batho Report). However, a full response is awaited.

It seems that the view of the Noise Advisory Council expressed in 1973 remains a valid goal to be attained:

"While noise cannot of course be totally eliminated, much can and should be done to reduce it. The law should be framed so as to provide practical and effective assistance to this end."

Control of Noise Through Planning

Official guidance on planning and noise was issued by the Department of the Environment in 1973 as DoE circular 10/73. The guidance given there relates to noise from transportation and industry and covers the building of new housing near to existing noise sources and the development of new sources near to existing housing. It laid down principles and specific sound level criteria, which the Secretary of State uses when taking planning decisions and which local authorities have used as a basis for their own policies to minimise noise annoyance. BS 4142 was the basis for assessment.

This document has served local authorities for nearly 20 years but is now superseded by a new document, Planning Policy Guidance: Planning and Noise PPGXX to be published early in 1993.

The document introduces the concept of noise exposure categories, ranging from category A, where noise is unlikely to be a significant factor, to category D, where noise is such an important factor that planning permission for an incompatible use should normally be refused in the absence of strong planning reasons to the contrary. It will be for the local planning authorities to determine what levels of noise they wish to attribute to each of these categories, and to define the type of development to be considered most sensitive to noise in the light of local circumstances and priorities. Thus much greater responsibility is placed with local authorities than before. The most appropriate means of assessing the likelihood of complaints from new development is BS 4142:1990, see chapter on Community Standards.

The document offers the following approaches to noise control at the planning stage:

Engineering: reduction of noise at source; improving sound insulation of sensitive buildings; screening by purpose built barriers

Layout: adequate distance between source and noise-sensitive building or area; screening by natural barriers or other buildings or non-critical rooms in a building

Administrative: limiting operating time of source; restricting activities allowed on the site; specifying an acceptable noise limit

The planning system is the most appropriate means of tackling potential noise problems from the outset. It cannot be used to deal with existing noise problems, for which the appropriate means are COPA and EPA. Nor is it appropriate to rely on planning controls to minimise the effect of noise, where building regulations can be used.

For large developments, the local authority or the Secretary of State may require the submission of a statement of the effects on the environment of the proposed developments. The effects include noise and vibration. The proposed developments are in the main restricted to such operations as:

oil refineries

power stations

chemical works

aerodromes

roads and railways

ports

waste disposal facilities

Other proposed developments which by size and location are likely to have a significant effect on the environment are also included. This requirement to make a submission arises out of EC Directive 85/337/EEC, implemented by Town and Country Planning (Assessment of Environmental Effects) Regulations 1988.

Noise at Work

The legislative framework is incorporated in the Noise at Work Regulations 1989, implementing EC Directive 86/188/EEC on the protection of workers from the risk of exposure to noise at work. Refer to the chapter on *Noise Regulations and Hearing Protection*. The EC has started work on a replacement directive, the draft of which is not yet published, but it seems probable that it will suggest a further tightening of the standards.

Noise from Mineral Extraction Sites

Mineral sites are subject to the other statutory controls outlined above, but more specifically they will need to meet the planning requirements of the Mineral Planning Guidance Note (MPG) Note 11, at present in draft form.

Aircraft Noise

Noise from airports can be dealt with under planning controls. However increasing pressure on manufacturers is coming from noise certification regulations, which require new aircraft to demonstrate compliance with limits agreed through the International Civil Aviation Organisation, specified in Chapter 3 of ICAO Environmental Recommendations.

Noise Control Through Limits on Machinery Noise

In many cases, particularly when taking action under COPA, the defence of best possible means can be used where the noise arises from industrial premises or construction sites.

Best possible means can include the use of equipment that meets the noise limits specified in the various EC Directives which have been issued. These have been implemented by the UK Government in the form of statutory instruments, a full list of which is incorporated into the chapter on standards, codes of practice and publications. The directives have been issued to comply with Article 100 of the Treaty of Rome, which relates to unequal competition. In some cases, the noise levels are phased in over a period of time and relate only to equipment placed on the market, and are therefore not retrospective. However, it seems that the defense of best possible means could only be sustained if the latest type of equipment were in use.

Certain of the directives specify the maximum noise levels, and define the method of test to ensure compliance. They have been drawn up in consultation with manufacturers, so the noise levels prescribed are attainable with existing technology and without being burdensomely expensive.

The machinery covered by the various directives to date includes:

> motor vehicles, including tractors, motor cycles and industrial trucks
>
> construction plant, including excavators, loaders, backhoe loaders and tower cranes
>
> generator sets, including welding generators
>
> concrete breakers
>
> compressors
>
> lawnmowers
>
> subsonic jets
>
> household appliances, including equipment for upkeep, cleaning, preparation and storing of foodstuffs and air conditioning
>
> machinery

The last two do not specify maximum noise levels. The Directive on Household Appliances specifies the method of test to be used when labelling for noise levels; the Machinery Directive requires noise level labelling when the sound pressure level exceeds 70 dB(A) at the workstation and sound power level labelling when the sound pressure level exceeds 85 dB(A); the Machinery Directive also requires labelling of hand held and hand guided machinery where the weighted vibration level exceeds 2.5 m/s^2.

SECTION 9

Chapter 1
Standards, Codes of Practice and Publications

British Standards with Corresponding International Standards

BS 848: Part 2: 1985	Fans for general purposes: methods of noise testing
BS 2497: Part 5: 1988 ISO 389	Specification for a standard reference zero using an acoustic coupler complying with BS 4668
BS 2497: Part 6: 1988 ISO 389/Add 1	Specification for a standard reference zero using an artificial ear complying with BS 4669
BS 2497: Part 7: 1988 ISO 389/Add 2	Specification for a standard reference zero at frequencies intermediate between those given in Parts 5 and 6
BS 2750:	Measurement of sound insulation in buildings and of building elements
BS 2750: Part 1: 1980 ISO 140/1	Recommendations for laboratories
BS 2750: Part 2: 1980 ISO 140/2	Statement of precision requirements
BS 2750: Part 3: 1980 ISO 140/3	Laboratory measurements of airborne sound insulation of building elements
BS 2750: Part 4: 1980 ISO 140/4	Field measurements of airborne sound insulation between rooms
BS 2750: Part 5: 1980 ISO 140/5	Field measurements of airborne sound insulation of facade elements and facades
BS 2750: Part 6: 1980 ISO 140/6	Laboratory measurements of impact sound insulation of floors
BS 2750: Part 7: 1980 ISO 140/7	Field measurements of impact sound insulation of floors
BS 2750: Part 8: 1980 ISO 140/8	Laboratory measurements of the reduction of transmitted impact noise by floor coverings on a standard floor
BS 2750: Part 9: 1987 ISO 140/9	Method for laboratory measurement of room-to-room airborne sound insulation of a suspended ceiling with a plenum above it
BS 3015: 1991 ISO 2041	Glossary of terms relating to mechanical vibration and shock
BS 3045: 1981 ISO 131	Method of expression of physical and subjective magnitudes of sound or noise in air

BS 3425: 1966 ISO 362	Method of measurement of noise emitted by motor vehicles
BS 3539:1986	Specification for sound level meters for the measurement of noise emitted by motor vehicles
BS 3593:1963(1986) ISO R266	Recommendation on preferred frequencies for acoustical measurements
BS 3638: 1987 ISO 354	Method for the measurement of sound absorption in a reverberation room
BS 3851: 1990 ISO 1925: 1990	Glossary of terms used in the mechanical balancing of rotating machinery
BS 3852:	Balancing machines
Part 1:1979 ISO 2953	Method of description and evaluation
BS 4142:1990 ISO 1996–1, ISO 1996–2 ISO 1996–3	Method of rating industrial noise affecting mixed residential and industrial areas
BS 4196	Sound power level of noise sources
BS 4196: Part 0: 1981 ISO 3740	Guide for the use of basic standards and for the preparation of noise test codes
BS 4196: Part 1: 1991 ISO 3741	Precision methods for determination of sound power levels for broadband sources in reverberation rooms
BS 4196: Part 2: 1991 ISO 3742	Precision methods for determination of sound power levels for discrete-frequency and narrow bands sources in reverberation rooms
BS 4196: Part 3: 1991 ISO 3743	Engineering methods for determination of sound power levels for sources in special reverberation test rooms
BS 4196: Part 4: 1981 ISO 3744	Engineering methods for the determination of sound power levels for sources in free-field conditions over a reflecting plane
BS 4196: Part 5: 1981 ISO 3745	Precision methods for the determination of sound power levels in anechoic and semi-anechoic rooms
BS 4196: Part 6: 1981 ISO 3746	Survey methods for determination of sound power levels of noise sources
BS 4196: Part 7: 1988 ISO 3747	Survey method for determination of sound power levels of noise sources using a reference sound source
BS 4196: Part 8: 1991 ISO 6926	Specification for the performance and calibration of reference sound sources
BS 4668: 1971 IEC 303	Specification for an acoustic coupler (IEC reference type) for calibration of earphones used in audiometry
BS 4669: 1971 IEC 318	Specification for an artificial ear of the wide band type for the calibration of earphones used in audiometry
BS 4675:	Mechanical vibration in rotating machinery
BS 4675: Part 1: 1976 ISO 2372	Basis for specifying evaluation standards for rotating machines with operating speeds from 10 to 200 revolutions per sec
BS 4675: Part 2: 1978 ISO 2954	Requirements for instruments for measuring vibration severity

BS 4718: 1971	Methods of test for silencers for air distribution systems.
BS 4727:	Glossary of electrotechnical, power, telecommunication and electronics, lighting and colour terms.
BS 4727: Part 3: Group 8: 1985 IEC 50(60), IEC 50(801)	Acoustics and electro-acoustics terminology
BS 4813: 1972	Method of measuring noise from machine tools excluding testing in anechoic chambers
BS 5108	Sound attenuation of hearing protectors
BS 5108: Part 1: 1991 ISO 4869–1	Subjective method of measurement
BS 5228:	Noise control on construction and open sites
BS 5228: Part 1: 1984	Code of practice for basic information and procedures for noise control
BS 5228: Part 2: 1984	Guide to noise control legislation for construction and demolition including road construction and maintenance
BS 5228: Part 3: 1984	Code of practice for noise control applicable to surface coal extraction by open cast methods
BS 5228: Part 4: 1986	Code of practice for noise control applicable to piling operations
BS 5265:	Mechanical balancing of rotating bodies
Part 2: 1981 ISO 5406	Methods for the mechanical balancing of flexible rotors
Part 3: 1984 ISO 5343	Recommendations for criteria for evaluating flexible rotor balance
BS 5330:1976 ISO 1999	Method of test for estimating the risk of hearing handicap due to noise exposure
BS 5331: 1976 ISO 2249	Method of test measurement of the physical properties of sonic bangs
BS 5512:1991 ISO 281	Method of calculating dynamic load ratings and rating life of rolling bearings
BS 5647: 1979 IEC 561	Specification for electro-acoustical measuring equipment for aircraft noise certification
BS 5721: 1979 ISO 537	Specification for frequency weighting for the measurement of aircraft noise (D– weighting)
BS 5727: 1979 ISO 3891	Method for describing aircraft noise heard on the ground
BS 5821	Methods for rating the sound insulation in buildings and of building elements
BS 5821: Part 1: 1984 ISO 717/1	Method for rating airborne sound insulation in buildings and of interior building elements
BS 5821: Part 2: 1984 ISO 717/2	Method for rating the impact sound insulation

BS 5821: Part 3: 1984 ISO 717/3	Method for rating the airborne sound insulation of facade elements and facades
BS 5944:	Measurement of airborne noise from hydraulic fluid power systems and components
BS 5944: Part 1: 1980 ISO 4412/1	Method of test for pumps
BS 5944: Part 2: 1980 ISO 4412/2	Method of test for motors
BS 5944: Part 3: 1980	Guide to the application of Part 1 and Part 2
BS 5944: Part 4: 1984	Method of determining sound power levels from valves controlling flow and pressure
BS 5944: Part 5: 1985	Simplified method of determining sound power levels from pumps using an anechoic chamber
BS 5966: 1980 IEC 645	Specification for audiometers
BS 5969: 1981 IEC 651	Specification for sound level meters
BS 6055: 1981 ISO 5008	Methods of measurement of whole-body vibration of operators of agricultural wheeled tractors and machinery
BS 6056: 1991 IEC 551	Determination of transformer and reactor sound levels
BS 6083	Hearing aids
BS 6083: Part 10: 1988 IEC 11810	Guide to hearing aid standards
BS 6086: 1981 ISO 5128	Method of measurement of noise inside motor vehicles
BS 6107	Rolling bearings: tolerances
Part 1: 1981 ISO 1132	Glossary of terms
Part 2: 1987 ISO 492	Specification for tolerances of radial bearings
Part 3: 1992 ISO 5753	Specification for radial internal clearance
BS 6140	Test equipment for generating vibration
BS 6140: Part 1: 1981 ISO 5344	Methods of describing characteristics of electrodynamic test equipment for generating vibration
BS 6140: Part 2: 1982 ISO 6070	Methods for describing characteristics of auxiliary tables for test equipment for generating vibration
BS 6177: 1982	Guide to selection and use of elastomeric bearings for vibration isolation of buildings
BS 6294:1982 ISO 7096	Method for measurement of body vibration transmitted to the operator of earth moving machinery
BS 6344	Industrial hearing protectors

Part 1: 1989	Specification for ear muffs
Part 2: 1988	Specification for ear plugs
BS 6402: 1983	Specification for personal sound exposure meters
BS 6414:1983 ISO 2017	Method for specifying characteristics of vibration and shock isolators
BS 6472: 1984	Guide to the evaluation of human exposure to vibration in buildings (1 Hz to 80 Hz)
BS 6611: 1985 ISO 6897	Guide to the evaluation of the response of occupants of fixed structures especially buildings and offshore especially buildings and offshore structures to low frequency horizontal motion (0.063 Hz to 1 Hz)
BS 6686	Methods for determination of airborne acoustical noise emitted by household and similar electrical appliances
BS 6686: Part 1: 1986 IEC 704–1	General requirements for testing
BS 6686:Part 2	Particular requirements
BS 6698:1986 IEC 804	Specification for integrating-averaging sound level meters
BS 6749:	Measurements and evaluation of vibration on rotating shafts
Part 1:1986 ISO 7919/1	Guide to general principles
BS 6805	Statistical methods for determining and verifying stated noise emission values of machinery and equipment
BS 6805 Part 1: 1987 ISO 7574/1	Glossary of terms
BS 6805 Part 2: 1987 ISO 7574/2, EN 27 574–2	Method for determining and verifying stated values for individual machines
BS 6805 Part 3: 1987 ISO 7574/3, EN 27 574–3	Method for determining and verifying stated values for batches of machines using a simple (transition) method
BS 6805 Part 4: 1987 ISO 7574/4, EN 27 574–4	Method for determining and verifying stated values for batches of machines
BS 6812	Airborne noise emitted by earth moving machinery
BS 6812: Part 1: 1987 ISO 6393	Method of measurement of exterior noise in a stationary test condition
BS 6812: Part 2: 1987 ISO 6394	Method of measurement at the operator's position in a stationary test condition
BS 6812: Part 3: 1991 ISO 6395	Method of measurement of exterior noise in dynamic test conditions
BS 6840: Part 16: 1989 IEC 268–16	Guide to the RASTI method for the objective rating of speech intelligibility in auditoria
BS 6841: 1987 ISO 2631–1	Guide to measurement and evaluation of human exposure to whole-body mechanical vibration and repeated shock
BS 6842: 1987 ISO 5349	Guide to the measurement and evaluation of human exposure to vibration transmitted to the hand

BS 6861:	Balance quality requirements of rigid rotors
BS 6861 Part 1:1987 ISO 1940/1	Method for determination of residual unbalance
BS 6916	Chain saws
BS 6916 Part 6: 1988 ISO 7182, EN 27182	Method of measurement of airborne noise at the operator's position
BS 6916 Part 8: ISO 7505	Method of measurement of hand transmitted vibration
BS 6955	Calibration of vibration and shock pickups
BS 6955 Part 0: 1988 ISO 5347–0	Guide to basic principles
BS 7025: 1988 ISO 6081	Method for preparation of test codes of engineering grade for measurements at the operator's or bystander's position of noise emitted by machinery
BS 7085: 1989 to	Guide to safety aspects of experiments in which people are exposed mechanical vibration and shock
BS 7129 ISO 5348	Recommendations for mechanical mounting of accelerometers for measuring mechanical vibration and shock
BS 7136	Brush saws
BS 7136: Part 1: 1989 ISO 7917	Method of measurement at the operator's position of emitted airborne noise
BS 7189: 1989 IEC 942	Specification for sound calibrators
BS 7285: 1990 ISO 8626	Method for describing characteristics of servo hydraulic test equipment for generating vibration
BS 7347: 1990 ISO 8568	Guide for characteristics and performance of mechanical shock testing machines
BS 7385	Evaluation and measurement for vibration in buildings
BS 7385: Part 1: 1990 ISO 4866	Guide for measurement of vibration and evaluation of their effects in buildings
BS 7527	Classification of environmental conditions
BS 7527:Part 1:1991 IEC 721–1	Environmental parameters and their severities
BS 7527: Section 2.6 : 1991 IEC 721–2–6	Earthquake vibration and shock
BS 8233:1987	Code of practice for sound insulation and noise reduction for buildings
BS AU 193a	Specification for replacement motor cycle and moped exhaust systems

Codes of Practice – British Standards

| CP 153: | Windows and rooflights |
| CP 153: Part 3:1972 | Sound insulation |

Codes of Practice – UK Government (HMSO)

The control of noise (Measurement and Registers) Regulations 1976

Code of practice on noise from ice cream van chimes 1982

Code of practice on noise from audible intruder alarms 1982

Code of practice on noise from model aircraft 1982

Code of practice on noise from construction and open sites 1984

DoE 10/73 Planning and noise. Shortly to be superseded by Planning Policy Guidance: Planning and Noise PPGXX

Building regulations 1985 SI 1065

Noise insulation regulations 1975 SI 1763

UK publications other than from government sources

BRE (Building Research Establishment) documents:

> Current Paper CP 6/73. Designing offices against road traffic noise

> Current Paper CP 20/78. Sound insulation performance between buildings built in the early 1970s

> Digest No 266. Thermal visual and acoustic requirements in buildings

> Digest No 293. Improving the sound insulation of party walls and floors

International Standards which have no BS equivalent at present

ISO 6258:1985	Nuclear power plants – design against seismic hazards
ISO 8662	Measurement of vibrations in hand-held power tools
Part 1	General (only part published at present)
Part 2	Chipping hammers, riveting hammers
Part 3	Rotary hammers and rock drills
Part 4	Grinding machines
Part 5	Breakers and hammers for construction
Part 6	Impact drills
Part 7	Impact wrenches
Part 8	Orbital sanders
Part 9	Rotary drilling tools

American National Standards

ANSI S series includes approximately 90 standards dealing withacoustics generally; many are similar to the corresponding ISOstandards. There are also ANSI series prepared in conjunction with AFBMA (Anti-friction Bearing Manufacturers), SAE (Society ofAutomotive Engineers), CAGI (Compressed Air and Gas Institute), ASME (American Society of Mechanical Engineers) and ASHRAE (American Society of Heating Refrigeration and Air Conditioning Engineers), ASTM (American Society for Testing Material). Some of the more important ones are included here:

ANSI	S1.4	Specification for sound level meters
	S1.13	Methods for the measurement of sound pressure levels
	S1.23	Method for the designation of sound power emitted by machinery and equipment
	S2.8	Guide for describing the characteristics of resilient mountings
	S2.61	Guide to the mechanical mounting of accelerometers
	S3.5	Methods for the calculation of the articulation index
	S3.14	Rating noise with respect to speech interference
	S3.19	Methods for the measurement of real ear protection of hearing protectors
	S3.28	Evaluation of the potential effect on human hearing of sounds with peak A-weighted sound pressure levels above 120 dB and peak C-weighted sound pressure levels below 140 dB
	S3.40	Guide for the measurement and evaluation of gloves which are used to reduce exposure to vibration transmitted to the hand
	S9.1	Guide for the selection of mechanical devices used in monitoring acceleration induced by shock
	S12.6	Method for the measurement of real-ear attenuation of hearing protectors
	S12.8	Methods for the determination of insertion loss of outdoor noise barriers
ASTM	C384	Test method for impedance and absorption of acoustical materials by the impedance tube method
	E90	Standard test method for laboratory measurement of airborne sound transmission loss of building materials
	E413	Classification for rating sound insulation
	E1124	Standard test method for free-field measurement of sound power level by the two surface method
	E336	Test method for the measurement of airborne sound insulation in buildings

European Community Directives on the approximation of laws of the Member States relating to noise. The UK implementation document, where available is given for each directive. There is a time delay between the issue of the directive and the implementation document.

70/157/EEC	Permissible sound level and the exhaust system of motor vehicles
77/212/EEC	Implemented by The Road Vehicles (Construction and Use)
81/334/EEC	Regulations 1986 S I 1986:1078
84/372/EEC	
84/424/EEC	
74/151/EEC	Tractors. Implemented by The Agricultural or Forestry Tractors and Tractor Components (Type Approval) Regulations 1979 SI 1979:221
77/311/EEC	Driver-perceived noise level of wheeled agricultural or forestry tractors Implemented by SI 1979:221
78/1015/EEC	Permissible sound level and exhaust systems of motor cycles

89/235/EEC	
79/113/EEC	Determination of the noise emission of construction plant and
81/1051/EEC	equipment. Implemented by SI 1985:1968 Construction Plant and
84/532/EEC	Equipment (Harmonisation of Noise Emission Standards) Regulations 1985
84/533/EEC	Acoustic power level admissible for motor compressors. Implemented
85/406/EEC	by SI 1985:1968
84/534/EEC	Acoustic power level admissible for tower cranes. Implemented by
87/405/EEC	SI 1985:1968
84/535/EEC	Acoustic power level admissible for welding generating sets.
85/407/EEC	Implemented by SI 1985:1968
84/536/EEC	Acoustic power level admissible for generating sets. Implemented
85/408/EEC	by SI 1985:1968
84/537/EEC	Acoustic power level admissible for hand-held concrete breakers and
85/409/EEC	pick hammers. Implemented by SI 1985: 1968
86/594/EEC	Airborne noise emitted by household appliances. Implemented by The Household Appliances (Noise Emission) Regulations 1990 SI 1990:161
86/662/EEC	Limitation of sound emission by hydraulic and rope-operated excavators, tractors with dozer equipment, loaders and backhoe loaders. Implemented by SI 1988:361 Noise – Construction Plant and Equipment (Harmonisation of Noise Emission Standards) 1988
86/663/EEC	Self powered industrial trucks
84/538/EEC	Acoustic power level admissible for lawn mowers. Implemented by
87/252/EEC	The Lawnmowers (Harmonisation of Noise Emission Standards)
88/180/EEC	Regulations 1986 and 1987. SI 1986:1795 and SI 1987:876
86/188/EEC	The protection of workers from risks related to exposure to noise at work. Implemented by SI 1989:1790 Noise at Work Regulations 1989
89/391/EEC	Improvements of health and safety at work
89/392/EEC	Machinery
89/629/EEC	The limitation of noise emission from civil subsonic jet aeroplanes Implemented by The Air Navigation (Noise Certification) Order 1990. SI 1990:1514
85/337/EEC	The assessment of the effects of certain public and private works on the environment. Implemented by Town and Country Planning (Assessment of Environmental Effects) Regulations 1988 European Standards (includes EC and EFTA)

European Standards with no corresponding BS

EN 352	Safety requirements and testing
EN 3521	Ear muffs
EN 3522	Ear plugs
EN 3523	Helmet mounted muffs
EN 3524	Level dependent muffs

| EN 3525 | Hearing protectors – safety requirements and testing, alternative performance |
| EN 458 | Recommendations for selection, use, care and maintenance of hearing protectors |

Note: These EN standards at the time of writing are preliminary and bear the designation prEN.

Pneurop Publications

(obtainable from British Compressed Air Society, 8 Leicester St London WC2H 7BN)

5604 (1969)	Measurement of sound from pneumatic equipment
6610 (1978)	Test procedure for the measurement of vibration from hand-held power driven grinding machines
66160 (1985)	Test procedure for the measurement of vibration from chipping hammers

Index

W

X

Index to Advertisers